HANS-JÜRGEN NITZ (ED.)

# THE EARLY MODERN WORLD-SYSTEM
# IN GEOGRAPHICAL PERSPECTIVE

# ERDKUNDLICHES WISSEN

SCHRIFTENREIHE FÜR FORSCHUNG UND PRAXIS
BEGRÜNDET VON EMIL MEYNEN
HERAUSGEGEBEN VON GERD KOHLHEPP
IN VERBINDUNG MIT FRED SCHOLZ

HEFT 110

FRANZ STEINER VERLAG STUTTGART
1993

# HANS-JÜRGEN NITZ (ED.)

# THE EARLY-MODERN WORLD-SYSTEM IN GEOGRAPHICAL PERSPECTIVE

FRANZ STEINER VERLAG STUTTGART

1993

Die Deutsche Bibliothek - CIP Einheitsaufnahme

**The early modern world system in geographical perspective /**
Hans-Jürgen Nitz (ed.). - Stuttgart : Steiner, 1993
   (Erdkundliches Wissen ; H. 110)
   ISBN 3-515-06094-4
NE: Nitz, Hans-Jürgen [Hrsg.]

# CONTENTS

Part VI  External Arenas

## Acknowledgements

This book owes its existence to the combined effort and co-operation of a number of institutions and individuals. First and foremost I would like to thank the "Deutsche Forschungsgemeinschaft" and the "Wissenschaftsministerium des Landes Niedersachsen". Their financial support made the conference at Göttingen-Reinhausen in April 1990 possible where the authors were able to present and discuss their contributions. In addition the "Deutsche Forschungsgemeinschaft" provided a generous grant towards the completion of this publication. As editor I would like to thank the authors whose contributions complied with my initial intentions allowing for this volume to follow a homogenous concept.

For their assistance with the final draft of this publication, I would like to thank the following persons: Julie Harris, Thomas Rolirad, Christoph Bergob (Göttingen), and Robin Butlin (Loughborough) for translations and corrections of style; Howard Davis and Tim Coles, Exeter, formerly Erasmus students at Göttingen University, for painstakingly correcting the contributions of those authors who are non-native speakers of English. Many thanks also to Irmhild Paulus for carefully typing a number of manuscripts, and to Erwin Höfer and colleagues for the necessary cartographic processing of the figures. For carrying out and assisting the computer-aided creation of the final draft my thanks go to Holger Paul, Wolfram Jäckel, and Wolfgang Aschauer.

My apologies go out to the readership due to the impossibility of providing an entirely uniform layout which would have caused a further delay of publication. In addition, I must add that not all of the diacritical marks in words of Polish origin could be reproduced accurately.

Last but not least I am indebted to the editors of "Erdkundliches Wissen" for incorporating the volume into their renowned series.

September 1992          H.-J. Nitz

## Contributors

Professor Robin A. Butlin
Department of Geography, Loughborough University of Technology, Loughborough, Leicestershire, LE11 3TU, Great Britain

Professor Salvatore Ciriacono
Instituto di Studi Storici e Politici,Università degli Studi "G. D'Annunzio", V. le Crucioli, 120, I-64100 Teramo, Italy

Professor A.J. Christopher
Department of Geography / University of Port Elizabeth, P.O.Box 1600, Port Elizabeth, 6000, South Africa

Priv.Doz. Dr. Dietrich Denecke
Geographisches Institut, Universität Göttingen, Goldschmidtstr.5, D-3400 Göttingen, Germany

Professor Robert A. Dodgshon
Institute of Earth Studies,University College of Wales, Penglais, Aberystwyth, Dyfed, SY23 3DB, Great Britain

Dr. Anna Dunin-Wasowiczowa
Instytut Historii, Polskiej Akademii Nauk (Polish Academy of Sciences), Rynek Str. Miasta 29, Warszawa, Poland

Professor Ursula Ewald
Geographisches Institut, Universität Heidelberg, Im Neuenheimer Feld 348, D-6900 Heidelberg, Germany

Docent Dr. Karl-Erik Frandsen
Historisk Institut, Københavns Universitet, Njalsgade 102, DK-2300 København S, Denmark

Professor Gerard A. Hoekveld
Faculteit der Ruimtelijke Wetenschappen, Rijksuniversiteit te Utrecht, Postbus 80115, NL-3508 Utrecht, The Netherlands

Professor Wolf-Dieter Hütteroth
Institut für Geographie, Universität Erlangen-Nürnberg, Kochstr.4, D-8520 Erlangen, Germany

Dr. Wolfram Jäckel
Geographisches Institut, Universität Göttingen, Goldschmidtstr.5, D-3400 Göttingen, Germany

Dr. Ian Layton
Department of Geography, University of Umeå, S-901 87 Umeå, Sweden

Professor Wilhelm Matzat
Geographisches Institut, Universität Bonn, Meckenheimer Allee 166, D-5300 Bonn, Germany

Professor Brian Murton
Department of Geography, University of Hawaii at Manoa, Porteus Hall 445, 2424 Maile Way, Honolulu, Hawaii 96822

Professor Hans-Jürgen Nitz
Geographisches Institut, Universität Göttingen, Goldschmidtstr.5, D-3400 Göttingen, Germany

Professor Jaques Pinard
Faculté des Lettres et des Sciences Humaines, Université de Limoges, Campus universitaire, rue Camille Guérin, F-870 366 Limoges-Vanteaux, France

Professor Arthur I. Ray
Department of History, The University of British Columbia, 1297-1873 East Mall, Vancouver, B.C., V6T 1W5, Canada

Dr. Pieter Saey
Seminarie voor Menselijke en Economische Aardrijkskunde (Department of Human and Economic Geography), Rijksuniversiteit Gent, Krijkslaan, 281, B-9000 Gent, Belgium

Professor Halina Szulc
Instytut Geografii, Polski Academii Nauk (Polish Academy of Sciences), ul. Krakowskie Przedmiescie 30, PL-00-927 Warszawa, Poland

Dr.Taeke Stol
Instituut voor Sociale Geografie, Universiteit van Amsterdam, Nieuwe Prinsengracht 130, NL-1018 VZ Amsterdam, The Netherlands

Dr.Tim Unwin
Department of Geography, Royal Holoway, University of London, Egham Hill, Egham, Surrey TW20 0EX, Great Britain

Dr. Anton Verhoeve
Seminarie voor Menselijke en Economische Aardrijkskunde (Department of Human and Economic Geography), Rijksuniversiteit Gent, Krijkslaan, 281, B-9000 Gent, Belgium

Dr. J.Malcolm Wagstaff
Department of Geography, University of Southampton, Highfield, Southampton SO9 5NH, Great Britain

Dr. David Watts
School of Geography and Earth Resources, University of Hull, Cottingham Road, Hull, HU6 7RX, Great Britain

Dr.Kevin Whelan
Newman Scholar, Department of Geography, University College, Belfield, Dublin 4, Ireland

Dr.Vera Zimányi
Magyar Tudományos Akadémia (Hungarian Academy of Sciences), Történettud. Intézete (Department of History), Uri u. 53 / Pf. 9, H-1250 Budapest, Hungary

# Introduction

Hans-Jürgen Nitz

The concept of an early modern capitalist world-system was introduced by the sociologist and historian I. Wallerstein and socio-economic historian F. Braudel in the 1970's. It is a concept with pronounced geographical dimensions and structures.

The economic system is centered around a core region which is equipped with a trade metropolis that leads and directs the world economy. It consists of those parts of northwestern Europe that have access to the Atlantic, the North, and Baltic Seas (the Netherlands, lower Rhine region, northern France, and England).

A second domain encompasses a peripheral zone dependent on the core in an economic-geographical sense. It provides raw materials, and stretches from Scandinavia (fish, wood, metal) through the Baltic region (grain, wood), to the Black Sea and Hungary (livestock). In the south it embraces southern Italy and the Iberian peninsula (sheep wool), and across the Atlantic the former colonies (plantation products).

The market forces and entrepreneurs of the capitalist core initiate and determine, based upon their demand, the extent to which raw materials are produced. Furthermore they set the prices for the goods, both at the locations where they are produced, as well as at the export market sites. In comparison to the finished products manufactured within the core and delivered to the periphery, the prices set for goods from the peripheries reflect a coerced and unequal exchange. In effect, it is a trade relationship based upon exploitation. This system displays a striking similarity between the economic role played out by the colonies that were subjugated in the 19th and 20th centuries and the role of today's developing countries as opposed to the industrialized states.

It is Wallerstein's and Braudel's primary goal to emphasize this historical continuity which has existed since the 16th century.

According to their model there is a third structural zone located between the core and the periphery - Wallerstein's 'semiperiphery'-, which Braudel refers to as the 'intermediate zone'. The term makes reference both to its geographical location as well as its intermediary position in terms of wealth and the degree of economic diversification.

The economic structure of the semiperiphery is more diverse than that of the periphery. Besides the labour-intensive production of grain (primarily consisting of barley for beer brewing) a developing 'proto-industry' within rural areas produces commercial textiles on a mass scale. These cheap, finished products are also accompanied by the production of half-finished goods (i.e. yarn and cloth), which are then converted to finished products in the core areas. Their production and export is organized on two levels: Firstly, by merchants from the core region who specialize in international trade, and secondly, through intermediary trade centres within the semiperipheries. Braudel accurately characterizes these centres as 'relay cities'.

It seems adequate to parallel the semiperipheral regions of the Early Modern Age with today's so-called 'Newly-Industrialized Countries'(NICs), and the role they play within the world economy. According to Wallerstein, this category includes countries rising in wealth as well as those on the decline. The latter were more specifically those regions which in the late Middle Ages functioned as a global, economic axis before northwestern Europe acquired that position - i.e. northern Italy, the trading centres of southern Germany, Flanders, as well as (if only temporarily) the chief Iberian ports of Sevilla/Cadiz, and Lisbon.

According to Wallerstein the reason for Spain's decline was its role as "at best a rather passive conveyor belt between the core countries and Spain's colonies" (Wallerstein II, 1980, p 185).

Whereas Wallerstein uses the term semiperiphery to designate a dependent zone, similar to that of the periphery, Braudel views this type of region as having greater similarities to that of the core due to the rank and functional position of countries such as France and the larger principalities of Germany, which he labels as 'splendid seconds'. As early as the 17th century these states began to emulate the economic and political powers of the core by adopting a mercantilistic policy.

The concept of a 'geo-economic macrostructure' of the world-system is for historical geographers concerned with the pre-industrial world a particularly fascinating model. This is largely due to the fact that it offers a meaningful analytical framework of that part of the world that has been dominated by Europe. It is namely a spatial system with an inter-regional and interdependent division of labour between its zones. This did not just begin to take shape with the Industrial Age but was already present in an earlier form since the 16th century.

Wallerstein's fundamental concept has been previously addressed by several geographers, however those concerned primarily with the recent world economy. To name a few examples: "The Global Economic System" by Ian Wallace (Boston 1990), and "Political Geography. World Economy, Nation State, and Locality" by P.I. Taylor (London 1985). C.P. Terlouw's recent book (published in autumn of 1992) on "The regional geography of the world-system" is intended as a critical discussion of Wallerstein's work and the pertaining literature. Terlouw concerns himself primarily with one weak point in Wallerstein's concept: The question of the spatial extension of the world system by incorporation of external arenas into its periphery. Otherwise the concept of an early modern world economy has remained to a large extent undebated by historical geographers. Thus, it is the aim of the authors in this volume to open up such a debate.

From a geographical perspective the Braudel/Wallerstein concept of a global system consisting of three parts tends to be too vague. This is especially the case when Wallerstein considers entire nation states as the elementary units of the system. At the same time he claims that the market forces of the world economy basically ignore national boundaries. Economic zones often transcend these boundaries to form larger functioning units within the framework of the world economy. This phenomenon is illustrated by the regions of the grain belt along the coastline of the Baltic states and along the rivers of eastern Central Europe, or the rural textile zone stretching from the Sauerland region of West Germany to

Silesia. Thus, it is important that the world economy is perceived on several different scales, below as well as beyond the boundaries of nation states.

In other words, a more accurate and detailed geographical analysis of the modern world economy is needed, while simultaneously incorporating geographical methods. Such an analysis must include finer distinctions and more accurate differentiations of the different economic and cultural regions that make up the three-zonal world-system.

In addition, the long term dynamics within this spatial system need to be clarified, including changes within the economic and cultural-geographical structures of each region. In particular, attention must be given to individual settlement, population, and social patterns, which develop in response to the specific functions of a region, and the spatial position it occupies within the greater system. A geographical analysis of change should also include the re-evaluation of the regional natural potential. All this, of course, in light of Braudel and Wallerstein's emphasis on the political, social, and commercial framework of their world-system concept.

The interpretation of a region's 'spatial position', meaning the relative and absolute distance of a specific region from the core - the core representing the chief market place of the European world-system -, is incomplete without considering the theory of J.H. von Thünen. Whereas Wallerstein failed to consider the theory at all, Braudel made but brief reference only to overlook its intrinsic value. The social historian viewed the theory as dealing too exclusively with economics.

Already in 1966, however, the German economic historian W. Abel drew attention to the fact that von Thünen's agrarian zones were indeed in existence in the early modern world economy. They were undoubtedly subjected to the principles of spatial organization that von Thünen described, while simultaneously under the influence of other forces at work as well. It is not surprising that von Thünen's observations have proven useful and accurate, for as the owner and manager of a large agrarian estate he was a late contemporary and immediate observer of the pre-industrial world economy. His theory was essentially based upon the costs of transportation to market places, at a time when the factor of 'distance-dependence' was of crucial importance. Employing this foundation as a starting point, von Thünen went on to deduce basic principles that characterized a world economy based upon a regional division of agricultural production.

Thünen's theoretical approach to spatial analysis which has long since been popular among geographers, is in my opinion superior to Wallerstein's concept, when examining the systemic character of the world economy with the evolution of its various functional zones.

A number of the studies in this volume analyze the applicability of von Thünen's principles. His theories have shown that the scale of a region is of great importance. Thus, it was no accident that von Thünen presented his theories through the use of an 'isolated state'.

Consequently, Nitz' contribution focuses on regions that parallel the size of von Thünen's spatial units. These units are continental agricultural zones of the

European world economy, which are centered around the world market metropolis (Antwerp in the 16th century, Amsterdam in the 17th century, and finally London) and other urban centres of the core region. These units represent a scale which corresponds best with von Thünen's premises. On the British Isles, Thünian rings around London were indeed already apparent as early as the 16th century. The same applies to the agrarian zones around the metropolises of northern Italy, as shown by Matzat in his article.

Because von Thünen worked with the homo-oeconomicus concept, his theory is unable to elaborate on the social disparities characteristic of a capitalist economy at the local, regional, and 'world-zone' levels (capitalism had even social repercussions on manorial feudalism). Wallerstein, on the other hand, corresponds a dominant form of social and labour system to each of the three zones of the world economy, and therefore fails to fully depict the socio-geographical reality of the world as well. He narrowly views a region's geographical location and its functional role as the determinant of labour conditions and the social reality. In contrast to this view a number of the authors in this volume, who concern themselves with regional studies, convincingly argue for a more differentiated view of the complex reality of various parts of the world-system.

After critically examining Wallerstein and Braudel's various works, the editor of this volume perceived the need to develop on the unique capacity of a geographical perspective in this context. This was carried out by bringing together a selected group of historical geographers and historians who place emphasis on geographical issues. Their aim was to illustrate the competence of historical geography in the inter-disciplinary analysis of the world-system, in theoretical and methodical approaches, and through regional case studies. In doing so, they attempted to demonstrate the advantages that a geographical perspective has, in connection with aspects of social and economic history, when analyzing regional sectors of the different zones of the world-system. Firstly, it sheds new light on several aspects of the world-system concept. Secondly, a regional approach aids in eliminating over-generalized, incomplete, and therefore inaccurate definitions made by Wallerstein about the main zones. Whereas Wallerstein considers complete nation states as the functional units of a certain zone, the authors here evaluate individual regions within states in order to determine their unique role as parts of a zone in the world economy. Thirdly, and lastly, it is the aim of this volume to make this scientific field more attractive not only to historical, but also economic geographers concerned with the present day world. An analysis of the pre-industrial world economy would benefit the economic geographer due to:

1. The resourceful information gained through the comparison of earlier economic stages with the present.
2. Its insight into long term processes of spatial change, i.e. expansions or contractions.
3. Conclusions that can be drawn from an examination of the effects of lawful processes in the field of economic geography. Thanks to the longevity of early modern capitalism ( which has existed for over three centuries) insights can be

gained into the effects of general spacio-economic laws like those established by von Thünen.

In an attempt to meet these goals and to join the debate on the world-system theory (which has already taken place for more than two decades in the fields of social and economic history), the editor of this volume invited scientists to a 1990 conference in Göttingen-Reinhausen. These scientists had each dealt with the world-system theme - in some cases but indirectly and not explicitly - within the framework of their own particular field of research.

Already in 1986, the editor was given the opportunity to present the topic of the early modern world economy, and its effect on the rural landscapes of Europe, as the focal point of the "Permanent European Conference for the Study of the Rural Landscape" (Nitz, 1987). One third of the authors of this volume already participated in this conference. In addition the editor dealt with the subject by expounding upon his ideas by way of several circulars. Various other personal correspondence led the editor to invite a larger team of competent scientists to prepare papers on specific themes which were intended to systematically cover the subject of the world-system. During a four day conference in 1990, the issues were presented, discussed and finally worked over in anticipation of this publication.

Hence, the present volume is intended as a comprehensive contribution by a team of historical geographers to the interdisciplinary analysis of the early modern world-system.

It is the aim of this introduction to emphasize the relevant conclusions of each of the contributions within the framework of the general subject. The publication follows a systematic order, with the contributions being grouped together as follows:

The three more general articles on the world-system concept critically examine the topic from a geographical perspective. Their aim is not to produce fundamental changes, but instead to further develop the concept of a modern world-system by adding a geographical dimension, thus making it more fruitful for both historical geography and history.

The structure of the main body of this work corresponds with the spatial structure of the world-system: the preliminary articles focus on the core, and are then followed by articles on the semiperiphery (respectively the intermediate zone), then the periphery, and lastly the 'external arenas'. The term external arena applies to those areas which were connected only loosely with the European world-system. The articles present deal with three such regions: The Near East (Ottoman Empire), India, and the Southeast Asian Archipelago.

One of the more general articles on the topic is R. Dodgshon's contribution which emphasizes that if the world economy is to be considered a system, it must have a 'systemic' structure, in which the various zones are interconnected by distinct processes. He is especially concerned with the way in which these processes organize and maintain the world-system in the form of a core-periphery structure. Furthermore, he points out that the spatial and quantitative growth of the world economy since the 17th century have changed the qualitative structure of the world-system, a topic left completely unaddressed by Wallerstein, and one which demands further attention.

It is Dodgshons's opinion that the constraints to the worldwide expansion of the European world economy were only overcome through the rapid urbanization and industrialization of the 19th century, through which a global world-system could be established.

G. Hoekveld sets out to combine the world-system theory with other geographical concepts of space that have long been in use. He begins with a theoretical overview of a number of geographical concepts that have thus far been applied to spatial structures at the local, regional and world economic levels. He emphasizes the idea of socio-economic evolution which is today widely accepted by geographers and forms the basis of Wallerstein and Braudel's concept.

This concept as well as theories adopted from social scientists have in return stimulated geographers to develop specific approaches in order to interpret spatial structures on all levels as the manifestation and result of social developments. Hoekveld, however, argues that a certain theory can only be applied to a certain spatial scale. Hence, the world-system theory applies to the world-system level, while certain regional theories apply to the regional level. Thus he warns against the 'top-down-determinism' implied by the world-system concept. On the local and regional levels the world-system is relevant only through external relations.

The last of the introductory articles, by H.-J. Nitz, draws on J.H. von Thünen's spatial theory which describes the geographical agrarian macrostructure as a subsystem of the economic world-system (other parts were the proto-industrial system and the trade system). Nitz emphasizes the importance of von Thünen's theory within the context of the world-system concept, in spite of the lack of recognition it has received from economic historians, with some exceptions such as W. Abel.

The author briefly outlines von Thünen's ideas and applies his principles and their effects to the social, rural and settlement structures at all spatial levels of the early modern world economy. In a selected section of northern Central Europe Nitz further elaborates on the systemic character of the von Thünen model (Dodgshon) and explores the external interactions that take place at the different spatial levels under study (Hoekveld).

The geographical structure of the European core, which is taken up in the second chapter is of interest concerning two aspects:
1. its internal differentiation into functional subzones, and
2. the regional changes that developed in response to the shift of the core in the late Middle Ages from northern Italy (with Venice as the main city) to the southern Netherlands (around Antwerp) and then to the northern Netherlands (centered around Amsterdam).

As T. Stol points out, it is necessary to make differentiations with respect to functions and intensity between sub-areas even within small regions like the northern Netherlands, which in the seventeenth century clearly belonged to the core. The core proper was located on the coast around Amsterdam, which served as the world trade metropolis. The area was highly urbanized, characterized by a high degree of proto-industrialization, and was equipped with an excellent traffic network of dense channels and waterways. It was surrounded by a first ring of rural areas, which were subjected to innovative impulses and capital investment by trade entrepreneurs who simultaneously expanded the channel network. The result was an increase in the production of, for example, cash-crops such as tobacco, and the exploitation of peat as fuel for the cities.

A second ring within the peripheral interior was made up of areas with less fertile soils. This region was not connected to the core by means of a channel network because that kind of investment was not considered profitable. The region's economy was for the most part based on agricultural subsistence but produced cattle for the core as well, and thus formed a part of the cattle breeding periphery that stretched from Holland, over northern Germany, to Denmark.

The following contributions analyse the social and economic development of northern Italy and the southern Netherlands since the late 16th century. Together with cities in southern Germany and along the Rhine river, they once represented a part of the 'dorsal spine' (Wallerstein) of the European economy in the late Middle Ages. Their succeeding development was characterized by Wallerstein as 'downward semiperipherilization'.

The editor, however, raised the question whether northern Italy could as well be seen as a region on the downturn that has maintained some core characteristics, and retained its functions as an important proto-industrial trade centre at the regional level with its own periphery.

The case of the southern (Spanish) Netherlands follows a similar suit. Once Amsterdam had replaced Antwerp as the world trade metropolis, the northern part of the Netherlands gained in importance whereas the South became the margin of the northwest European core. A careful spatial analysis is presented by A. Verhoeve and P. Saey in order to clarify whether this region is still predominantly characterized by core structures, or whether it ought to be defined as a semiperiphery.

In his analysis of northern Italy, Matzat argues that in the early modern era this region formed the core of its own economic spatial system, with a periphery that extended all the way down to Sicily. A shift in the gravity centre of the European world economy to the northwest undoubtedly caused a depression in its production and trade of urban manufactured goods.

Ciriacono does confirm in his article on northern Italy's Venetian province Wallerstein's hypothesis with respect to the changes in manufacturing. Intense growth becomes apparent in the countryside as the urban economy stagnates. Rural growth is caused by low wages and a rise in population, which not only encourages proto-industrialization but also allows for the establishment of progressive large capitalist estates that operate on wage labour, as shown by Matzat in the case of Lombardy. This last point clearly diverges from Wallerstein's view of a semiperiphery, in that he defines a semiperiphery's agricultural sector as characterized by share-cropping. In the case of northern Italy the traditional system of share-cropping, however, is replaced by wage labour.

This and the evolution of a progressive large-scale capitalist agriculture are according to Wallerstein characteristic of a core region.

Ciriacono does not address the issue of northern Italy's semiperipheral role dependent on forces within the northern European trade centres.

Evaluating the findings of the two authors the editor views early modern northern Italy as occupying a position somewhere between a core and a semiperiphery.

Similar conclusions are presented by Saey and Verhoeve with respect to the second former core sphere, which was reduced to an inferior position, i.e. the southern Netherlands. In contrast to the growth of the dynamic core region in the North, the South's urban sector stagnated during the 17th century, due to a crisis in the world economy. Rural regions in northern Italy, Flanders, and Wallonia were seen to flourish simultaneously. Agrarian sectors as well as regional division of labour patterns underwent further diversification. The result was an intensification of land use and labour accompanied by strong economic growth in these regions. It is debatable whether the entire development is to be perceived as negative only because of the stagnation in larger cities.

It is even more important, however, to consider whether the industrial regions in the southern Netherlands were becoming dependent on trade interests of the North, or whether they were able to remain independent.

In their discussion of the question, Saey and Verhoeve consider whether the transportation costs of agricultural products influenced the structure of rural areas in the southern Netherlands as von Thünen's theory would suggest. They conclude based on their research that this was not the case. Due to the high density of urban centres, and the density of the rural manufacturing population, the farms were located in the immediate neighbourhood of the market places. Transportation costs were therefore minimal. Hence, other local factors such as soil quality and the size of farms played a much more influential role in the development of spatial structures in these rural areas.

Furthermore, Saey and Verhoeve argue that the proto-industrial regions of the southern Netherlands were more advanced than, for example, those of the semiperipheral regions of Germany. This may be assumed from the fact that their proto-industrial products (predominantly textiles) were sold directly from the producer to the local merchants. Germany, on the other hand, operated under the putting-out-system, which was almost entirely dependent on foreign capital.

Because Wallerstein realized that his spatial category of 'semiperiphery' was too vague , he developed sub-categories for regions like northern Italy, and the southern Netherlands, which he referred to as a part of the 'inner semiperiphery' due to their having undergone 'downward semiperipherilization'. Although no longer a part of the core, these regions were nonetheless highly developed in terms of agriculture and industry.

Those regions, on the other hand, that were incorporated into the evolving European world-system in the 16th century, such as northern Central Europe and parts of France, qualify as 'rising semiperipheries' in spite of the fact that they share factors with the 'inner semiperipheries' (such as the existence of a rural industry, or rapid population growth).

These regions, for example interior central Germany, became part of the European world-system when capitalist entrepreneurs from southern Germany and later from England started to develop and exploit them, for instance Saxony, northern Bohemia, and Silesia, in a fashion that economic historians such as Aubin and Kunze (1940) refer to as 'economic colonialism'. Students of rural industry have repeatedly emphasized that the lack of natural potential in these regions was the most important factor for their development, because the road to an economy based on grain exports was closed.

This attempt to distinguish between the declining but still highly developed semiperipheries and those developing and on the rise, can, for the latter, be tested by a comparison with the contributions of Butlin and Pinard who focus on the proto-industrial regions of England and western France.

Wallerstein views the rural proto-industrial regions of England at the beginning of the 16th century as belonging to those semiperipheral regions in the early phases of industrial mass production, which together formed a part of the European world-system. Butlin however disagrees. Instead, his findings show that in the 16th century rural industrial regions were mostly controlled by the national market, and that it was not until the 17th century that international market forces became strong enough to partly integrate the rural manufacturing regions of England.

In order to completely understand, however, the degree to which this was the case, additional comparative regional studies are required by historical-economic geographers. These studies need to cover the quantity and quality of exports from the different regions to the internal market, as well as those between these regions and other European countries.

Furthermore, Butlin's study undermines Wallerstein's method of equating entire states with a certain role in the world-system. Butlin emphasizes the need for a regional approach in the case of England. Through such an approach it would soon become clear, that London and the surrounding regions of southern England, which London economically dominated, formed a part of the northwest European core. Farther outlying regions, with economies based on agriculture and proto-industrial mass production, functioned more as typical semiperipheries or peripheries.

Pinard, on the other hand, agrees with Wallerstein when examining the coastal region of western France, which was obviously a part of the European semiperiphery. The region serves as a good example of the effects of international economic expansion during the 16th and 17th centuries. Rural areas that had previously been largely self-sufficient now became subjected to a strong international influence. This process led to the development of an intense regional division of labour as well as specialization in industries and crafts.

In contrast to the semiperipheral regions in Central Europe, and eastern and southern France, western France had the advantage of being located near the ports of the Atlantic which allowed for direct access to world trade. In addition, imported raw materials from the colonies were converted to finished products in these ports.

Pinard's observations may serve as a challenge for comparative studies of coastal and land-locked semiperipheral regions.

It may be adequate to identify another new type of semiperiphery in the case of the mining regions, which are the subject of Denecke's article. This is an 'extra-zonal' type of region even found on the margins of the peripheries, such as the mining regions of Upper Hungary (Slovakia), the Austrian Alps or Peru. Characteristically they contain highly developed industrial and urban economies with dense populations, and a relatively advanced traffic infrastructure. These factors result in a strong demand for foodstuffs which gives way to the development of a ring of agricultural regions which deliver products to the urban centres. Thus, these metal producing regions have a structural and functional character similar to that of the proto-industrial semiperipheries.

The role of the mining regions in the world economy is difficult to identify. D. Denecke, who is an expert on the historical geography of the metal economy, points out that an economic-geographical analysis only begins with an examination of the production sites, which were operated by a large, skilled work force that was very mobile and worked according to free wage labour. In additon, attention must be given to trade routes, trade centres, and sources of capital ( especially with respect to copper and silver production). After considering all these factors, Denecke arrives at the conclusion that the copper and precious metals sector in the 16th and 17th centuries was an autonomous sub-system of the European world economy. Only a handful of capital centres directed the mines' production, as well as the processing, transport and trade. They initiated the exploration and exploitation of new sites, and set the levels of supply and demand. Special fairs in the large trading cities of the core and semiperiphery functioned as the centres of this distributive network. Wallerstein's spatial terminology therefore provides a somewhat inadequate portrayal of the operations of the highly complex metal economy, and its very specific market network.

The seven contributions on the European periphery focus on one-sidedly specialized, regional economies stretching from Scandinavia to Hungary and - in the northwest- to Ireland. Producing for the world market, these economies are based on a small number of raw, organic products such as grains, cattle, wine, lumber and tar. They produce and rely for the most part on single products, which makes them exceedingly vulnerable in times of economic crisis. The articles attempt to analyze developments in their social, economic, and settlement patterns while taking into account the dynamics of economic and political changes in these resource peripheries.

A. Dunin-Wasowiczowa and H. Sculz examine regions in eastern Central Europe that are dominated by the classic large landed estates. The primary crop produced in these regions was corn for bread, which was later exported to the international market. In the case of Silesia it was used to support its own textile regions.

Focusing on the regions of Pomerania and Silesia, Sculz analyzes the different stages of development that rural villages underwent in the process of estate formation by the feudal lords. Starting with the peasant villages in the Middle Ages, which included a small estate, Sculz traces their transformation to a stage, in which these settlements were dominated by large, single estates, to which a hamlet of dependent day-labourers was attached.

A. Dunin-Wasowiczowa's work for the historical Atlas of Poland has made her quite familiar with the issues of geographical differentiation. In her contribution to this volume, she provides an historical analysis of Poland's economic geography in the 16th and 17th centuries. In contrast to regions of the interior, which are dependent on local production and small-scale exchange, Dunin-Wasowiczowa argues that the dominant form of economic space is the grain exporting regions that extend as belts along navigable rivers, a spatial pattern, which emerged under the limiting conditions of transport costs.

Another such economic region, having been thoroughly examined by Polish scholars, exemplifies perfectly the theory of von Thünen as well. It is the beef cattle zone located in the southeast of the Polish Kingdom (the grand principality of Polish Lithuania, respectively); it stretches out as far as the Ukraina. The forests and forest steppes, which were still in a semi-natural condition at the time, served as pasture land for oxen that were delivered to areas as far as the Rhine river.

The articles by K.E. Frandsen and V. Zimanyi deal with the same type of specialized periphery. Frandsen compares the mast oxen regions of southeastern Europe to a region in Denmark, which was dominated by the breeding of lean oxen. The former had climatic advantages which resulted in sufficient pasture land during all four seasons, and raised a species of a strong steppe-cattle which thrived under the existing natural conditions. Denmark, on the other hand, was located on the northern margin of the continental beef cattle breeding zone and was therefore forced to apply the costly method of stall feeding in winter. This disadvantage, however, was compensated by the fact that the markets of the Rhineland and the Netherlands were located within close proximity (less than 1000 kilometers as opposed to more than 2000 kilometers). In addition, Denmark had the advantage of a regional division of labour in co-operation with the fattening regions located west of the lower Elbe river (see the article by H.-J. Nitz).

Frandsen focuses particularly on the effect on economic-geographical structures once the Danish cattle breeding regions produced predominantly for the West European meat market in the 16th century. He analyzes the development of two royal estates in eastern Denmark, where surrounding peasant villages as part of the manors were forced into a peculiar form of unequal economic symbiosis with the former. They had to breed and care for the cattle for the first four to five years. Then the estates bought them by pre-emption and handled stall feeding in winter until the oxen were sold to the West. Frandsen's analyses are based on very detailed sources, which show the ups and downs of the cattle economy caused by the uncertain conditions of the export market.

V. Zimanyi studies the steppe pasture economy of regions in southeastern Europe. In the Danube-Tisza plain of Hungary and the western regions of southern Romania (Walachia), a 'monoculture' of mast oxen predominated, which allocated Hungary a position in the European world economy. That the country formed a part of the Ottoman Empire in the 16th century did not seem to interfere with its cattle exports to southern Germany and Venice. This underlines what Wallerstein and Braudel point out: the European economy at the time had an international scope indeed.

Of particular interest from a geographical standpoint is the spatial structure and organization of the pasture economy, with its complementing agricultural sector in the expansive Pußta areas of 'giant' townlike villages. The same applies as well for course changes apparent in the driving routes of the oxen, which developed in response to changes in the trade network with Central Europe and Venice.

Zimanyi's findings demonstrate the accuracy of some of von Thünen's deductions. For instance, he argued that areas located far off from market sites are forced to produce goods of relatively high value, such as meat, instead of less valuable ones, such as grain. Thus, the high cost of transportation is compensated through a reduction in the costs of production, for a pastoral economy with large herds is not labour intensive and also based upon rather extensive land use. The production of agrarian products of an extremely high value, however, can also be profitable under a system of more intensive land use. The Tokayer gourmet wines of Hungary, for example, were produced and processed according to labour intensive methods but were delivered to areas as far off as Silesia and Poland.

According to Wallerstein, 'coerced labour' was one of the most important factors in the formation of the European peripheries. Zimanyi, however, disagrees. Instead, she argues that in the case of the oxen and wine economies, wage and family labour were used both on burgher's as well as peasant property.

Hungary, on the other hand, was definitely dominated by coerced labour conditions on the grain producing estates. Because these estates were located far from the western world market, the produce was delivered only to interior markets of the wine-growing, cattle producing, and mining regions. Zimanyi's observations raise questions concerning those factors involved which led to a particular form of labour system. Because grain is a labour intensive product but of relatively low value (as compared with meat or wine), the costs of production needed to be kept at a minimum in order to compensate for the transportation costs over long distances (especially land transport in Hungary). Thus, unpaid coerced labour provided a means for cutting back on costs, and was therefore the dominant form of labour system in the grain periphery. While coerced labour guaranteed a great amount of work days - since six days of work per week were demanded - the quality of labour was relatively low. In contrast to these conditions, only free wage labour could provide the care, expertise and responsibility which goes into wine-growing or the cattle sectors. The same is true for the highly specialized lumber, mining, and glassworks economies, which were partly located in resource peripheries and were operated under wage labour for the same reasons.

Whelan expounds on the role of Ireland in the world economy. Ireland functioned as an economic colony of the English core, with London as the metropolis. In comparison to other peripheries, Ireland's role in the British sector of the world economy was uniquely flexible. Dependent on England's changing economic interests, Ireland was forced into ever new peripheral roles. In the 16th century it delivered live cattle to the towns of the English core, until the Cattle Act put an end to this type of trade. The Cattle Act aimed to protect the producers of the English semiperiphery from Irish competition after they had introduced a new type of convertible husbandry, which resulted in a significant increase of beef production. Thus, Ireland had to switch to the new task of providing salted meats and butter for the British fleet, and the Caribbean plantation colonies.

An economic-geographical interpretation of the spatial structure of the Irish periphery depicts grazing areas around the meat processing export centres, characterized by concentric zones of breeding and fattening. Whelan compares the rural labour conditions with those in eastern continental Europe, where estates were worked on by peasants through means of coerced labour. Ireland paralleled these standards with its own socially and economically degraded population with the exception that its peasant population had made the transformation over to extremely poor cottiers who earned minimal wages.

Wine is a commodity that has hardly been dealt with by historical geographers and historians. T. Unwin is an historical geographer whose expertise covers the wine economy, and he therefore integrates wine into the debate. As early as the Middle Ages wines were traded over long distances, if but in limited quantities, as a luxury good. According to Unwin, the growth of the European economy was accompanied by a boom in the consumption of wines, particularly in the most urbanized parts of the core. A diversity in incomes further led to a branching out of the wine market, with higher and costlier sorts for the well-to-do, and cheaper qualities for the masses. With remarkable improvements in trade networks and traffic conditions, various wine-growing regions were then able to specialize their products, with different sorts of wine aimed at different sectors of the market.

Already in 1953, Karl Heinz Schröder described the pronounced expansion in wine production. The market had opened up to the consumption of wine on mass levels, which could be witnessed by an increase in quantities, as well as in the expansion of wine-growing regions in spatial terms. Some regions specialized in the production of costlier wines of higher quality (for example, Champagne in the Bordeaux region) where skilled wage labour and considerable amounts of bourgeois capital went into the cultivation and processing of wines.

Unwin's study demonstrates that the wine regions further away from market centres tended to produce labour intensive, sweet gourmet varieties intended for wealthy aristocrats and the bourgeois elites of Europe. This applies to Hungary, most parts of the Mediterranean, and the Isles of the south Atlantic (which were then opened up for the first time, such as Madeira and Teneriffa). There once again the question arises whether the distance to market sites is the key factor, which determines the type of wine production employed within a region. Whether this or other criteria would allow for the integration of certain types of wine-growing regions into the zonal model of the world economy, is as yet unaddressed by Unwin.

Von Thünen's 'isolated state' model of a market economy based on the production of natural resources, surprisingly depicts the lumber economy as located near market sites. Von Thünen's calculations were based on the assumption that natural forests no longer existed. In his days - at the beginning of the 19th century - wood could already be produced by advanced methods, in forest plantations. This became a necessity since already in the early modern period the primary forests had been recklessly and short-sightedly exploited for centuries, which shut out the possibilities of sustainable regeneration of the forest resources. Nevertheless the production of wood as a fuel could not expand beyond short distances from market places, since the costs of transportation would not be met by the prices of the end products otherwise.

Of great interest in the context of the early modern international European market is the lumber production intended for building and manufacturing purposes. This included beams, deals, boards, and planks for houses, carriages, carts, boxes, barrels, and most importantly shipbuilding. The shipbuilding industry was the main consumer of tar and pitch as well, which were produced from wood. In addition, potash, which is derived from wooden ashes, was a necessary commodity in the production of other materials, such as glass, soap, gun powder, and cloth dyes. Because the lumber supplies of the densely populated core and semiperipheries could not keep up with demand, the forests of the sparsely populated peripheries were soon integrated into the world economy. Scandinavia, the southern coastal regions of the Baltic Sea, and the catchment areas of Central Europe that were close to rivers were soon drawn upon for their supply of wood products, both semi-finished goods (beams, etc.) and raw materials (trunks). This opening up of resource peripheries in order to meet the demands of countries within the core, was organized by merchant capital, which continuously expanded the regional division of production in the world economy. Dutch merchants were for the most part at the centre of these activities.

The great distances between the sites of production and the core markets resulted in high costs of transportation. Hence, the export of wood could only then be profitable if exploited forest regions were located along streams and rivers, on which rafts could flow down to the North and Baltic Seas from where ships would transport the wood to its final destination.

Drawing on detailed, historical statistical sources, I.G. Layton examines the changing spatial structure of an export economy based on naval stores (tar) and timber. His study focuses on the forest regions of Sweden and Finland (at a time when Finland was part of Sweden). Layton analyzes changes in the long-term spatial structures in response to the dynamics of external demands and of the regions, namely the exhaustion of resources, the growth of demand, and the subsequent rise in prices. Layton further cites the innovation of the 'single frame fine blade sawmill' as an important factor for change in spatial structures. Developed by the Dutch sawmill industry, and later introduced in Sweden by Dutch wood wholesalers, this innovation enabled the lumber industry to push the formerly dominating tar industries into the country's interior.

In addition, Layton argues that the economic policy of the Swedish state had a pronounced impact on spatial dynamics, in an attempt to provide charcoal for its iron and copper industries. These industries dominated northern and central Sweden until the 18th century and prevented tar and sawmill industries from entering these regions. Following a slump in the Swedish metal economy in the 18th century, which was largely caused by the state of the world economy, these conditions were seen to change. Layton thus describes how a resource periphery became integrated into the world economy, and the latter's effects on the temporal dynamics of spatial change in the periphery.

The peripheries of three characteristic sectors of the transatlantic colonies are presented and analyzed in this volume, with Mexico representing the Spanish colonies, the Caribbean Islands as classic plantation colonies, and the Indian interior in the northernmost region of North America as suppliers of furs. While there is no disagreement that the Caribbean and the Indian interior functioned as peripheries, historians still passionately debate the role of the Spanish colonies. Wallerstein and Braudel define them as a part of the world economy's periphery. This is due to the function they served as producers of raw materials, in particular silver from Mexico and Peru. They further argue that large agrarian estates - or 'Haciendas' - can be perceived as the counterparts of the manors in the peripheries of eastern Central Europe. In the colonies, however, they had the task of supporting the metal economy and the towns of the colonial administration.

In contrast to Wallerstein and Braudel, other theories have proposed the idea that the colonial provinces of Spain functioned as outposts under direct control of the royal administration. Still yet is the theory that these colonies formed a 'New Spain' economically independent of the motherland and only loosely connected with the European world economy. Under Wallerstein's defined categories, this last interpretation would make the colonies a part of the 'external arena'. This position is also held by U. Ewald in her discussion of Mexico.

In order to facilitate arriving at a conclusion as to which function the Spanish colonies served, and in anticipation of J.M. Wagstaff and W. Hütteroth's discussion of the similarly complex role of the Near East (the Ottoman Empire and other regions that belong into this category are discussed in the final chapter of this volume), the editor would like to make some general comments on Wallerstein's concept of 'periphery' and 'external arena'. Wallerstein defines them as follows:

> "The periphery of a world-economy is that geographical sector of it wherein production is primarily of lower-ranking goods (that is, goods whose labor is less well rewarded) but which is an integral part of the overall system of the division of labor, because the commodities involved are essential for daily use. The external arena of a world-economy consists of those other world-systems with which a given world-economy has some kind of trade relationship, based

primarily on the exchange of preciosities, what was sometimes called the 'rich
trades'."(Wallerstein I, pp. 302)

According to Wallerstein, the 'external arena' of the Early Modern Age
consists only of:
a. 'world-systems' that are politically homogeneous world empires, such as the
   Ottoman or Russian Empires, or
b. world economies, formed by states interconnected through very intense
   international trade.

In both cases, it is a given that some form of core-periphery structure exists.
In the case of the Asian world economy (Wallerstein) - according to Braudel there
are three Asian world economies: the Chinese, Southeast Asian and Indian world
economies [Braudel, p. 21] - both authors have not proven the existence of a core-
periphery structure. The two last contributions of this volume tackle this topic
from an historical-geographical perspective.

For the discussion of the following articles as well as for future research the
assumption should be made that the European world-system's 'external arena'
could also include states and tribal territories which did not belong to world
empires or world economies. Indirectly Wallerstein also made this assumption in
the case of West Africa, which he considered part of the external arena
(Wallerstein III, p. 187). European trading posts set up within various strong local
states bartered for ivory, gold, malaguette-pepper, and most importantly slaves, in
exchange for their manufactured goods, especially rifles.

The same would also hold true for the North American Indian tribes
(discussed in the article by A.J.Ray) who exchanged furs in return for
manufactured goods offered by European traders. According to Wallerstein,
however, the members of the external arena should have been "resistant (perhaps
culturally?) to importing manufactured goods in return" (Wallerstein III, p. 167),
which is true for the Asian states and empires but obviously not for tribal societies
of Africa and North America. Furthermore, as slaves served as an 'export
commodity', Wallerstein's restriction to 'preciosities' and 'luxuries' is essentially
insufficient. This is particularly the case, when he defines these goods as:

"dispensable at moments of contraction, and consequently not central to the
functioning of the economic system".(Wallerstein I, p. 307)

Slaves, however, were undoubtedly indispensable in order for plantation
peripheries to function as well as in times of recession and crisis in the European
world economy. Furthermore, economic slumps did not curb demands for such
products as pepper from Asia, and furs from Siberia and North America, as
opposed to preciosities such as silk from China, and high quality cotton fabrics
from India.

Thus, Wallerstein considered pepper, for example, as a 'semi-necessity'
(Wallerstein I, p. 333) as it was used in Europe even to preserve meats. In fact,
the habitual consumption of sugar and tobacco among the middle and upper classes
transformed these former luxury goods into 'semi-necessities' as well. Cochenille-
dye and cocoa, on the other hand, remained expensive luxuries from the Spanish
colonies throughout the 16th century.

Here again, Wallerstein has failed to systematically integrate von Thünen's laws into his own theory. Von Thünen has accurately pointed out that only products of high value can be exported to distant markets, and yet still turn a profit although voyages as in the case of Asia often lasted two or more years. Slaves are an example of such products, as they were costly enough to cover high transportation costs overseas. In addition, the distinction between the terms 'semi-necessity' and 'preciosity' is fluid with commodities such as sugar, tobacco, and pepper at the lower end of the scale, while silk, furs, and finer spices were located at the upper end. Thus they cannot indicate whether the regions that produced them were parts of peripheries or external arenas.

One of Wallerstein's insightful observations deals with the effect the world market has on regions that are in close contact with its operations. Once a region functions as a periphery in the world-system, its economic, social, and political structures are seen to undergo basic transformations, which are also observable geographically. This holds true for tribal societies as well, which hunted the commodities that they traded to the world market (furs, slaves) instead of producing them. The external arenas, on the other hand, maintain their autonomous structure since their commercial contact with the world market is much less intense, for example the Russian and Ottoman Empires, which remained relatively independent until the 18th century.

But these Empires did not function in their entirety as either a periphery or an external arena. Certain regions of the Empires, however, played different roles for the European world-system. Russia's northwestern territory, for example, conducted trade with the outside world through rivers that flowed into the Baltic Sea transporting hemp, flax, and lumber for ship building. Thus it formed a part of the Baltic periphery. Starting in 1553, Russia also exported an ever increasing number of furs over the port of Archangelsk.

R. Reynolds argues that the transformation of Siberia into an economic colony of the Russian Empire occurred as a result of the insatiable demands of the European market, which was made possible through the co operation of the merchant family Stroganov with the Cossacks (1967, p. 450). The result was undoubtedly, that the indigenous population underwent similarly drastic changes in their way of life as can be observed among the North American Indians, who were incorporated by the English and French due to demands for valuable furs (as shown in the article by A.J.Ray). Hence, with Siberia becoming the economic periphery of the Russian world empire, as Wallerstein claims, it can as well be argued that Siberia was incorporated at the same time as a 'fur periphery' of the European 'world economy', since a large part of the furs delivered to Russia was later only re-exported to Western Europe. Hence, Siberia can be seen as the Eurasian counterpart of the Indian 'fur periphery' in North America.

This argument, however, can only be reconciled with Wallerstein's concept of the periphery if such hunted commodities such as slaves and furs are to be equated with "cash-crops, agricultural or analogue forms of primary sector production" (Wallerstein III, p. 138). If Wallerstein accepts timber and dye-wood, which are collected from natural forests, why not furs and slaves? Instead these are only accepted at the first stage of transition when an external arena becomes incorporated.

> "Slave raiding took on importance initially when West Africa was in the
> external arena, and continued (even grew in importance) as a mode of
> incorporation". (Wallerstein III, p. 164)

But tribal, hunting, and collecting economies as such could not enter the status of the genuine periphery because

> "incorporation involved the integration of the production sphere into the
> commodity chains of the capitalist world-economy." (Wallerstein III, pp.
> 166).

This is followed up with the development of a new state structure which either evolves independently as a consequence of economic changes, or is set up through European colonialism. In effect, the newly formed structure as part of the "interstate system" ensures the flow of the cash-crops into the capitalist world market (Wallerstein III, p. 189).

From an economic-geographical standpoint, this view of the peripherilization of a region is too limited. It does not account for what actually happened in those regions whose economies were based solely on the delivery of 'resources' (i.e. slaves, furs, or spices) to the European world economy. This undoubtedly led to the restructuring of their political, social, and economic conditions, as well as the pattern of their settlements. In West Africa, for example, it was only through the use of firearms acquired by way of exchanges with European traders, that strong state machineries were eventually set up. These in turn organized the systematic slave raids and trade which caused anarchy and disorder in the areas where slaves were hunted. This development not only caused changes in the political order of the weaker tribes but similarly led to a complete restructuring of their economic and settlement patterns, in which villages were seen to retreat from the open plains to more easily defendable hill sites; that in turn led to changes in land use. As A.J.Ray shows in his article, the evolution of the North American Indian fur trade clearly resulted in the formation of politically dominating tribes, which controlled those of the interior who supplied furs. This development parallels that of West Africa. It therefore may be proposed, that the intensity of such structural and at the same time spatial changes ought to be taken as an indicator for establishing the categories of periphery and external arena; what kind of cash commodity caused the changes seems to matter much less.

These remarks were formulated by the editor during the final revision of this volume, once the contributions had been critically compared with Wallerstein's theory for the last time.

U. Ewald is a geographical historian who specializes on colonial Mexico. In her study on Spanish America, Ewald focuses on the elements involved which enabled a colony like Mexico to establish an autonomous national economy relatively early in its history. She depicts how this development came about by focusing on political and economic-geographical factors. Ewald cites Mexico's size and intense regional differentiation as factors which not only promoted economic independence, but also prevented its full integration as a peripheral zone into the European world economy, and its one-sided, colonial exploitation by a mercantilistic mother country.

Hence, Mexico qualifies merely as an external arena for the European world economy, based upon its geographical and economic structure. Although external trade relations took place, their influence on the economy remained small, which in turn allowed for it to develop according to its own laws.

Mexico is yet another country which parallels von Thünen's spatial model. Distance is the organizing principle, which determines the spatial structures of the intensity of land use as they exist around the country's capital.

But could this status of relative independence also be maintained by the gulf fringe of Central America? U.Ewald does not make this inquiry. According to F.Braudel, G.Sandner (1985), and others, however, the extensive coastal regions of these provinces, alongside their small size, allowed for their rapid exploitation by European powers. Above all it was smuggling by overseas merchants, which opened up these regions as an economic periphery to the European world economy. Smuggled goods are estimated as far exceeding official imports and exports both in terms of production as well as demand. Only those regions which lay far off the coast, and were difficult to penetrate, remained external arenas, as in the case of Mexico, which according to U. Ewald "was to a large degree...not part of the world-system".

D. Watts examines the transatlantic 'periphery proper', focusing his study on the Caribbean plantation zone. According to Watts, the Spanish colonization was characterized by the destruction of the most important resources, and the annihilation of the indigenous labour force. As opposed to those of the English and the French, the economies set up by the Spanish were of a solely exploitative nature, without any attempt made at mobilizing the region's potentials through investment or the setting up of a trade network.

It is still a matter of debate whether even this form of ruthless exploitation - namely, the total destruction of a region's resources - can be seen as characteristic of peripheries. The same can be witnessed in Europe where the primary forests were destroyed for lumber, tar, and potash, with only the devastated heath-land remaining. Those peripheries characterized by fur 'resources' are yet another example of recklessly exploited areas.

The agrarian peripheries demonstrate a different development. The Caribbean, for example, witnessed a successful establishment of a plantation periphery due to an influx of capital (mainly Dutch) and colonizers who produced exotic goods, which were in great demand on the world market. They started out with tobacco, and cotton, and finally sugar, which produced the greatest profits.

Watts attributes the successful development of this agrarian periphery to the innovative nature of the planter societies who were willing to experiment. In addition, the capitalist planters who managed their own estates re-invested a great share of the profits so that new plantations could be founded, thereby expanding the realm of production. During the initial phase of expansion, the core provided the starting-capital for the plantations but did not further interfere with the planter economy except by way of the indirect effect of its growing demand.

The fur periphery underwent an entirely different process, with its resources eventually exhausted due to severe exploitation. The particular spatial structure that evolved in the 17th century as a result of the combined efforts of the North American Indians and the European fur traders is depicted by the Canadian historian A.J.Ray.

For all purposes, these peripheries paralleled those of northern Russia and Siberia. They contained 'primitive' populations characterized by subsistent economies, which underwent significant transformation once they were integrated into the world economy.

According to A.J.Ray, certain tribes along the frontier regions where European trading posts were located, instead of being passive victims of exploitation took on an active role as intermediaries between Indians of the remote interior and the Europeans. They monopolized trade, to a great extent by blocking contacts between the two groups, purchased the furs, and transported them over long distances (up to 1800 kilometers) to the trading posts.

Ray argues that the expansion of this tripartite economic system into the interior of western Canada was propelled mostly by the Indians. Growing competition between French and English fur trading companies was skillfully used by the Indians to their advantage. It was not growing demand in Europe but the Indians' effort to actively increase supply, which extended the commercial frontier westward across half the continent.

Ray further examines the much debated issue of the effect that trade with Europeans had on the traditional barter economy of the Indians, which once warranted them as autonomous and self-sufficient. He argues that the introduction of fire-arms and household appliances resulted in a preference for European goods. Exchange grew steadily in conjunction with the increase in trading posts. The result was the eventual overexploitation of natural resources which still increased the dependency on European traders.

The final segment of this volume covers a sphere which once formed the external arena of Asia. The articles analyze these regions at a time before they had become peripheral zones of the European world economy. Employing their historical-geographical expertise, the authors focus on the structural and functional roles of these regions as external arenas and on the question whether they became incorporated before the industrial age, as Wallerstein and Braudel claim in the case of the Ottoman Empire.

European commodities commanded little demand within the highly developed economic regions of the Asian market. As a maritime trade network already operated, the Europeans had little to offer these regions as woolen fabrics and metal wares had long since been in existence, and Asian silk and cotton proved far superior to European linen. Thus, the Europeans could only pay with minted silver and gold in exchange for Asian luxury goods. Courageous European seafarers made several attempts to intervene in the inter-Asian trade systems, and finally gained some influence on production. European trading companies were indeed successful in establishing direct control in several areas where spices were cultivated and cotton fabrics produced.

I.M. Wagstaff and W.D. Hütteroth are two well-known experts on the historical geography of the Near East. In a critical controversy they contrast the interpretation presented by the Wallerstein group with their own findings concerning the spatial structure of production and the extent to which trade relations existed within the Ottoman Empire.

Wagstaff and Hütteroth outline, for the first time, the economic-geographical structures of a 'world empire', a subject that until now economic historians have unfortunately hardly dealt with. According to Wagstaff and Hütteroth, the dominating metropolis of Istanbul and the surrounding region was once the core of an economically independent empire complemented by additional provincial centres (such as Cairo in Egypt) of regions, which had once been politically and economically autonomous.

If Wallerstein's use of the vague and abstract term 'world empire' is to acquire clarity and accuracy, further examples of world empires within the early modern world will have to be studied to reveal their geographical structures of core and peripheries. The same applies to the 'non-capitalist world economies' in the case of East Asia and the region of the Indian Ocean.

Wagstaff compares trade within the Ottoman Empire (between the provinces and large trade centres) with that which existed between the Empire and Europe. His results demonstrate that before 1800 foreign trade with Europe accounted for no more than 10%-20% of the entire trade. This - in his opinion - rather marginal percentage  contradicts assumptions that as early as the 18th century the Near East fulfilled peripheral functions for the European world economy.

W.D. Hütteroth analyzes a second aspect of the Ottoman Empire, namely the spatial structure of its economy. At the height of its power, around 1600, the Empire was economically independent and of impressive size. For political reasons the central government designed a spatial structure that possessed two core categories and several peripheries of different ranks, which operated under a centralized 'redistributive-tributary' economy. The system lacked a powerful economic elite in the agrarian areas, like, for example, the manorial gentry in eastern Central Europe. The result was that no part of the population was responsive to attempts by outside traders to incorporate these regions into the European resource periphery. Thus, the system was immune to outside influence. Only the traditional exports of luxury goods over the large trading centres were maintained under Ottoman rule, which characterizes the classic situation of an external arena.

Hütteroth also addresses the issue raised by Braudel and Wallerstein whether the Ottoman Empire acquired peripheral functions before the development of industrial capitalism and the rapid expansion of the world economy in the 19th century, in terms of the growth in resource bases and markets. He examines the changes that took place in the structure of the economic geography of the Ottoman Empire at that time.

According to Hütteroth, the most important reason for the late incorporation of the Ottoman Empire was a cultural resistance amongst the indigenous population due to the traditional ideology of Islamic superiority.

The development of the large 'Ciftlik' estates is another economic-geographical aspect, which both authors address. Based on research by the German geographer R. Busch-Zantner (1938), Wallerstein and Braudel perceive the Ciftlik estates as an indicator that the Ottoman Empire had indeed some peripheral functions within the European world economy. They argue that the Ciftlik estates were built up by productive and innovative sectors of the upper classes, who cultivated cotton and grain destined for the external European market on a large scale. Wagstaff and Hütteroth, on the other hand, argue that the role of the Ciftlik-economy as an export sector has always been overestimated.

Ottoman centralized rule proved more and more inept at controlling the margins of the Empire, so that the most prosperous coastal regions and ports, some of which Hütteroth identifies, were then able to export their goods to Europe. An accurate account of the composition, origin, and volume of these exports is unavailable at present due to a lack of research. This is therefore an important challenge to historical geographers who want to contribute to a better understanding of the transformation of parts of the late Ottoman Empire into peripheries of the European world economy.

The last three contributions deal with the spatial structure of the external arena in the East Indian sphere. The Dutch United East Indian Company (VOC) had successfully developed a network of overseas trade in the Southeast Asian Archipelago, in which the Cape Colony served as a replenishment-base half-way between Europe and East India (studied by A.I. Christopher).

In accordance with strategic trade principles the VOC designed a system of fortified trading posts in southern Asia which monopolized commerce (this network is analyzed by W. Jäckel).

B. Murton finally presents an example of the cotton textile regions of India's interior. Their considerable exports connected them with the maritime network of long-distance trade, that had been dominated by European trading powers since the 16th century.

A. Christopher demonstrates, as do other authors of this volume, that a territorial unit cannot be defined as a homogeneous entity with regard to its role in the world economy. He focuses on the example of the Cape Colony, founded by the Dutch at the beginning of the 17th century. As a result of the European trade expansion to India and East Asia, a waystation was needed as a supply-base for the trading vessels of the VOC which was provided by the agricultural region around Cape Town. After 1700 the colony was 'modernized' with the help of the colonial administration. Pronounced spatial expansion was not necessary. In the southwest of the colony the administration introduced a plantation economy based on large landowners. A 'landed gentry', the elite of the rural areas, organized this economy and profited from it. Slaves were brought into the region in order to produce wine, fruit, and wheat. Surplus crops were even exported. This development strengthened the role of the Cape region as an element of the VOC's infrastructure, as part of its 'supply-periphery' so to speak.

Christopher describes in detail how the interior, which originally had not been part of the colony at all, pursued a different path of development with respect to the world economy. It is similar to the squatter-expansion of settler colonies in North America. Due to the growth in population, the peasantry of the Cape Colony was forced to advance into the interior in order to colonize new land. Because the demand of the Cape Town market was limited, and the distance separating the settlers and the market increased, a self-sufficient pasture economy soon developed, composed largely of isolated farms. In addition, travelling merchants organized a small-scale exchange of skins and hides in order to meet the necessities of daily life.

With respect to the world economy, only the immediate neighbourhood of Cape Town with its agrarian supply-sector can be regarded as a functional part of the spatial infrastructure of the VOC. Hence, this region forms a part of the external arena.

The 'superfluous' population, namely the people not needed for the functioning of the supply-base, crossed the border of that functional space, thus leaving the external arena of the world economy. They became comparable to the self-sufficient, indigenous tribes, a fact that is well illustrated by their self-perception as 'Afrikaaner', and by their semi-nomadic way of life. This region was integrated into the rapidly expanding world economy no sooner than in the 19th century.

W. Jäckel's contribution demonstrates the need for more geographical studies on the VOC trading ports and the development of their spatial structures in the East Indian Archipelago.

According to Wallerstein, the incorporation of the external arena into the resource periphery of the world economy occurs automatically as a result of the laws of capitalism. In the long run, profits from productive exploitation (as in the peripheries), rather than speculative trade (as with external arenas), provide the only solid base for success in the capitalist world economy.

Jäckel does not doubt that the description of these processes is empirically correct. He considers it necessary, however, to identify the exact concrete conditions that cause a change in economic strategies within an historically given situation - in this case the strategies of the VOC, other trading groups, and those European states which stood behind them. The main task is to find out what exact mechanism of change brought on this thoroughgoing transformation of the economic structure, which at the same time led to profound spatial changes.

Jäckel's research aims at answering this question, a question of immense importance for the colonial world of the Archipelago. He begins by examining the fundamental difference between the British and the Dutch trading strategies in the external arena. Whereas the former utilize indigenous trading centres alongside traditional forms of barter, the VOC created its own network of trading ports while simultaneously eliminating the indigenous centres of commerce.

Jäckel considers it possible to calculate whether this investment of capital into a new infrastructure of trading ports paid off. The development of the settlement patterns of these trading posts can serve as a geographical indicator of whether such an investment was in the long run profitable or not.

Another issue raised by Jäckel is closely connected with the problem of incorporation mentioned above. The aim of creating a new infrastructure of trading ports was to secure a trade monopoly. Jäckel inquires to what extent this strategy led to the almost unintentional conquest of the hinterlands of the trading ports, that had formerly been indigenous principalities. The conquests were necessary in order to provide control and stability for the commercial centres. The result was regional instead of punctiform colonization. This process in turn led to the creation of the plantation system, the profits of which were needed to cover costs incurred through controlling and maintaining the security of and stability within the colony. This is perhaps the very mechanism that caused the final incorporation of this part of the external arena into the resource periphery of the world economy.

B. Murton's contribution on "Commercial Manufacturing in Southeastern India" is an important counterpart to the economic-geographical analysis of the network of trading centres in the external arena of Asia. It provides a concrete description of the spatial structure of those manufacturing regions of India, that delivered by far the greatest amount of commodities within the international maritime trade between the Persian Gulf and the Southeast Asian Archipelago. The historical documents of Southeast India (Coromandel) have been evaluated thoroughly, which allows for a differentiated examination of the spatial and social organization of its productive and commercial structure. The general conditions were favourable in the 15th and 16th centuries and in some respects very similar to those in Europe. The rapid growth of the population caused internal colonization, particularly in those regions suited best for the cultivation of cotton. Commerce flourished, and starting in the 17th century the principalities developed a fiscally well-organized economic policy, with mercantilistic elements intended to promote the economy.

Murton refers to Salem, twenty days from the coast, as an example of a region whose economy was based on cotton weaving. Salem possessed proto-industrial features similar to those found at the same time in parts of the interior of Europe. Surprisingly, only about 10% of the overall production was exported, and that mostly to southeast Asia. Only after the 17th century did exports to Europe increase. For a time the amount of imports from India even posed a threat to the English textile industry.

In contrast to Braudel and Wallerstein's interpretation, southern Asia cannot be perceived as one continuous world economy. Instead, as Murton argues, there is rather the co-existence of several regional economic systems. The cores of these regions are densely populated coastal areas characterized by intense commerce and by textile industries; yet they lack a dominating, centralized, trade metropolis, nor is there a 'stock market for information'. In short, there is no city with a role similar to that of Amsterdam and London.

Hence, a prominent task for future research in this area is the identification of factors and processes that prevented these 'mini-world-systems' (Murton) to develop into an all-embracing capitalist world economy, in spite of highly developed systems in finance, trade, and industry. These areas were comparable to such advanced European regions as Flanders, and northern Italy in the Middle Ages. Perhaps the distances between these regions were too great for the successful development of trade and communication, or maybe a lack of navigable rivers caused resource peripheries to be inaccessible.

It will be the task of historical geographers to answer some of these questions; for the comparative analysis and evaluation of locations there are valuable geographical methods at hand that ought to be applied to contribute to the interdisciplinary, scientific examination of this highly interesting theme.

This is similarly true for all the other themes covered by the authors of this volume. It is to be hoped that the geographical perspective may contribute to a deeper understanding of the spatial structure and the long-term dynamics of the early modern world-system.

## References

Abel, W. (1966): Agrarkrisen und Agrarkonjukturen. Eine Geschichte der Land- und Ernährungswirtschaft Mitteleuropas seit dem hohen Mittelalter. Stuttgart

Aubin, G. und A.Kunze (1940): Leinenerzeugung und Leinenabsatz im östlichen Mitteldeutschland zur Zeit der Zunftskäufe. Stuttgart

Braudel, F. (1979): Civilisation matérielle, économie et capitalisme, XVe-XIIIe siècle. Le temps du monde. Paris

Busch-Zantner, R. (1938): Agrarverfassung, Gesellschaft und Siedlung Südosteuropas in besonderer Berücksichtigung der Türkenzeit. Leipzig

Nitz, H.-J. (ed.,1987): The Medieval and Early Modern Rural Landscape of Europe under the Impact of the Commercial Economy. Göttingen

Reynolds, R.L. (1967): Europe Emerges. Madison

Sandner, G. (1985): Zentralamerika und der Ferne Karibische Westen. Konjunkturen, Krisen und Konflikte 1503-1984. Stuttgart

Schröder, K.H. (1953): Weinbau und Siedlung in Württemberg. Forschungen zur deutschen Landeskunde 73. Remagen

Terlouw, C.P. (1992): The regional geography of the world system. External arena, periphery, semiperiphery, core. Nederlandse Geografische Studies 144, Utrecht.

Thünen, J.H. (1826/1842/1875): Der Isolierte Staat in Beziehung auf Landwirtschaft und Nationalökonomie. Berlin. New edition based on the 1875 edition, Darmstadt 1966

Wallerstein, I. (1974): The Modern World-System (I): Capitalist Agriculture and the Origins of the European World-Economy in the Sixteenth Century. New York/London

Wallerstein, I. (1980) The Modern World-System (II): Mercantilism and the Consolidation of the European World-Economy, 1600-1750. New York/London

Wallerstein, I. (1989): The Modern World-System (III): The Second Era of Great Expansion of the Capitalist World-Economy, 1730-1840s. New York/London

# Part I    Critical Discussion of the Concept

# 1

# The Early Modern World-System: A Critique of its Inner Dynamics

Robert A. Dodgshon

## 1.    Introduction

Both as an interpretation of historical change since 1450 and as a statement about the ordering of the contemporary capitalist world, Immanuel Wallersteins's world system model is distinguished by its aggregative and synthetic approach, its concern for what he calls "totalities".[1] At issue here is not simply the question of whether the transformation of Europe can be viewed at a larger scale, but whether it has to be if its dynamics are to be properly explicated. For Wallerstein, the task of generalizing upwards to see the problem in terms of the wider global system is a matter of necessity not convenience. Herein lies the essence of his approach and its essential difference with those interpretations of European history since 1450 which stress the separateness of local or national systems and their individual capacity for change through modernization. At a time when post-modernist thinking favours more disaggregated, less all-embracing concepts of interpretation, the success of Wallerstein's argument - its appeal to so-many different types of specialists - has been based on a quite contrary approach, one that bundles the economic, social and political dimensions of Europe's post 1450-development into a single scheme and makes their collective character contingent on the role played by regions within the wider world system.

Though the world-system model can be broken down into its constituent parts, and seen as the experience of particular localities in relation to an emergent "world-system", any rounded debate of its worth and utility must at some point address the debate over its meaning as a "totality" and the core assumptions about how its different parts or roles interacted. I want to direct my comments at four aspects. First, I want to look at how the world-system has been defined in terms of roles and in terms of the processes binding these roles together. I will pay particular attention to establishing whether these processes had needs, thresholds or implications which have so far been understated. Second, I want to look closely at how its constitutive processes organized the world-system around a core-periphery structure. At issue here is more than a review of unequal exchange, or the exploitive relations that are threaded through the various economic, social and political interactions that activate the world system. We need also to specify how these processes of unequal exchange structure themselves, functionally and geographically, into a core-periphery system. We are dealing here with the very essence of Wallerstein's argument. Particular countries or regions may change roles but the roles themselves - core, periphery and their mid-way condition, the so-called semi-periphery - and their ordering of relations through unequal exchange remain fixed. If there is always a core and a periphery, then we are faced with a problem of morphogenesis, of understanding the system's structure-maintaining processes. Third, I want to re-examine how such a system can override political structures, a necessary subjection if there is to be a system-wide allocation of roles. My fourth and final point of argument is that if we take a more strictly-systemic view of the world-system, it has implications for its chronology. Having written three splendidly-argued and richly illustrated volumes on the subject, and numerous essays, Wallerstein has right to claim that this is the aspect on which he has most to say. Yet some see his argument as confronting but not entirely overcoming the inevitable conflicts engendered by a model which relies on both synchronic and diachronic analyses, with his stress on the horizontal space-integrating processes of the world system raising conceptual and interpretative difficulties when we ask how he steers his system chronologically through the events of the late seventeenth, eighteenth and nineteenth centuries.[2]

I want to make it clear that I see these questions as questions about the systemic character of the world-system. I make no apology for this. Wallerstein himself defines his world-system in these terms. Admittedly, he softens the approach by talking about his system in largely qualitative terms and, surprisingly, at no point does he reduce it to diagrammatic form. Yet in every other way, there can be no mistaking the fact that he uses the term "system" in a meaningful rather than figurative way, defining it as a structured system of relations that has roles, interactions, needs and constraints. What I want to argue is that by tightening its systemic definition, particularly in relation to the four questions that I have posed, we are better able to see its strengths and weaknesses as an interpretation.

## 2.    The Systemic Character of a World-System: Its Implications

I have tried to represent the systemic character of a capitalist world-system in three ways. First, we can define it in structural or static terms, that is, as a set of stable, spatially-discrete but functionally inter-linked roles (core, periphery and semi-periphery), each distinguishable through fixed and stable characteristics, the core by trade in manufactured goods, democracies, free labour systems and high wages and the periphery by the converse, trade in primary products (both exhaustible and reproducible resources), totalitarian or authoritarian regimes, coerced labour systems and low wages (see Fig. 1).

Second, we can see it as an operating or functioning system, by defining the various interactions, their constituent flows (goods, capital, information, profit) and the mechanisms which serve to direct them. As flows and mechanisms that give shape and stability to the system, we can expect this aspect of the system to

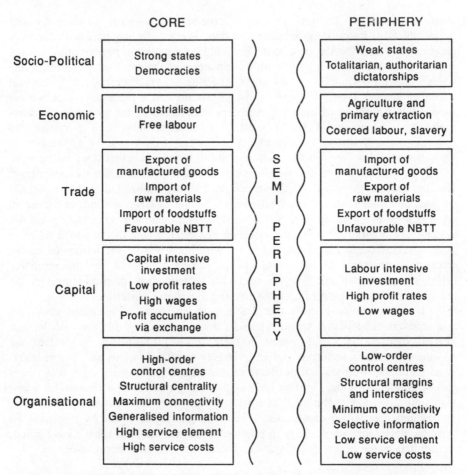

*Fig.1: What should a world-system look like? A static or structural view*

be bound by definite constraints and thresholds and to have been regulated within set limits by negative feedback loops (see Fig. 2).

Third, we can define it as a dynamic system, one that works to continually expand to embrace all resources and opportunities until it becomes a truly global system. At issue here is the question of how we accommodate this expansionary tendency by detailing the positive feedback loops that made it possible.

If by the term world-system we really do mean a system - something greater than the sum of its parts and not simply a loosely-coupled aggregate of parts[3] - then we have to accept that we cannot be said to have defined such a system convincingly until we have commenced the task of not only specifying, but also of quantifying the world-system as an operating or functioning system. Personally, I

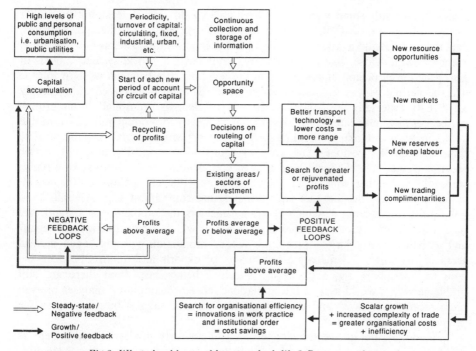

Fig.2: What should a world-system look like? Processes of operation

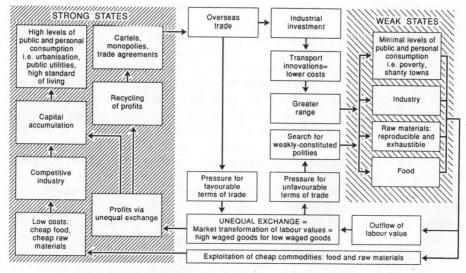

Fig.3: What should a world-system look like? Its dynamics of operation

think we are still some way from being able to do this at least for its earlier phases. Moreover, as O'Brien has demonstrated, bringing together the statistical data that is currently available on early trade points towards conclusions that are at variance with those insisted upon by Wallerstein's world-system model of European development. It suggests that the level of trade between western Europe and the periphery was smallscale before 1640. Though it increased quite sharply after 1640, especially the Atlantic traffic, it still appears "marginal" to the growth of the core even as late as 1830 with the periphery providing only 27% of the core's imports and taking only 14% of its exports.[4] Moreover, at no point prior to this date can the periphery be seen as a critical source of profit, one whose absence would have changed the nature of core development.[5] By way of a reply, we could argue that what is in dispute is simply the chronology of the world-system rather than the fact that one ultimately has come into being. Afterall, 1830 - the most recent point at which O'Brien considers the problem - is a significant date. More than any other, it marks the point at which rapid urbanization began[6] and at which really large, integrated factory production systems began in lead areas of the core[7]. To reanalyse the budget of exports and imports between the core and periphery by, say, 1913, reveals a better case[8], but there are still problems with a world-system interpretation. Indeed, quantitative analyses of even more recent trade flows and prevailing terms of trade suggest these interpretative problems to be persistent.[9]

However, before we dismiss the idea of an early modern world-system as anachronistic, we should probe closer into how Wallerstein constructs the world-system. What matters most to his argument are the generative processes developed around capitalist forms of unequal exchange and their three antinomies: economy/polity, supply/demand and capital/labour.[10] Any structure to the world system, its differentiation between core and periphery, has no independent meaning other than as a projection of these generative processes. In other words, once we establish their operation, we can deduce the existence - latent or actual - of the world-system as a structure. The problem with this approach is that it strips the world-system of any critical parameters or thresholds because its structure is not scale dependent. If such a world-system was simply an aggregate of interacting roles, there would be no difficulty with this approach. But once we talk about the whole being greater than the sum of its parts, we are in effect saying that the way in which core and periphery interacted with each other is more important than intra-actions within each sector. The difficulties which any insistence on this sort of condition raise for the world-system model is highlighted by the extent to which past analyses have concluded that most industrial countries of the core have tended to trade more with each other than with non-industrial countries or the periphery.[11] Clearly, how we phrase the strategic importance of linkages demands very careful specification as to their scale and value. I have indicated the range of possible measures in Tab. 1. The choice on hand underlines just how much specification still needs to be added to the world system model if it is to be seen to work as a system.

Tab. 1: Ways of defining interdependence

a.  51% = Absolute degree of interdependence of all trading flows
b.  Strategic variable = Trade in a particular commodity or raw material critical for either
    importing or exporting country
c.  Structural interdependence = complimentarity, factor substitution
d.  Growth opportunities externalised = Export linkages constitute prime means of growth
    owing to saturation, exhaustion or inadequacy of home/export demand
e.  Conditionality = Trade links contribute to development in a significant or critical way
f.  Connectivity = Linkages exist but their presence/absence does not necessarily affect
    development
g.  Orientation of dependence = Core has some form of dependency on periphery or vice-versa
    or both
h.  Content of relations = Can be defined in terms of either profit, GNP, volume, food, raw
    materials and/or manufactured goods

A further consequence of this need to see the world-system in more systemic terms is that it counters those arguments which try to reduce its formation and operation to a simple cause-effect sequence by prioritizing relations, making one particular variable (eg. class relations) deterministic of the total set of relations. I would not agree that such cause-effect formulations can cope more easily with the essential time element of human interactions. The world-system needs the reciprocation and feedback that comes from a complex, interactive system, one in which particular variables (eg. class relations, the free market) may be critical but are not the only ones capable of changing system behaviour.[12] Arguably, the fact that the world-system is structured around different spaces or roles suggests that the operation of variables, their balance one with another and their capacity for determining system behaviour, needs to be adjusted between these spaces or roles.

If we concede structure to the world-system then we also must concede a degree of homeostasis or goal behaviour. By that I simply mean that there must be control functions built into the world-system, including negative feedback loops, which acted to preserve this structure. For the present, I mention such control functions simply to distinguish them from the positive feedback loops which act to continually expand (and adjust?) the system (see Fig. 3). If we survey the literature on how such positive feedback operated, most analyses root its expansion in the core. The world-system is nothing if not core driven. Attempts to explain how the needs of the core can inflate the world-system emphasize three possibilities. First, increasing levels of urbanization and industrialization are seen as sustainable only if regions of food production can be incorporated into the system, so that any expansion of the core's role necessarily means an expansion of the periphery's role.[13] A slight but crucial variant stresses the need for cheap sources of food if labour costs were to be kept low and profit rates maintained. Second, there is the problem of capital's constant quest to expand itself, to over-accumulate. When faced with a decline in profit rates, or the averaging out of rates, it seeks via technological innovation and/or the exploitation of still cheaper reserves of labour to rejuvenate its profit rates, to renew disequilibrium.[14] Such conditions were more easily realized on the periphery of established systems, so that - seen over time - capital has a tendency to move outwards from its core areas to the periphery. Initially, such shifts mainly involved what I would call the inner periphery, or the many marginal areas of the core.[15] By the nineteenth century though, it had broadened to embrace the periphery of the world-system at large,

providing opportunities for what Harvey has called a "spatial fix" for capital
during times of recession.[16] Third, and the factor favoured by Wallerstein, capital
faced a constant tension between its problems of supply and demand. If levels of
production were to be expanded, then new markets had to be found. For this
reason, capital seeks to constantly expand into the periphery.[17]

        Each of these different forms of positive feedback employs a different
process or dynamic. However, they have this in common: they work to
incorporate peripheral areas whether we define their peripherality relatively
(=inner periphery) or absolutely(=the outer periphery). Their compatibility as
interpretations of how the world-system may intensify and expand is, I think,
easier to understand if we accept that, like any other system, the world-system has
to be seen in terms of its organizational strengths and weaknesses. To put this
another way, what may matter most for change is what we can call "free-floating
resources" or "redundancy", resources that are not yet embedded within a
system.[18] In any organized system, these are more likely to occur on its edges, or
least organized parts. In reply to those who see my point as too mechanistic, I
would reply by saying that the role of the periphery as a source of change is
already well established in European history precisely because it offers greater
scope for change. Some authors, like Rokkan and Skocpol have couched the point
in socio-political terms, seeing the periphery as providing "tactical space" within
which "marginal elites" and others could institute change more easily.[19] Others
see it in terms of economic opportunity. Throughout the late medieval and early
modern period, we can find instances of old established urban centres of
production stagnating and new, innovative rural centres developing on the edges of
the system. To use Olson's words, "the longer an area has had stable freedom of
organization, the more growth-retarding organizations it will accumulate".[20] In
others words, irrespective of any precise input we attribute to economic or socio-
political factors, there may well be organizational constraints at work that direct
change towards peripheral areas, co-opting them into an ever larger system.


## 3.    The "World-System" as a Core - Periphery Structure

        If there is a central, organizing process to the capitalist "world-system then it
lies in the unequal exchange of goods and commodities between the core and
periphery of the world-system". Wallerstein defines unequal exchange along two
axes: the exchange of high waged products for low waged products and the
exchange of industrial or manufactured goods for basic raw materials, including
both industrial raw materials and staple foods like grain.[21] By either definition,
unequal exchange is seen as creating a greater flow of resource and profit from the
periphery to the core than vice versa.[22] In short, it is the operation of unequal
exchange that makes the world-system exploitive. A number of questions follow
on from this stress on unequal exchange. In particular, we need to ask why terms
of trade within a capitalist world-system should be unequal. Is it to do with the
compound value of manufactured goods compared with primary products? Is it to
do with who controlled trade or with the capitalist relations through which trade

was organized? Many discussions make the assumption that once long-distance trade existed, then it necessarily involved unequal exchange as if long-distance trade and exploitive relations were mutualities. The organization needed to overcome the time-space displacements involved certainly admitted a potent role for merchants and exchange value. Yet in the context of the "long sixteenth century" such an assumption needs to be handled with care. The low density and selective nature of early trade routes, plus their very considerable search and capitalization costs, made them highly speculative.[23] Had the risks not been further reduced through the granting of monopolies, many trade routes would have developed much more slowly. The returns on early voyages makes this clear.[24] Moreover, many were only sustained under the political and military protection of dominant states, a social cost which is easily ignored when calculating gross returns for core areas.[25] The sum effect of these factors means that we must be cautious before assuming that profit invariably flowed from the periphery to the core. Indeed, in so far as long distance trade is concerned, some have actually speculated on whether the net flow of capital between the core and periphery in relation to the exchange of primary products and manufactured goods was so clearly in favour of the former during the "long sixteenth century".[26] Clearly, there is a need to establish a more exact budgeting of capital flows at this point. Though by way of a warning, we should appreciate that attempts to establish Net Barter Terms of Trade (=NBTT), the only true measure of inequalities in trade, have acknowledged the severe difficulties facing such an exercise even when using data for later, better documented periods.[27]

For this latter reason alone, we possibly can make more headway by asking why - in principle - we would expect trade to have been based on unequal exchange. If we bundle Wallerstein's own views and those of his critics together, we are faced with arguments that range over the economic, social, and political dimensions of unequal exchange. Wallerstein himself sees it as developed through a combination of, first, class structures and the different forms of labour control fostered by them, second, through the relative strengths and weaknesses of the various state or political systems that make up the world-system and the degree to which these strengths and weaknesses can affect their relations to the world-system, and third, the workings of a burgeoning international market.[28] Whilst not necessarily quibbling with the ingredients that he feeds into the debate, his critics have questioned the way in which he prioritizes these relations. For Wallerstein, the role played by a state within the world-system shapes its internal character both as regards its political structures and its prevalent system of labour control: thus, regions in the core tend towards democratic polities and systems of labour control based on personal freedom, whilst peripheral regions tend towards totalitarian states and coerced systems of labour control. In reply, Wallerstein's critics would argue that the only mechanism by which such roles could be allocated is through the market system. After all, only by seeing the role of the market as determinate can one justify its description as a capitalist world-system. In effect, Wallerstein is seen as offering a global version of Polanyi's argument, seeing society as embedded in the economy and allocative markets as playing an instrumental

role.[29] Such a view though, over-simplifies Wallerstein's argument. By stressing
the distinction between strong and weak states, he is effectively saying that
unequal exchange works itself out through differences in market power, both
between groups and between states.[30] Like frictionless space, he argues, the ideal
of a "free market" is never actually attained, its workings being constantly
thwarted by the self-interest of cartels, monopolies and governments.[31] Even this
qualification, though, would not satisfy critics like R. Brenner, for whom class
relations are paramount rather than accessory.[32]

I do not myself feel that this conflict between a world-system driven by
market forces and one shaped around by prevailing class relations is easily
resolved. However, I would argue that two extra points need to be fed into the
debate. The first concerns the role of the market mechanism. If markets are to be
allocative, then they must be free or self-regulating, able thereby to allocate
resources to those uses or users able or prepared to bid the most. The problem
with invoking the free market as an organizing force behind the world-system is
that it does not operate in the way the structure of the world-system would
suppose. When we actually survey how free or freer markets came into being,
there is a strong case for arguing that they tended to emerge by default rather than
by a formal act of de-regulation. Whatismore, when we ask where they emerged,
we find a different geography to what we might expect. Traditional centres of
trade remained largely regulated throughout the early modern period. In London,
gilds still tried to exercise control over their members down to the early nineteenth
century, long after non-regulated trade had overwhelmed their control of the
market.[33] Likewise, when we look at the expansion of long-distance overseas
trade, we find that - far from being the source of free trade - it was used merely to
extend regulation, with the position of the early trading companies being protected
by a welter of monopolies and regulations.[34] Between these two extremes of
regulation - the local and international - there emerged over the early modern
period patterns of freer trade in what I have called the inner periphery or those
marginal areas which witnessed the growth of rural industry. Set beyond the
geographical limits of old centres of industry, the production systems of these
newer areas can be said to have pioneered freer forms of trade.[35] By comparison,
long-distance international trade did not shift into free trade until the nineteenth
century. Even then, it has been argued that when countries like Britain accepted
the case for free trade, it was not because she had accepted the equity or efficiency
of free markets but because it was assumed that she could now compete just as
effectively in world markets beyond her immediate control(=the informal empire)
as she could in those within her formal Empire.[36] With such a delayed appearance
on the international stage, it is difficult to see how free markets in themselves
could affect the core-periphery structure of the world-system prior to the mid-
nineteenth century.

My second point is that there is much we can learn from the economic
debate over unequal exchange. Quite apart from emphasis on class relations and
market power, we need to take into account the basic differences that exist
between the market for industrial goods and that for primary products and these

differences affect the Net Barter Terms of Trade between countries specializing in such products. The argument was first put by Prebisch. Writing in 1950, he questioned the standard assumption by economists that, over time, we can expect a progressive improvement in NBTT. Instead, he argued, NBTT had declined. His initial explanation for this was that in the manufacturing countries of the core, gains in productivity were distributed through higher incomes whereas in developing countries of the periphery, gains in productivity have been distributed through lower prices.[37] In subsequent work, he tightened the argument further by pointing out how the relatively inelastic demand for food coupled with the presence of cheap labour has tended to manifest itself more in price decreases than in quantity increases. Though developed in relation to the last 100 years or so, Prebisch's argument clearly has implications for earlier periods.

## 4.    Is the World-System an Autonomous Level of Control?

A contentious aspect of Wallerstein's argument is his insistence that the world-system can allocate roles that are charged with socio-political as well as economic meaning. Some writers, such as Skocpol, find this particular suggestion so unpalatable that they feel it sufficient to reject the argument in its entirety.[38] We can hardly temper such criticism by downgrading the importance of the point at issue. If the world-system model is to have any meaning as a system that is greater than the sum of its parts, then it must have force as a system. In other words, there must be some part of its interpretation which explains how this ability to shape its parts operates. In fact, Wallerstein's own position here cannot be said to support a system greater than the sum of its parts. The world-system extends beyond the power of polities, he argues, because no one state is able to control the total system.[39] Yet having freed the system from state control, he reins its potential autonomy back in again by seeing the market power of strong states as instrumental in shaping the system through unequal exchange.[40] This, of course, leaves us with the question of how such a system manages to co-opt weaker states into it. Logically, such co-option works through some form of colonial power - colony relationship (sometimes backed by military coercion) or through the co-option of local ruling elites into the system in return for a share of the benefits, these elites then sanctioning investment and the re-orientation of local economies.[41] Either way, we can hardly talk of the system acting as if it was an autonomous level of control. Rather is it simply a system that serves the interests of capital and elite groups who opt into it, dragging space with them.

However, I want to re-phrase the question and ask whether the world-system exerts an influence largely through what economic geographers would call its externalities. By such externalities, I have in mind savings or benefits attached to the system as a whole and which make it more profitable as an integrated or world-system than a disaggregated one. Obviously, larger systems will offer a greater range of trading complementarities and a greater range of opportunities for buying cheap and selling dear. Over and above such benefits though, I want to suggest that a unified or singular world-system that functions as such offers states

considerable savings in terms of organization notably through savings in the
gathering and handling of information.

I think we must separate out organization or structure, and its capacity for
handling information, from the processes surrounding unequal exchange and their
material embodiment in flows of commodities and capital.[42] There are two
features about the nature of organization that most concern me. First, the
organization of the world-system, the range over which it must collect and process
information, represents its opportunity space. The more the world-system can
enlarge and integrate this opportunity space, the more likely it is to allocate
resources and capital efficiently within it. We need only compare the early
speculative voyages by trading companies like the VOC or English East India
Company with the regular timetabled voyages of the nineteenth century to
appreciate how this enlarged opportunity space mattered. My second point is one
forcibly made by Arrow, the economist. Organization in any production system is
easily taken for granted yet represents an enormously costly investment. These
costs are controlled by the merging of previously independent systems and by
adjustments towards a more linearised and hierarchical system.[43] If, for instance,
we have a system of 100 ports and associated hinterlands, a diffused matrix-like
pattern of trade would involve over 10,000 connections. If we multiply the
number of ports and hinterlands ten-fold to 1000, the number of such connections
increases a 100-fold to a million. Clearly, any incremental growth of the world-
system led to an exponential growth in the demands made on its organization. As
with other sorts of transaction cost, we can expect constant shifts to simply the
problem.[44] In view of this problem, it is hardly surprising that the world-system
has been increasingly organized on a core-periphery basis or that it has been
centered around dominant information-clearing nodes like Amsterdam and
London.[45] Their dominant organizational role within the world-system made them
ideal as transshipment ports, handling far more commodities than they actually
consumed. Likewise, both helped pioneer the joint-stock company, bringing
together the myriad flows of investment capital with information about the
potentially myriad flows of commodities and investment opportunities. The point
which I want to distill from these observations is this: The growing ability of the
world-system to organize and co-ordinate information on a global scale provided
huge savings in transaction costs. Such savings would be a powerful inducement
for even weak states to join in with it, rather than build a rival set of linkages.
Indeed, the considerable costs involved in building the information needs of the
world-system coupled with the gains to be made from centralizing such
information are likely to have been a powerful source of stability for the system,
or negative feedback, just as the need to add to its network of information must
have been one of its more important positive feedback loops.

## 5.   The Chronology of World-System Development

Seeing the world-system in systemic terms has implications for its
chronology. In his first volume of the Modern World-System, Wallerstein

indicated that earlier world-systems had existed. Since its publication, others have expanded on the point, defining them as any socio-economic system that involved dependencies between local and extra-local space.[46] Mindful of this broadening debate, Wallerstein has become careful to stress the individuality of the capitalist world-system as it developed over the long sixteenth century.[47] Though his first volume treats the emergence of the world-system in a qualified way, later discussions are less equivocal: the capitalist world-system being seen as in existence by 1640.[48] We must be clear in our mind how Wallerstein has made an important choice here. The history of the world-system is effectively seen by him in epigenetic terms, with its operational character taking shape quickly so that most of its history - meaning its history since 1640 - has consisted simply of a smallscale system becoming a largescale system, as new areas were absorbed, the adult version being simply a projection of the child. In effect, to restate a point which I made earlier, Wallerstein's world-system is a paradox, a geographical system with no scale dependencies, a system that apparently does not need to adjust organizational in response to its scalar success. The scale at which it operates does not seem to matter.

I want to argue that this sort of view works against its interpretation as a system. For reasons I hinted at a few moments ago, there is a considerable difference between the incipient world-system of ca. 1500 and that of the mid-seventeenth century but there is an even greater magnitude of difference between the mid-seventeenth and late nineteenth century. I know of no analysis of organizational change which does not stress the importance of qualitative change as a primary response to sustained quantitative change.[49] I would add that for any spatial system, the increase of scalar stress is especially acute. By establishing a chronology which does not draw out how the world-system responded to these problems, we weaken the case for its interpretation as a system. As an alternative to the somewhat compressed chronology offered by Wallerstein, we need to consider the possibility that the world-system has come together more slowly. We need to build into our interpretation how it slowly overcame a succession of scalar thresholds, as it passed from, first, a diffused pattern of extended trade to, second, a pattern with more co-ordination and third, to a system whose whole was greater than the sum of its parts. Needless to say, for dating the system, I see this second transition as the more critical. My own personal view is that it is a transition that is easier to explain in relation to the rapid urbanization and industrialization of nineteenth-century Europe than the scalar stresses posed by the "long sixteenth century".

## Notes

1.  See, for example, I. Wallerstein, The Rise and Future Demise of the World Capitalist System: Concepts for Comparative Analysis. Comparative Studies in Society and History 16 (1974), pp.389-91.
2.  Wallerstein is fully aware of the standard criticisms directed at functional or equilibrium analysis and its capacity for change. See, for example, his comment that "the use of any concept is a capturing in fixed form of some continuing pattern" but we need concepts "with all the limitations that any reification, however slight, implies". I. Wallerstein, The Politics of the World- Economy. The States, the Movements and the Civilizations, Cambridge 1984, p.14.
3.  A comparable analysis occurs in R.L. Bach, On the Holism of a World-Systems Perspective. In: T.K. Hopkins and I. Wallerstein (eds.), Processes of the World-System, 111, Political Economy of the World-System Annuals, Beverly Hills 1980, pp.289-310. Note particularly his coment that it "is quite possible and more consistent with the emphasis on a world-system to claim that the whole possesses properties of its own", p.294.
4.  P. O'Brien, European Economic Development: The Contribution of the Periphery. Economic History Review, 2nd series XXXV (1982), p.4; P. Bairoch, Geographical Structure and Trade Balance of European Foreign Trade. Jnl. of European Economic History 111(1974), pp.560-1.
5.  Ibid., pp.4 and 18.
6.  A. F. Weber, The Growth of Cities in the Nineteenth Century. A Study in Statistics, New York 1899.
7.  K. Tribe, Genealogies of Capitalism, London 1981, pp. 114-15.
8.  See the broad-based analysis by Kuznets which looks at changes in international trade from the early/mid nineteenth century down to the mid-twentieth century but which uses 1913 as one of its benchmarks. Its conclusions suggests "old established" countries (=core) strengthened their dependence on foreign trade down to 1913 but the same cannot be said of what he calls newer or young developed countries eg. USA. Overall the trends raise all sorts questions for a world-system model. Thus, whilst manufactured goods formed a high percentage (over 50%) of exports in countries like Britain, Germany, Belgium and France by the mid/late nineteenth century, a country like the USA provides us with this anomaly. Looked at in 1913, it possibly had the highest per capita level of income in the world, yet it exported in value more primary products than manufactured goods! See S. Kuznets, Quantitative Aspects of the Economic Growth of Nations: X. Level and Structure of Foreign Trade: Long-Term Trends. Economic Development and Cultural Change 15 (1967), pp.1-140.
9.  See, for example, comments by R. J. Johnston, The World Trade System. Some Enquiries into its Spatial Structure, London 1976, especially pp.111-20; D.B. Freeman, International Trade, Migration, and Capital Flows. University of Chicago Dept of Geography Research Paper No. 146, Chicago 1973, pp.88-96; Kuznets, op. cit., pp.1-140.
10. Wallerstein, The Capitalist World-Economy, Cambridge 1979, p.272 but see also his reference to two basic dichotomies on p.162.
11. O'Brien, op. cit., p.4; Johnston, op. cit., p.120; Kuznets, op. cit, p.56.
12. In fact, contrary to those who see systems theory as incapable of handling the sequential nature of human activity (eg. D. Gregory, The Ideology of Control: Systems Theory and Geography. Tijdschrift voor Economische en Sociale Geografie LXX1 (1980), pp.327-42), some see its ability to handle the time-space distanciation of processes as one of its merits, see for example, G. Bateson, Mind and Nature, London 1980, p.119.
13. It is worth noting here that when political economists in Britain began talking about broad regional specialization within the world-economy during the 1830s and 40s, the periphery

was seen as the supplier of food. Some justified the export of capital during the nineteenth century because it would lead to cheap food, see G.S.L. Tucker, Progress and Profits in British Economic Thought, Cambridge 1960, p.178. Amongst modern writers, the "ghost acreage" afforded by the Discoveries is well-reviewed in E.L. Jones, The European Miracle, Cambridge 1981, chapter 4. His point, summarized on p.82 is that the Discoveries raised the amount of land available on a per capita basis to the average European from 24 acres to 148 acres. Of course, Wallerstein also stresses the extent to which food is a major commodity flow between the periphery and core, see for example, The Modern World-System. Capitalist Agriculture and the Origins of the European World-Economy in the Sixteenth-Century, New York 1974, p.100. Yet arguably, a country like Britain did not become "dependent" on food imports until the 1860s, see R. Peet, Influences of the British Market on Agriculture and Related Economic Development in Europe before 1860. Transactions of the Institute of British Geographers 56 (1972), p.4.

14.  This of course is the Schumpeterian view, see J. Schumpeter, Business Cycles: A Theoretical, Historical and Statistical Analysis of the Capitalist Process, New York 1939. The need to preserve disequilibrium, to maintain differences, is stressed in A.O. Hirschman, The Strategy of Economic Development, New Haven 1958, p.66.

15.  The penetration of mercantile capital into the marginal areas of north-west Europe, or the livestock-woodland and heath areas, during the sixteenth and seventeenth centuries is now a well- researched theme. Its ecological character is brought out studies in by E.L. Jones, The Agricultural Origins of Industry. Past and Present 39 (1968). pp.58-71; C. Lis and H. Soly, Poverty and Capitalism in Pre-Industrial Europe, Brighton 1979, pp.144-53; J. Thirsk, The Rural Economy of England, London 1984, chaps. X1-X111: F. Mendels, Proto-Industrialization: The First Phase of the Industrialization Process. Jnl. of Economic History 32(1972), pp.247-8; H. Kellenbenz, Rural Industries in the West from the End of the Middle Ages to the Eighteenth century. In: P. Earle (ed.), Essays in European Economic History, Oxford 1974, pp.48-72; P. Kriedte, The Origins, the Agrarian Context and the Conditions in the World Market. In: P. Kriedte, H. Medick and J. Schlumbohm (eds.), Industrialization before Industrialization, Cambridge 1981, esp. pp.13-33; R.A.B. Houston and K. Snell, Proto-Industrialization? Cottage Industry, Social Change and the Industrial Revolution. Historical Jnl 27(1984), pp.473-92; R.A. Butlin, Early Industrialization in Europe: Concepts and Problems. Geographical Jnl. 152(1986), pp.1-8. Further references to the distinct ecology of much proto-industrialization can be found in the volume on Industrialization et Desindustrialization, Annales. E.S.C. 39 (1984), no.5. especially the papers by A. Dewerpe, pp.896-914 and F. Mendels, pp.977-1008.

16.  D. Harvey, The Spatial Fix - Hegel, Von Thünen and Marx. Antipode 13 (1981), pp.1-12.

17.  Wallerstein, Capitalist World Economy, pp.159-60 and 278. Each "wave of expansion has revitalized demand", ibid., p.278. See also, his The Modern World-System 111. The Second Era of Great Expansion of the Capitalist World-Economy 1730-1840s, New York 1989, pp.57-126 which discusses how Britain gained the upper hand in world markets 1763-1815.

18.  G. Bateson, Steps to an Ecology of Mind: Collected Essays in Anthropology, Psychiatry, Evolution and Epistomology, London 1972, pp.285 and 511; S.N. Eisenstadt, The Political Systems of Empires, New York 1963, p.317. Compare Wallerstein's use of "tactical space" when he talks about change in the "interstices of the world-system", in Politics of the World-Economy, p.56.

19.  S. Rokkan, Dimensions of State Formation and Nation-Building: A Possible Paradigm for Research on Variations within Europe. In: C. Tilly (ed.), The Formation of National States in Europe, Princeton 1975, p.596; T. Skocpol, France, Russia and China: A Structural Analysis of Social Revolutions. Comparative Studies in Society and History 18(1976), p.202.

20.    M. Olson, The Rise and Decline of Nations, New Haven 1982, pp.71-3.

21.    Wallerstein, The Modern World-System, vol. 1.

22.    Eg. Wallerstein, Politics of the World-Economy, p.15.

23.    Some of these points are well-made in D. Mackay, In the Wake of Cook. Exploration, Science and Empire, 1780-1801, New York 1985. As regards capitalisation, the first ten voyages of the English East India Company averaged almost 50,000 £. See K. N. Chaudhuri, The English East India Company. London 1965, p.209.

24.    According to ibid., p.212, the early voyages gave returns of 24 1/2% if one takes into account how long stock ran for, at a time when interest rates in London were around 8-10%, but compare this with the general conclusion reached by O'Brien, op. cit, pp.7-9 and the general point made by Davis that whilst the English India Company and the Vereenigde Oost-Indische Compagnie or Dutch East India Company (=VOC) were particularly profitable others, like the West Indische Compagnie or Dutch West India Company, the English African Companies, the Spanish Caracas Company, notoriously the Scottish Darien Scheme tell a different story, R. Davis, The Rise of the Atlantic Economies, London 1973, p.238.

25.    This was true of both the Dutch and English trading companies. See, for example, I. Wallerstein, Dutch Hegemony in the Seventeenth-Century World-Economy. In: M. Amard(ed.), Dutch Capitalism and World Capitalism. Capitalisme Hollandais et Capitalisme Mondiale, Cambridge 1982, p.117.

26.    R. Brenner, The Origins of Capitalist Development: A Critique of Neo-Smithian Marxism. New Left Review 104(1977), p.84.

27.    R. Ballance, Trade Performance as an Indicator of Comparative Advantage. In: D. Greenaway(ed.), Economic Development and International Trade, London 1988, p.7.

28.    See, for example, Wallerstein, Capitalist World-Economy, p.149.

29.    For a statement of Polany's views, see K. Polanyi, The Great Transformation, Boston 1957.

30.    Eg. Wallerstein, Rise and Demise of the World Capitalist System, p.401.

31.    Wallerstein, Capitalist World-Economy, p.149.

32.    Brenner, Origins of Capitalist Development, pp.25-92 esp. p.39.

33.    J.R. Kellett, The Breakdown of Gild and Corporation Control over the Handicraft and Retail Trade in London. Economic History Review, 2nd series X (1958), pp.381-94.

34.    For exemplification, see E. Coorneart, European Economic Institutions and the New World: The Chartered Companies. In: E.E. Rich and C.H. Wilson (eds.), Cambridge Economic History of Europe, IV, Europe in the Sixteenth and Seventeenth Centuries, Cambridge 1967, pp.223-74 and C.H. Wilson, Trade, Society and State, ibid., pp.487-575.

35.    See, for example, H. Kisch, Growth Deterrents of a Medieval Heritage: the Aachen-area Woollen Trades before 1790. Jnl. of Economic History XXIV (1964), pp.517-37; Kriedte, op. cit., p.22; J. Langton and G. Hoppe, Town and Country in the Development of Early Modern Europe. Historical Geography Research Group, Norwich 1983. Wallerstein makes similar points in his Capitalist World-Economy, p.46.

36.    J. Gallagher and R. Robinson, The Imperialism of Free Trade. Economic History Review 2nd series VI(1953), pp.1-15; P.J. Cain, Economic Foundations of British Overseas Expansion, London 1980, p.11.

37.    H. Singer, The Distribution of Gains between Investing and Borrowing Countries. American Economic Review 40(1950), pp.473-85; H. Singer, The Distribution of Gains from Trade Revisited. Jnl. of Development Studies 11(1974-5), pp.376-82. See also, J. Spraos, The Inequalising Trade? A Study of Traditional North/South Specialization in the Context of Terms of Trade Concept, Oxford 1983 and D. Sapsford, The Debate of Trends in Terms of Trade. In: Greenaway (ed.), op. cit., pp.117-131.

38.    T. Skocpol, Wallerstein's World Capitalist System: A Theoretical and Historical Critique. American Jnl. of Sociology 82 (1977), pp.1075-90.

39.  Wallerstein, Capitalist World-Economy, p.275; Future of the World-Economy, p.174; and Politics of the World-Economy, pp.13-14.

40.  Wallerstein, Rise and Demise of the Capitalist World-system, p.401. Yet compare his statement that the "major social institutions of the capitalist world-economy - the states, the classes, the peoples, and the households - are all shaped (even created) by the ongoing workings of the world-economy", Politics of the World Economy, p.14.

41.  Examples of "incorporation" can be seen in I. Wallerstein, The Modern World-System II: Mercantilism and the Consolidation of the European World-Economy, 1600-1750, New York 1980, p.47-8. But note, in his Modern World-System III, p.129 Wallerstein writes "Incorporation into the capitalist world-economy was never at the initiative of those being incorporated".

42.  Compare Bach, op. cit., p.296.

43.  K. Arrow, The Limits of Organization, New York 1974, p.54. See also K. Flannery, The Cultural Evolution of Civilizations. In: P.J. Richardson and J. McEvoy (eds.), Human Ecology: An Environmental Approach, North Sciutate 1976, pp.96-118; K.W. Deutsch, On Social Communication and the Metropolis. In: A.G. Smith (ed.), Communication and Culture, New York 1966, pp.386-96. Compare the comments by Wallerstein in The Modern World-system II: Mercantilism and the Consolidation of the European World-Economy 1600-1750, New York 1980, pp.55-56. He is certainly conscious of the inescapable fact that the costs generated by a world-systems' organization are high and that these costs have to be serviced by those most directly involved in its organization.

44.  D. North, Structure and Change in Economic History, New York 1981, p.31. For an explicit example of how innovations in organization can confer gains, see N. Steensgaard, The Dutch East India Company as an Institutional Innovation. In: M. Aymard (ed.), op. cit., pp.235-57.

45.  For a review of Amsterdam as a centre of information processing, see W.S. Smith, The Function of Commercial Centers in the Modernization of European Capitalism: Amsterdam as an Information Exchange in the Seventeenth Century. Jnl. of Economic History XLIV (1984), pp.985-1005.

46.  Wallerstein, The Modern World-System I, p.348. For a review of earlier systems, see K. Ekholm, On the Limitations of Civilization: the Structure and Dynamics of Global Systems. Dialectical Anthropology 5 (1980), pp.155-66.

47.  Wallerstein, Capitalist World-Economy, pp.37 and 158-60.

48.  Ibid, p.37 and 166.

49.  See, for example, R. L. Carneiro, A Reappraisal of the Roles of Technology and Organization in the Origin of Civilisation". American Antiquity 39 pp.179-86; B.A. Segraves, Central Elements in the Construction of a General Theory of the Evolution of Societal Complexity. In: C. Renfrew, M.J. Rowlands and B.A. Segraves (eds.), Theory and Explanation in Archaeology, New York 1982, pp.287-300.

# 2

# World-System Theory: Implications for Historical and Regional Geography

## Gerard A. Hoekveld

Pour nous, l'oekoumène reste, comme pour les Anciens la terre habitée, mais
la terre habitée avec ses annexes et ses marges, l'air d'extension du genre
humain, qui tend à se confondre avec la surface du globe.
Max. Sorre, 1952, p. 438.

## 1.    Introduction

World-system theories matter in both regional and historical geography.
Taylor (1988, p.262) claims that, in contrast to the traditional geographer's
'natural regions', "the world-systems analyst's regions are historical, they are
created, exist for some period of time and come to an end". According to Smith
(1989, p.231) a world-system approach offers "the possibility of grasping the
simultaneity of global and regional processes operating within an historically
unfolding framework". During the phase of 'spatial analysis', geographers lost
sight of the interrelationship between historical processes and their arealization that
formed the basis of much geographical work, particularly at a macro-scale. World-
system theory brings this nearly lost dimension of geography back to the center.
That should be applauded, particularly by historical and regional geographers, but
they also ought to be aware that it has been brought back at a price.

This paper is an attempt to assess the value of its come-back. First, some
comparisons are made between a couple of traditional geographical world views
and that of the modern world-system theory, as framed by Braudel, Wallerstein and
others. Then consideration is given to some of the methodological limitations of
the regional and historical geographer's point of view. Finally, some conclusions
are drawn on how geographers should work with the world-system theory.

## 2.    The World-System is Part of the Traditional Geographical Heritage

This statement could be illustrated by many examples; a few salient ones will
suffice for our purposes. In the nineteenth century, founding fathers of our
discipline in Germany and France developed a world view based on the bond
between history and geography. Often they even intended to emancipate
geography from its status of history's servant (Ratzel 1909, pp.54-55). "Solchen
Auffassungen gegenüber ist mit Entschiedenheit zu betonen, daß die Geographie
zunächst die Erforschung und Beschreibung ohne Rücksicht auf Menschliches und

Geschichtliches zur Aufgabe hat und daß die selbständige Lösung dieser Aufgabe voran zu gehen hat der gemeinsamen Arbeit mit der Geschichte auf anthropogeographischen Felde". But Ratzel continued immediately: "Beide sind freilich unzertrennlich". These forefathers also had a world-wide view; Ratzel, for instance, said that world history should be world-encompassing (1909, p.55): "Die Weltgeschichte muß erdumfassend sein". In that respect geographers had two leading concepts: "humanity" and "ecumene". The notion of humanity had lost its normative aspect in geography. It came to designate the world population, as differentiated according to its races, peoples and (levels, stages or other types of) cultures. These cultures were often described in ethnographic terms, but these descriptions were always organized by the way a material subsistence was wrung from the ecological environment. Ecumene, the inhabited earth, was studied from a holistic viewpoint, or, as Ratzel phrased it, "die hologische Erdansicht". Partly, this holism was expressed in the study of the intermingling of the forces of nature and human self-actualization in a certain region ("vertical relationships"); in part, it was expressed in "horizontal relationships", or interactions over the earth's surface. To cite Ratzel again, his concept of situation ("Lage") was circumscribed as the size and form of an area. This implies that a situation consists of areally distributed characteristics of phenomena on, or of, the surface of the earth; it also implies interaction, or interrelationships, between human groups (Ratzel 1909, p.137). Ratzel thought that "Lage" was a determinant of "Raum" (the site) in being its spatial context. The ecumene was seen by German and French geographers alike as the setting of these contexts.

The opposition between land and sea in the geographer's view of the world was very important, at least since Carl Ritter's time. Vallaux (1908, Part I, p.6) even spoke of an "oecumène maritime" and believed that "civilization went from the fluvial phase to the Mediterranean phase and then to the oceanic phase" (Vallaux 1908, p.415). He distinguished even a class of "mediterraneans", a.o. the Caribbean sea, the Indonesian seas and the Mediterranean (1908, p.5). That opposition, present in all major works about general geography of the time, dominated political geography (Mackinder, 1904) and work in economic geography of commerce and transport. For instance, Brunhes and Vallaux (1921, p.493) pointed out that the sea governs Europe's history. F. von Richthofen wanted to base a geography of trade and commerce on causal and genetic, or historical, methods (1883, p.63). He viewed the world as a mosaic of regions ("Siedlungen") where different peoples exploit the natural resources under the natural conditions to which they were forced to adapt, producing different products that are exchanged by commerce via traffic. He suggested the formation of a world-economy in the seventeenth century (von Richthofen 1908, p.203, p.279). The economic geographer Eckert (1905, p.161) suggested that the development of modern means of transport has brought together the parts of the world that are equipped with different gifts of nature. A general completion has become possible, just like economic satisfaction of the demand of the globe. Eckert (1905, p.172) divided the formation of the world-economy in three phases: "eine terrestrisch-littorale Periode, die ozeanische Periode und die ozeanischterrestrische Periode". In his book, a traditional geography of goods ("Warengeographie"), he offered an economic geography of the world-economy, world trade and world traffic (Weltwirtschaft, Welthandel und Weltverkehr). (Implicitly he used the economic theoryof comparative (production cost) advantages). In 1930 Dieterich and Leiter (1930, p.259) concluded that the world had become one market:

> "In the time of steam navigation, the railways, and industry based on the steam engine, the Europeanization forced its way into the hearts of the other continents; the mass emigration and the exchange of mass produced goods began; the economic and, recently, spiritual connections with Europe became tight. World commerce, world economy, world culture and world politics are terms coined in the past few decades."(translation G.A.H.)

The world-system of the economic geographers was thus based on a world-wide division of labour maintained by trade; resulting in a seaborne process of world-Europeanization. The word "Europeanization" was used by Ratzel (Anthropogeographie, Teil II, 1912, p.191) in the sense of demographic expansion and dispersion of Europeans over the globe. Although they did not use the concept of capitalism, they were interested in the functioning of monetary systems and the flow of international capital (Eckert 1905, pp.168-172, Demangeon 1920, Brunhes & Vallaux 1921, p.684). They were completely aware of the oppositions in that system between developed and underdeveloped countries. Dieterich and Leiter (1930, p.79) speak about the major antagonism between countries producing raw materials, enforced by coercive economic bonds with the consumer countries, which is a consequence of economic imperialism. The same analysis, even the same wording, can be found in Demangeon (1920, 1932) and Brunhes & Vallaux (1921, p.684), and also of the different situations of center and periphery. However, they were not the first, for Ratzel (1909, p.141) had already elaborated the relationships between center and periphery. He saw them as a result of the "Lage" with respect to the seas and continents.

The main traditional geographical views of the world-system may be summarized as follows:
1) as a global system of the interrelated realms of nature, inclusive of humanity (an idea picked up again by the Club of Rome);
2) as a changing mosaic of peoples;
3) as a rather stable mosaic of regions or landscapes (primarily defined by "internal vertical relations");
4) as a system based on diffusion, sequence of occupancy, and interactions of peoples with specific cultures and ways of life;
5) as a political interactive system (of states);
6) as a commercial interactive system.
(The last two systems, 5 and 6, were often considered interdependent, as Müller-Wille (1966) has shown.)

Each of these six traditional ways of viewing the world has its modern version(s). For example, the first category has been revived by the Club of Rome and by E. Laszlo (1978), who has compiled essays on different types of world-systems. For world-system approaches in political geography see Van der Wusten (1987). One may conclude that all the ingredients of the modern world-system theory of Braudel, Wallerstein and others were already present (division of labour and dependency; opposition between the West and the non-Europeanized countries; center-periphery ideas; the historical evolution of a world-system via European dominated maritime commerce, i.e., the capitalistic economy. Why, then, did the geographers not assemble these bits and pieces into a world-system theory as Wallerstein and some others did? The answer may be that the development of geography as a discipline with a clear identity (always, of course, recognizing, that this identity was different in geography's various "schools" and in successive periods) weakened its bonds with other disciplines studying society.

The emancipation from history has already been mentioned; we now know its high price. The separate development of geography and sociology is another consequence of the formation of geographical schools. In Germany, important factors leading to the isolation of the discipline may include deterministic trends in the pre-war period, landscape geography as elaborated by the "Schlüter school", and the fact that German geographers remained aloof from the *Methodenstreit*.

In France, the "Vidal-Durkheim debate" at the beginning of the century may have had the same effect, notwithstanding cooperation in the "Annales". "Durkheim had a unitary view of the social sciences, with sociology synthesizing the elements of the social sciences" (Berdoulay 1978, p.78).

> "Thus as opposed to the Durkheimians, who tried to establish rigorous concepts that they thought were the only ones capable of capturing the reality of the subject matter at hand, the Vidalians were imbued with the idea of the relativity and the conventional character of concepts and theories. These were viewed as heuristic devices to approach the study of spatial relationships between phenomena" (Berdoulay 1978, p.84).

After the period of 'spatial analysis', marxist geographers re-established the link from geography to history and the social sciences within their own ideological framework. This closed a period of mutual neglect. We will not go into the current process whereby marxist geographers are disengaging themselves from such ideological rigidities as economism and class struggle. (See the discussion on "reconsidering social theory" in the journal Society and Space ,1987, 5.) Harvey still takes an orthodox point of view, but most others do not any longer.

Eisenstadt and Shachar (1987, p.67) state that systems that are basic to human life are open. "No human population is confined within any single such system, but rather in a multiplicity of only partly coalescing organizations, collectivities and systems. Such systems and the division of labour entailed are constructed by special social actors, by different carriers; ideological, power and material components are always interwoven in the process of such construction.

The pluralism in geography occurring after the period of spatial analysis, and leading to neo-positivist, humanistic and neo-marxist geography, seems legitimated by the post-modernistic geography (B. Becker, 1990, Cooke 1990) although that pluralism is still of a high level of abstraction. Therefore it is difficult to build bridges between post-modernistic theoretical reflection on

geography and the practise of geographical research. The fact, however, that
human geography considers itself now to be a social science, irrespective of this
pluralism, confronts us with the methodological basis of the social sciences and
requires us to rethink the geographical use of theories. This has to do with the
specific methodological characteristics of geography and with the specific ways in
which geographers attach relevance to the historical evolution of society. We
should consider this last aspect first.


## 3.    The Relevance of Historical Evolution in Geography

Nitz (1984) demonstrated the power of a historical-social science approach
for the study of settlement geography. Newson (1976, p.239) identified cultural
evolution as a basic concept for human and historical geography. It is "a valuable
mode of explanation in the analysis of cultures and cultural change and since the
presence of a culture implies an area, this applies to spatial patterns too". It is also
a key concept because it offers "a much needed holistic framework for human
geography". Dodghson (1987, p.xx) formulated it concisely: "the principles which
govern the organization of human spatial order have changed in step with the
substantive or qualitative changes that have occurred in the organization of society
itself". Friedman and Rowlands (1977, p.241) state that there is "a multi-linear
evolutionary trajectory in which variant pathways are generated by the constraints
imposed by particular local conditions". Newson surveyed an array of
evolutionary schemes and typologies of stages used by geographers (e.g. that well
known one of Bobek). Friedman and Rowlands suggest, however, "epigenetic"
models, models that depict "the structure of processes and not of institutions".
They name, for example, the tribal system (strongly dominated by ritual
interaction and positioning), "prestige good systems", the "Asiatic State" (strongly
dominated by a class structure leaning on a ritual center and external exchange),
"urban-commercial states" and "the world-system". A recent example of a stadial
concept in modern regional development on a global and regional scale is the
study of Gritsai and Treivish (1990, p.66). These authors use the four Kondratieff
long cycles and combine them with a "core and (semi) periphery" typology of
regions.

There seems to be a strong tendency to work along three lines. One line is
the study of the evolution within cultures, or cultural realms, allowing for the
specificity of the historical trajectory of that particular culture, e.g., the European
or the Japanese evolution. The development of communication (in a technical
sense as well as in the sense of ideas, norms and values) is an essential part of this
evolution. A second line is the study of the interactions between these cultures.
Migrations and sequential occupancy, diffusion of innovations, assimilation and,
finally,"Europeanization and contra-movements" form the basis of this line. A
third line is to investigate the fundamental structures of processes within these
different developments. These structures might be described in terms of

generalizations. Braudel's and Wallerstein's ideas about the world-system fit into this line.

There are at least two areas of agreement on these structures. The first is that distributional and market relations govern exchange, which is a prerequisite for existence of every cohabitational group. This idea has a long history in economics and ethnology. It has been superbly elaborated in recent years and synthesized by Polanyi and, especially, Wallerstein, who elucidated that market relations culminate in capitalism conquering the world (see also Dodgshon 1987, pp.193). The second area of agreement is that there are centers and peripheries. The best example of this reasoning is given by Ekholm, who believes that the relation between center and periphery is a structural continuity between the systems of all epochs and cultures. (There is no geographical continuity, however, in the distribution between the center-periphery systems in the different periods.) Ekholm describes a global system as being composed of two hierarchical levels. The lower level he calls the "social formation", or "the production and local social structure". The higher level is the set of external relations linking such local sub-units. "The logical interdependence of the two levels should be evident: Production can only be local, exchange can only take place between local production processes" (Ekholm, 1981, p.246). Archaeologists and anthropologists have shown that local production has been organized into larger economic networks since the first civilizations. Thus geographical interdependency has been a crucial factor (p.251). Ekholm stated that the control over these external relations, as a resource, is exercised by "the centre" that has access to it. This is in direct contradiction to "a periphery" that has, by definition, no access. Development should therefore affect both levels, the "local and the external relations". The centre-periphery concept has become a completely accepted tool, not only in theorizing about societal evolution and diffusion, but also in planning and regional development theory. It has become a concept that indicates a form of societal integration expressing itself also in areal complementarity. At the same time as the centre has mostly been operationalized by cities, it has forged a link between the conceptualization of societal integration and urbanization.

Dodgshon (1987, p.361) has brought together various strands of thinking about general cultural evolution; societal integration; the growth of societal complexity and the concomitant change and control; and, finally, space. These are integrated in his theory of societal change from without: "the early stages of societal evolution can be seen in terms of whole systems leap-frogging each other toward greater complexity. With the rise of complex hierarchical world-systems we approach the possibility of a single, linear course of development at a system level. Once we couple the evolutionary implications of the drift towards a single, incorporated system with the radical changes which have taken place in the bases of societal integration, then clearly, we are dealing with a form of evolution in which some of the vital processes have changed both qualitatively and quantitatively." Dodgshon believes that the freedom to adapt by changes in societal integration is greater in the peripheries than in the core.

The discussions referred to above emphasize explaining spatial order in a changing society. The integration of society is propelled by changes in local production and external relations (Ekholm), and by increasing societal complexity and scale (Newson, Dodgshon). In other words, traditional geographers saw spatial order as a product of ecological adaptation of societies within the framework of areally complementary production systems. Modern geographers think that spatial order arises from an interaction of certain processes of societal differentiation and specialization within frameworks of external relations organized and controlled by centers. For geographers, this means that 'spatial order' must be conceptualized in such a way that the 'mechanisms' of societal change can be phrased in terms of explanatory concepts and theories. These theories should explain the changing spatial distribution of phenomena or changing areal characteristics of the territories under study.

At present there are several options. The first is based on internally generated change in the controlling centres (Eisenstadt & Shachar 1987, p.67). The second is based on externally generated change, either from an expanded system, culminating in a poly-centered world-system, or from a periphery that is adapting itself to changed environments. The third is based on a combination of the first two possibilities. The second possibility in particular is combined with theories of urbanization and of the formation of the (nation-) state and world-system. But these very theories and the ensuing empirical research have shown that urbanization, state formation, and world-system development are long-lived processes, changing in form and comprehensiveness. Therefore these theories have a partial impact and a more or less restricted ability to explain the spatial order. This leads to a dilemma. When the geographer concentrates on the mechanisms or agents of change, he loses sight of autonomous, often stable systems, and the explanations of their behavior. When concentrating on equilibrium and its explanations, the geographer loses sight of the processes of change. In both cases his explanations are only partial.

## 4.    The Limitations and Strength of Geographical Methodology

As shown above, spatial order may be perceived in terms of societal evolution, be that linear, bifurcated, or anagenetic. Does this viewpoint force geographers to adopt a specific level of analysis? Does it necessarily lead to 'the most appropriate spatial framework' to deploy the mechanism of societal integration, namely the world-system? Wallerstein, Hopkins, Bergesen and other adherents of the world-system theory would answer this question in the affirmative. Geographers would deny it. Geographers may concede that the world-system theory offers the right way to study spatial order. Yet they should not be committed to analyze the world's differentiation at the units of analysis pertaining to that theory: states; the three combinations of states in core, semi-periphery, and periphery; and the within-state centres and peripheries (Wallerstein 1980, 1988, Wallerstein & Hopkins 1980). That would force geography back into the mold of organistic essentialism. This approach prescribes the regional units of analysis on the grounds that the geographer must not construct a region on the basis of his interests. Instead, the units of analysis must help him to discover the essence, the unique individuality resulting from the local confrontation of general forces (formerly these were considered to be "laws") and the unique site & situation characteristics. Geography has been liberated from that vicious circle by spatial analysis and should never return to it.

There are two kinds of geography that implicate different methodological elaborations of the relationship between society and space. Topical, or thematic human geographies, e.g. economic or political geography are focussed on the spatial characteristics of some categories of phenomena, and the possible causal relationships between those spatial, and other, characteristics of these phenomena. They may use the theories, research methodology and techniques adopted from the neighbouring sciences. Then they work with different units and levels of analysis. The first of these is the level of the "elements" or "members"; the second that of systems combining in some way or another these elements. The elements' level is characterized by three subtypes. The first subtype is the so-called individual level, at which activities are understood through knowledge of attitudes, values, motives and perceptions of individuals. The second subtype level refers to groups and organizations as elements of sets, or classes, or as systems. The third subtype concerns aggregated categories of the two lower levels. This means that "levels of analysis" are concerned with different elements, or configurations, of a societal nature when used in thematical human geography. In regional geography, however, we can use the same distinction between levels of analysis; namely elements, sets or classes, and systems, but then these levels denote individual regions, classes or configurations of regions. The last are regional systems formed by subregions that are interrelated with each other. (These interrelationships distinguish a system from an aggregate or a set). A geographical system may have the same spatial appearance as an aggregate of areas, namely the mosaic form. One might say that such methodologic treatment of areas has parallels in the other levels of analysis: individuals and single regions, aggregates and (probably not adjacent) mosaics of regions, organizations, or groups, and regional systems.

The problem of regional geography is even more complicated. The societal elements (e.g. inhabitants of cities) are taken to be characteristics of the area (e.g. its population and population density or degree of urbanization). Often there are 'jumps' in the description. One can start by mapping the distributions (spatial characteristics of localized elements). Then the distribution pattern that is representative of the area may be transformed into a property of the area, e.g. a distribution of factories may be transformed into a map of the density of factories in different subareas. These jumps are often necessary when one changes the scaleof the observed areas.

Every level of analysis may contribute to the explanation. For example, the structure of an aggregate, like the market as an arena of demand and supply-activities, may be a constraint on localized individuals and their activities. To quote Warf (1988, p.183), "at successively larger spatial scales - those encompassing larger territories progressively removed from the immediate context of localities - the explanatory utility of theories centered around conscious action declines; it is analytically more efficient to explain such broad, 'macro-level' issues on the basis of theories pertaining to the nature of production, trade, and the dynamics of the world-system than it is to invoke the factors relevant at more local scales". Neither a top-down determinism nor a bottom-up voluntarism is postulated by geographers, because every level of analysis must be theorized in terms compatible with that level. The foregoing does not implicate, however, a straight correlation between the levels of analysis and the scales of the elements of regional and historical geographical analysis. Because the societal characteristics have become properties of the areas studied by regional geographers, the method of analysis is determined by the level of the societal characteristics, even in regional geography. When one changes the scale (e.g., when a smaller area is subsumed in a larger area), then the phenomena in the smaller area no longer form the whole set or class of phenomena under study but reappear as a mere subclass, along with other sub-classes that together form the new class of phenomena of the larger area. (See Hoekveld-Meijer, 1990, p.151 and Hoekveld-Meijer and Hoekveld, 1991).

This necessity is manifest in the thesaurus of theories used by geographers. Geography uses three types of theories. The first consists of theories explaining locations and distributions of elements, be they households, firms or other establishments. The second type is formed by theories explaining interaction (e.g., migration, trade, flows of communication, control, etc.). The third one comprises theories about areal or ecological structure, e.g. the von Thünen theory, Christaller's model, the Burgess model, Friedmann's centre-periphery theories, etc. These theories are suited to an explanation at a certain level, with specific units of analysis at pertinent areal scales of societal organization. Organizations and groups may also be seen as systems, e.g. composed of individuals with specific relationships. They are studied on the second level, and the individuals in that group are not elements of a sector class, but members of a system. Regional and, mostly, historical geography, in contrast to the topical geographical subdisciplines, studies areas and not societal elements. These societal elements and their relations, then, are characteristics or properties of regions and they are only studied as such. Again there are different levels of analysis, but of quite another nature (Hoekveld-Meijer, 1990, p.190). Firstly an area, or a region, may be viewed as an isolated unit, or as an element of a set or class of areas or regions, and as such an element of an aggregate of different regions. It may also be seen as a geographical system, a differentiated, but in some ways integrated region (Hoekveld and Hoekveld-Meijer 1991). The theories that geographers use are intended to explain the spatial or areal characteristics of the phenomena. This means that geographers may - and even should try to elucidate geographically influences on societal processes. Yet principally, they need theories that explain spatial or areal characteristics by societal processes. Although these processes provide explanations, they are not the ultimate object of geographical query.

In Figure 1, the areal and historical-societal-evolutionary perspectives in geography are combined. The historical perspective is shown in the upper half of

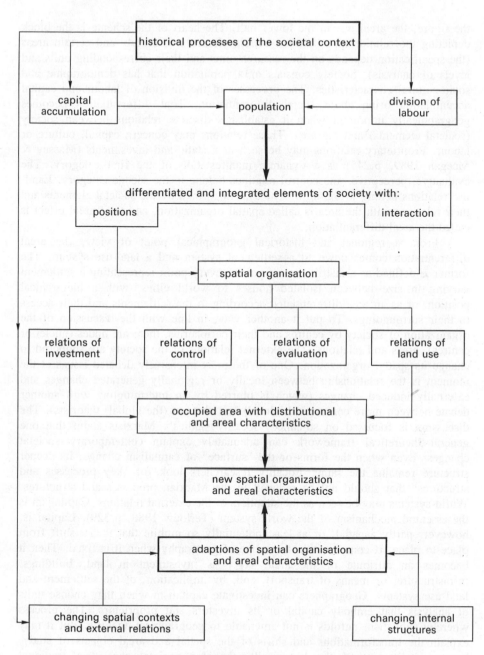

*Fig.1: Relationships between society and space: a conceptual scheme.*

the figure, the areal one in the lower half. The heart of the scheme is the block depicting the relationships between certain elements of society and certain areas (the specification depends on the research aims and their corresponding units and levels of analysis). Society consists of a population that has demographic and socio-cultural characteristics. The processes of the division of labour and capital accumulation bring about social differentiation. That differentiation becomes geographically important when it establishes discrete relations between society (societal elements) and the area. These relations may concern capital, culture or labour. Proprietary relations may be seen as a static and investments (Massey & Meegan 1989, p.245) as a dynamic manifestation of the first category. The evaluative, perceptive, and control relations belong to the second category. Land-use relations belong to the third. The system formed by the societal elements and their relations with the area is called spatial organization, and their areal effect is called the areal differentiation.

From a regional and historical geographical point of view, the areal differentiation comes down to a settlement system and a land-use system. The former is defined as a distribution of points, every point representing a settlement varying in size between isolated houses to world cities, with a hierarchical position, or as an area differentiated according to the settlements and their access to their surroundings. To put it another way, in line with the discussion of the organization of society by centres and their peripheries, there are nodes (nuclei or centres), areas and relations. The external relations of the society are supposed to change its spatial organization. One of the most intensively debated issues at the moment is the relationship between locally or regionally generated changes and externally induced changes (which is blurred by an intermingling with another debate between more concrete (the local) and abstract (the global) theorists). The discussion is focussed on so-called "locality studies". Marxists claim that one general theoretical framework can adequately explain contemporary societal changes. Even when the forms or the "surface" of capitalism change, its deeper structure remains the same. Locality researchers look for "key processes and structures" that should not be reduced to the Marxian processes and structures. World-systems may be seen as the suppliers of the external relations. Capitalism is the essential mechanism of the world-system (Terlouw 1988, p.329). Capital is, however, partly aspatial, or at least potentially so mobile that it can shift from place to place. It comes within the scope of geography when it is fixed. Then it becomes an attribute of the area, e.g., as investment in land, buildings, infrastructure, or means of transport, and, by implication, of the settlement and land-use systems. Geographers can investigate capitalism when they choose units of analysis that embody capital or its investors and proprietors. The process whereby capitalism unfolds is not amenable to geographical analysis. But it may explain the transformations and shifts of the spatial and areal objects of study, because it is part of the historically changing societal context of regional development.

We may conclude, then, that geography's limitations lie in its commitment to the spatial or areal aspect of society. Geography has to be receptive to theories that formulate processes of societal development, but it lacks the apparatus to develop these theories on the foundations of geographical experience alone. One strength of the discipline is that it can test these theories when they must have predictable spatial outcomes. Space undergoes a "metamorphosis" when the scales and levels of analysis are changed. This is the key to another strength: geography is not bound to stick with a theory that was formulated at a prior level of analysis. Geographers have to explain at different levels. This may be illustrated by describing the core of much geographical theorizing: the theories of regional development. These explain the transformations of regional systems and are, therefore, at the heart of regional and historical geography.

## 5.   Regional Development

The literature on regional development theories is overwhelming. An overview is beyond the scope of this paper. We may distinguish between two broad post-war periods of theory forming. The first one is focused on the regional economy. Refined pre-war theories based on the development of regional sectors or elements and their interactions were combined in several theories: export-base theories; Myrdal's theory of the circular causation of economic growth; growth pole theories; and centre-periphery theories. At a lower level of analysis, these theories could be combined with location theories (von Thünen, Christaller, Weber, and Lösch) and with theories of sectoral dynamics (e.g., institutionalist theories, the product cycle model, and the oligopoly-profit cycle theory; see Markusen 1985, p. 21). The planner's praxis brought the insight that economic theories do not suffice and that non-economic factors should be included. Fischer (1973, p. 62) concluded after a thorough overview of the theories that regional development is determined by economic and non-economic (social, political, infrastructural) factors. An important factor is the historically grown starting-point of regional development; this is determined by the settlement system and the internal division of labour. Fischer considers the external prime movers as given and the (potential) adaptability to these external forces as the key to regional development. The determining aspects of that adaptability may be very different in nature, however, varying from an antiquated transport system or an unresponsive population to a specific religion or ethnically based economic attitude.

In the eighties theory forming focussed on a search for regional society. Martin's admirable review (1989) distinguishes several competing theoretical frameworks. The first is the debate among proponents of Marxist locality studies . At the same time, long-wave theories of economic development were prevalent. Among these, the theory of Mandel is important. Like Fischer, Mandel stresses the significance of autonomous radical restructuring of the socio-political and geographical environment for the upswing from stagnation to expansion. Martin also presents Gordon's theory of the internal dynamic of the social and institutional structure, which explains how a once well-adapted regional system can become dysfunctional. He calls attention to its parallel, Olson's theory of institutional ageing. Martin also discusses the theories of change from "Fordist accumulation" to "flexible accumulation" based on recent changes in industrial organization and production.

Lash and Urry's theory of "the end of organized capitalism" (1987, p. 16) is also based on these recent changes. First there was a liberal period during which "large-scale collapsing empires, built around dynastic rulers or world religions saw the emergence of weak nation-states", "growth of substantial commercial cities as well as the expansion of new urban centres in rural areas" and growth of tiny pockets of industry. In the subsequent period of organized capitalism, nation-states dominated parts of the world by colonization and imperialism, developing distinct regional economies around growing centres and major inequalities between industrial and non-industrial states and regions. In the third period, that of disorganized capitalism, Lash and Urry describe the international division of labour, the multinationals, the separation and internationalization of finance and industrial capital, and the development of the "service classes".

Theories of urbanization take a very special place in regional development theory. Although seen as an outcome of deeper lying societal evolution, urbanization may play an intermediary role in regional development because of its internal spatial structuring and "metabolic" processes. Eisenstadt and Shachar (1987, p.278) compared medieval European urbanization with that in other continents. They observed a structural pluralism characterized by steadily increasing structural differentiation and constantly changing boundaries of collectivities, units, and frameworks. They defined the specific European configurations of "centrality and concentration" in the following terms: a multiplicity of centres, a high degree of permeation of the periphery by the centre, relatively little overlap of the boundaries between classes, ethnic & religious entities, a high degree of autonomy of groups & strata and of access of these groups to the centres of society, and, lastly, a relatively high level of autonomy of the legal system. Finally, a new paradigm has grown around structuration theory, originally formulated by the sociologist Giddens. In that theory at least three levels of analysis are necessary because of the deeper interwoven and phenomenal structures that come together in man's socio-spatial activities ("structures", "practices", and "agency"; see Wolch & Dear 1989). Regional structuring is a new frontier for that theory.

From the foregoing birds-eye view of contemporary discussions, we may, firstly, conclude that even though many authors now reject a monolithic all-encompassing Marxist theory, they stick to the contention that theory should guide research on historical and current developments. Another, second, conclusion is that theories endeavor to explain regional development by placing it in the theoretical context of an extra-regional (now global) division of labour, which entails giving much attention to external relations. At the same time they put it in the holistic context of development of society by the nation-state. Earlier we mentioned that these theories are split in two camps. The oldest, now losing ground, are the economistic Marxist theories that cling to "the universals that Marxist research produces" (Harvey 1987, 376). The more recent theories acknowledge the central position of capitalism's development in societal and regional structures and changes. Nevertheless, they admit that not only top (capitalistic world-system development) down (regional development) exists, but that there are also spheres with a certain autonomous development. They even assume that the local configuration may create regional or locally specific developments "e.g. the creative activities of certain key-persons" or "actors" so that bottom-up explanations may be feasible. These may start with an analysis at the individual level.

Finally, we may conclude that the phrasing of modern regional development theories does not prescribe particular areal units of analysis or define them as the most relevant ones. This would be the case if the world-system theory were the only valid explanatory instrument, prescribing the following units: core; semi-periphery; periphery and outer arena; and, at a lower level, states and the parts of the states that are differently, totally or partially, incorporated. By way of mental exercise, a typology of regions was designed (Figure 2), based on the notions of

| Nodality | Relations | | |
|----------|-----------|---|---|
| | Predominant internal relations | Predominant external relations | Interlaced internal and external relations |
| No predominant node | 1.  Autarchic rural region | 2.  Bounded specialized region | 3.  Interlaced rural region |
| One predominant node | 4.  City-hinterland region | 5.  Linked specialized city-hinterland region | 6.  Metropolitan regions with urban fields |
| Multi-nodal | 7.  City-hinterland region in network | 8.  Specialized city-hinterland mosaic regions | 9.  Interlaced rural-urban mosaic regions (macro-urban domains) |

Empirical types of regions by theoretical category:

1. Autarchic isolated regions: Alpine Valleys, pionier fringes; backward rural regions with a very high degree of autarchy and rudimentary formation of nodes
2. Large scale farm agricultural rural storeroom regions; small-scale farm specialized agricultural regions; industrialized rural regions; recreational rural regions
3. Inter-metropolitan peripheries
4. Von Thünen like central cityoriented rural regions
5. Trade-city-hinterland regions; land-river trade-city hinterland regions; traffic nodes
6. Capital cities, worldcities
7. Small market-city rural regions
8. Trade-axis city-hinterland regions; industrial city regions, modern service/industrial city regions
9. Becoming forms of macro-urban domain regions

*Fig.2:Types of regions according to nodality and external relations*

centre-periphery and external relations, concepts which define the scale of these areal units of analysis. These logical constructs may be translated into empirical ones. In Figure 3, the empirical referents are, arbitrarily worded and tentatively placed in the context of European historical evolution (see the appendix). These typologies demonstrate that it is perfectly possible to work at areal levels of analysis other than those implicated by world-system theory. While this last typology retains its basic idea of external relations by exchange, centrality, and core, it does so without ideological ramifications, thus leaving room for "intermediate theories". Theoretically, completely different theories may be applicable. These include the following, among others: Von Thünen, Christaller, growth pole theories, and polarization-reversal theories; these may apply

| Medieval Phase | Mercantilistic Capitalism | Industrial Capitalism | Neo-Industrial Capitalism |
|---|---|---|---|
| autarchic isolated rural regions | idem | – | – |
| nearly autarchic rural backyard regions | idem | idem | – |
| large-scale agricultural rural store-room and mining regions | idem | idem | idem |
| region-based city-hinterland regions | idem | idem | – |
| landriver trade axis based city-hinterland regions | idem | – | – |
| maritime trade-city-based (hinterland) regions | idem | – | – |
| | industrialized rural regions | – | industrialized rural regions |
| | capital city regions | idem | idem |
| | rural market-city regions | idem | idem |
| | | industrial city regions | old industrial city regions |
| | | small-scale agricultural regions | idem |
| | | traffic-node regions | idem |
| | | | modern service industrial city regions |
| | | | world-city regions and urban fields |
| | | | recreational rural regions |
| | | | intermetropolitan peripheries |
| | | | urban domains (macro-urban domains) |

*Fig.3:Types of regions according to their externally related structures*
*in successive historical phases*
*(scale based on the relevant center and its environment)*

respectively to city-hinterland regions and Vance's (1970) colonization-motherland model, city-hinterland mosaic regions in a network (von Thünen and Christaller), specialized city mosaic regions (growthpole theories), modern interlaced rural-urban mosaics, and interlinked city-hinterlands in a trade-axis region (polarization-reversal theories).

## 6.    Geography and the World-System Theory

Traditional geographical world-views had rather direct, predominantly descriptive concepts with which to analyze the workings of the world as a system. When the unity of the discipline crumbled in the aftermath of specialization in topical geographies, many explanatory theories came into use. These theories were methodologically adapted to the specific phenomena studied, by defining relevant units and levels of analysis, contexts, concepts, and conditions. The "grand theories" of our classical ancestors like Vidal de la Blache, Hettner, Herbertson, and Ratzel were considered out of date. The same fate befell the Weberian and Parsonian grand theories of sociology and those of classical economy. Neo-marxism brought an attractive (because holistic) alternative in the shape of Wallerstein's world-system theory. It synthesizes many processes that are interdependent but have been conceptualized in isolation, due to the impermeability of the boundaries between disciplines. This synthesis is generating fascinating empirical research. Nevertheless, the theoretical discussions in the wake of these studies bring to light certain conditions relevant to the geographer's use of that theory. By way of conclusion, these conditions are summarized in the following points.

Geography is bound to spatial or areal aspects of societal phenomena. This means that geographers are interested in societal evolution or development in so far it can throw light on the development of the spatial configurations of that society and its relevant areas. As the "mechanism" of that evolution is not essentially geographical in nature, geography cannot contribute to the understanding of that "mechanism". Nevertheless, geographers can use it as an explanatory instrument that may be supplemented by regional (historical and areal) specificities. World-system theory may help explain the development of geographical external relations, nodality, and the spatial context of areas or regional societies. Via these external relations, world-system theory may elucidate the areal and spatial structuring of these societies.

When using world-system theory, geographers should be aware of the implications of working with different levels of analysis and different areal scales. Theories designed for a particular level of analysis may be inappropriate at other levels, as the many ecological fallacies demonstrate. Geographers need to explore different explanations and theories at different levels (and units) of analysis. This may not be impeded by a theory of the world-system that allows only one deep structural explanation, to be applied at all levels and scales. The different types of regions distinguished in the typologies of Figures 2 and 3 show that very different theories may be necessary.

The incorporation of societies into the world-system has been a long process. This implies that incorporation was "completed" for some elements and not for others. Regional development during these periods is only partially explained by the world-system (Hall 1986, p.399).

If the thesis of "disorganized" modern capitalism is true, and if there are autonomous dimensions in the development of sectors of society, then there is a reason to use different levels ofanalysis. This would allow us to test the power of the world-system theory in determing external relations in combination with the theories in use at levels of analysis other than global.

Notwithstanding the aforementioned restrictions, the world-system theory offers a useful heuristic framework. It has two advantages over traditional views of the world. First, it combines traditional geographical interaction (exchange), which takes place by way of flows of commodities, people and ideas, with societal evolution to enlarge the scope of geography. Second, it is suited to an institutional or organizational mode of analysis whereby the abstract mechanisms of surplus accumulation are embedded in units of analysis other than the tripartite world and its states, with their areal manifestations. "The missing concept culture" (Worsley 1984, p.41) can be revived as well as the multi-dimensionality of the world. The classic concept "ecumene" has acquired a new relational connotation.

### Appendix: A Typology of Regions According to their Structures of External Relations

The development of capitalism has led to the incorporation of areas into an ever expanding geographical system. That system has been described by Wallerstein as a triad, consisting of a core, semi-periphery, and periphery. At the sub-state scale, other typologies should be developed to reflect the nature of regions' external relations to the wider system as well as their nodality. Figure 2 presents such a typology.

In every phase, the changed societal structures generate new societal elements and a new spatial organization. These manifest themselves at a lower level of analysis as period-specific or predominant. For example, in the Middle Ages, important elements of society included the church, the guild, the nobility, long-distance trade, the manorial rural organization, and the feudal organization. In the era of mercantile capitalism, these important elements were the commercial maritime ports, the urban "bourgeoisies", the strengthening state with its court, garrisons, decentralized industry, proto-industrial manufacturing, and country houses. The industrial-capitalistic phase was characterized by railroads, mines, large-scale factories, hotels, post offices, department stores and shopping districts, suburbia, and rural co-operative societies. The neo-industrial phase saw the growth of airports, motorways, multinationals, high-tech firms, mass recreation, specialized agriculture, and so on. In every period these new elements had to fit into the spatial organization and its areal structures, and to replace the obsolete residue of the earlier phases. This leads to new and different spatial distributions, dependencies and areal characteristics and thus to new regional structures.

The scheme presented in Figure 3 names the different types of regions that may occur in each phase. Some types of regions persist in the following phase. That does not imply, however, that their "representatives" stay in the same areas in the next period. For example, a large-scale agricultural storeroom rural area may be transformed by demographic pressure or a land reform into a small-scale agricultural area. Traffic flows may be diverted, changing trade axis-based city-hinterland regions into region-based city-hinterland regions. Seventeenth-century industrialized rural regions may again become agricultural rural regions in the nineteenth century, and so forth.

In the typology, the external relations differ with every sort of region. Of course, the autarchic isolated rural regions, e.g., in mountainous areas or isolated islands, disappear. They often change into rural backwaters only loosely connected with the rest of the world. Often these regions develop as rather autarchic Christallerian rural regions organized into a pattern of small market towns. The same may be the case with regions organized by a powerful isolated city. In the Middle Ages, long-distance trade established strings of cities along coastal or continental trade routes (Kellenbenz 1976). Their hinterlands had to adapt to that wider horizon. Just like in Antiquity, when the Black Sea region and Sicily were the granaries of the Roman Empire, frontiers of medieval colonization (e.g., the Baltic) developed into storerooms, producers of agricultural or mineral raw materials. In the era of mercantile capitalism the networks of trade cities were disrupted by the increased maritime orientation. In some respects the land-river-based trade-city regions became elongations of the maritime ports that were able to survive the competition.

The proto-industrialization of rural areas was mentioned above.The capitals had a major impact on their regions, attracting infrastructure and wealth and thus becoming big markets. In the period of industrial capitalism, new regions came into being. The industrialization of rural regions did not cease until well into the twentieth century. Agricultural areas had to adapt to world markets and as such became smaller or larger in scale as dictated by these markets. The railway changed the webs of trade but gave prominence to traffic regions, particularly those combining overseas and continental traffic.

In the modern era, many of the once-booming industrial city regions became old, deteriorated areas. But new city regions with a modern sectorial mix (service/industrial) have appeared, either sprouting out of the old urban regions or developing out of rural regions. Some cities attained a particularly strong position in world trade and became first or second echelon world cities, generating huge urban fields around them. Some rural regions have specialized in recreation. Between big metropolitan areas there may exist extensive inter-metropolitan peripheries with many smaller and, perhaps, a few larger nodes. In modern times, the urbanization of the whole society makes it difficult to distinguish between rural and urban. The ensuing labelling of spatial and areal characteristics is often ambiguous. Therefore, the term "macro-urban domain" (Van Paassen 1965, p.19) is introduced.

Already at the beginning of the sixties, Van Paassen (1965) envisaged these as incipient constellations and termed them "macro-urban domains". It denotes the interdependent configurations of rural and urban nodes that are shaping modern urban and rural relations and thus regional patterns.

## References

Becker, J. (1990): Postmoderne Modernisierung der Sozialgeographie? In: Geographische Zeitschrift 78, pp.15-23

Berdouley, V. (1978): The Vidal-Durkheim debate. In: Ley D. and M.S. Samuels( eds.): Humanistic Geography, Prospects and Problems, London, pp.77-90

Brunhes, J. & C. Vallaux (1921): La geographie et l'histoire, geographie de la paix et de la guerre sur terre et sur mer, Paris

Cooke, PH. (1990): Modern urban theory in question. In: Transactions of the Institute of British Geographers, NS.15 pp.331-343

Demangeon, A. (1920): Le déclin de l'Europe, Paris

Demangeon, A. (1932): Les aspects nouveaux de l'économie internationale. In: Annales de Geographie 41, pp. 1-21, 113-130

Dieterich, B. & H. Leiter (1930): Produktion, Verkehr und Handel. In: Andree-Heiderich-Sieger : Geographie des Welthandels. Eine wirtschaftsgeographische Weltbeschreibung, 4.Aufl., Wien

Dodgshon, R.A. (1987): The European Past: Social Evolution and Spatial Order, London

Eckert, M. (1905): Grundriß der Handelsgeographie, 1. Band, Allgemeine Wirtschafts- und Verkehrsgeographie, Leipzig

Eisenstadt, S.N. und A. Shachar (1987): Society, Culture and Urbanization, Newbury Park

Ekholm, K. (1981): On the structure and dynamics of global systems. in: Kahn, J.S. und J.R. Llobera (eds.): The Anthropology of Pre-capitalist Societies, London, pp. 241-261

Fischer, G. (1973): Praxisorientierte Theorie der Regionalforschung, Tübingen

Friedman, J. & M.J. Rowlands (1977): Notes towards an epigenetic model of the evolution of "civilisation". In: Friedman J. & M.J. Rowlands, eds., The Evolution of Social Systems, London, pp.201-276

Gritsai, O. and A. Treivish (1990): Stadial concept of regional development: the dynamics of core and periphery, a theoretical discussion. In: Geographische Zeitschrift 78, pp.65-77

Gritsai, o. and A. Treivish (1990): Stadial concepts of regional development; Centre and periphery in Europe. In: Geographische Zeitschrift 78, 3, pp.137-149

Hall, T.D. (1986): Incorporation in the world-system; toward a critique. In: American Sociological Review 51, pp.390-402

Harvey, D. (1987): Three myths in search of a reality in urban studies in the issue "Reconsidering social theory, a debate". In: Environment and Planning D (Society and Space): vol. 5, pp.367-376

Hoekveld-Meijer, G. (1990): Metamorphosis, how spatial facts change into classes of geographical regions. In: Johnston R.J., J.Hauer & G.A. Hoekveld ( eds.): Regional Geography: Current Developments and Future Prospects, London (forthcoming)

Hoekveld, G. and G.Hoekveld-Meijer (1991): Regional development, spatial and societal contexts: key-concepts. In: regional geographic methodology. Series REGIS-studies, no 9, publicated by the Geographical Institute of the State University at Utrecht, and the Centre for the Geography of Education of the Free University at Amsterdam, Utrecht/Amsterdam

Kellenbenz, H. (1976): The Rise of the European Community. An Economic History of Continental Europe 1500-1750, London

Lash, S. & J. Urry (1987): The End of Organized Capitalism. Cambridge

Laszlo, E. (1973): Uses and misuses of world-system models. In: Laszlo, E.(ed.): The World-System, Models, Norms, Applications, New York, pp.3-17

Mackinder, H. (1904): The geographical pivot of history. Reproduced in: Kasperson R.E., E. Kasperson & J.V. Minghi( eds.) (1969): The Structure of Political Geography, Chicago, pp.161-169

Markusen, A.R. (1985): Profit Cycles, Oligopoly and Regional Development, London

Martin, R. (1989): The reorganization of regional theory: alternative perspectives on the changing capitalist space economy. In: Geoforum vol. 20, pp.187-201

Massey, D. & R. Meegan (1989): Spati al divisions of labour in Britain. In: Gregory D. & R. Walford (eds.): Horizons in Human Geography, London, pp.244-257

Müller-Wille, W.C. (1966): Politisch-geographische Leitbilder, reale Lebensräume und globale Spannungsfelder. In: GeographischeZeitschrift 54, pp.17-38

Newson, L. (1976): Cultural evolution: a basic concept for human and historical geography. In: Journal of Historical Geography 2, pp.239-255

Nitz, H.J. (1984): Siedlungsgeographie als historisch-gesellschaftswissenschaftliche Prozeß-forschung. In: Geographische Rundschau 36, pp.162-169

Paassen, C. Van (1965): Over vormverandering in de sociale geografie. Groningen

Ratzel, F. (1909): Anthropogeographie, Erster Teil, Grundzüge der Anwendung der Erdkunde auf die Geschichte, Stuttgart, 2.Aufl.1909 (1. Auflage 1882). Anthropogeographie, Zweiter Teil, Die Geographische Verbreitung des Menschen, 2 Aufl. Stuttgart 1912 (1. Auflage 1891)

Richthofen, F. Freiherr von (1883): Aufgaben und Methoden der heutigen Geographie, Akademische Antrittsrede, gehalten in der Aula der Universität Leipzig am 27 April 1883, Leipzig

Richthofen, F. Freiherr von (1908): Vorlesungen über allgemeine Siedlungs- und Verkehrsgeographie, Bearb. und herausgegeben von O. Schlüter, Berlin

Smith, G.E. (1989): Privilege and place in Soviet society. In: Gregory, D. & R. Walford (eds.): Horizons in Human Geography, London, pp.320-340

Sorre, M. (1952): Les fondements de la geographie humaine, T. III, l'habitat; conclusion générale, Paris

Taylor, P.J. (1989): The error of developmentalism in human geography. In: Gregory, D. & R. Walford (eds.): Horizons in Human Geography, London, pp.303-319

Taylor, P.J. (1988): World-systems analysis and regional geography. In: Professional Geographer, vol. 40, pp.259-265

Terlouw, C.P. (1988): De ecumene als wereldsysteem: een overzicht van de wereldsysteemtheorie van Wallerstein. In: De Aardrijkskunde, 4, pp.321-341

Vallaux, C. (1908): La mer, populations maritimes, migrations, pôches, commerce, domination de la mer, Paris.Vance, J.E. (1970): The Merchant's World, Englewood Cliffs

Wallerstein, I. & T.K. Hopkins (1980): The future of the world economy. In: Hopkins, T.K. & I. Wallerstein (eds.): Processes of the World-System, Beverly Hills, pp.167-180

Wallerstein, I. (1980): Imperialism and development. In: Bergesen, A. (ed.): Studies of the Modern World-System, New York, pp.13-23

Wallerstein, I. (1988): The inventions of time-space realities: towards an understanding of our historical systems. In: Geography 7, pp.289-295

Warf, B. (1988): Locality studies. In:Urban Geography, 10, pp.178-185

Wolch, J. & M. Dear (1989): How territory shapes social life. In: Wolch J. & M. Dear (eds.): The Power of Geography. How Territory Shapes Social Life, Boston, pp.3-18

Worsley, P. (1984): The Three Worlds, Culture and World Development, London

Wusten, H. Van der (1987): Op zoek naar de juiste mate van vergruizing: de bruikbaarheid van de wereldsysteem-benaderingen in de politieke geografie. In: Boer, W.Ph.G. et al., Regio's in wereldcontext, Meppel, pp.20-28.

# 3

# The European World-System: A von Thünen Interpretation of its Eastern Continental Sector

Hans-Jürgen Nitz

## 1. The von Thünen-System - Introduction of the Concept and its Application to the Early Modern European World-System

In the concept of I. Wallerstein[1] and F. Braudel[2] the early modern European world-system included large parts of Europe as well as the transatlantic colonies. The core as the steering centre of the whole system was constituted by the Netherlands and England around London. It functioned as the main region of consumption, of manufacturing and of trading with the other parts of the world-system, with Amsterdam (earlier Antwerp and later London) as main entrepôt. These other parts were dependent on the entrepreneural mercantile economy of the core: it were the traders and shippers of the core who organized the imports from as well as the exports to the dependent parts which were exploited through unequal exchange. It is because of this economic dependence that Wallerstein and Braudel have termed the dependent regions as peripheries of the world-system. While Wallerstein discerns two ranks of peripheries -the semiperiphery and the "true" periphery- Braudel stresses the fact that the former has as well many characteristics similar to those of the core economy, including regions of a rather high level of development. Its peripheral rank is marked by the fact that its trade is dominated by colonies of foreign traders.[3] Because of these ambivalent economic characteristics, the present author prefers the term "intermediate zone".

Another approach to the functional spatial structure of the early modern European world-system is offered by Johann Heinrich von Thünen, who lived around 1800 as a contemporary of its late phase. As a practitioning big farmer (of the Mecklenburg region) who supplied grain to the regional and to the world market, and as a keen observer of agriculture and forestry, of transportation and marketing elsewhere in Europe, he searched into the economic conditions of production in their dependence on variable market prices. He paid special attention to transportation costs which rise with growing distance from the marketplace and therefore work as a reduction of the price paid at the farm. Von Thünen concluded that there must exist a limit of distance beyond which grain cannot be profitably sold to an external market because of prohibitive transportation costs. Thus the market-grain region would form a concentric circle around the central market. He tested his ideas on his own farm with many years of detailed book-keeping and also based them on informations from other regions and finally cast them into a general location theory in his well-known book of 1826 - "The Isolated State".[4]

Within this circle of market-grain production the ring closest to the market has the least transportation costs which permits the farmers to save capital and invest it in production in the form of labour and manure. The more distant the location of a farm from the market, the less investment of this kind is possible: intensity of farming has to be reduced by necessity, i.e. the farmers have to adapt by choosing less intensive landuse systems. This results in the well known sequence of three concentric market-grain zones (Fig. 1): In the grain zone nearest to the central market the farmers have to apply the most intensive landuse system of the time - "Fruchtwechselwirtschaft" - , i.e. permanent arable with grain and fodder crops alternating annually, with labour-intensive stall feeding all the year round which results in a maximum amount of dung and, consequently, high yields. This is a very narrow zone of small areal extension.

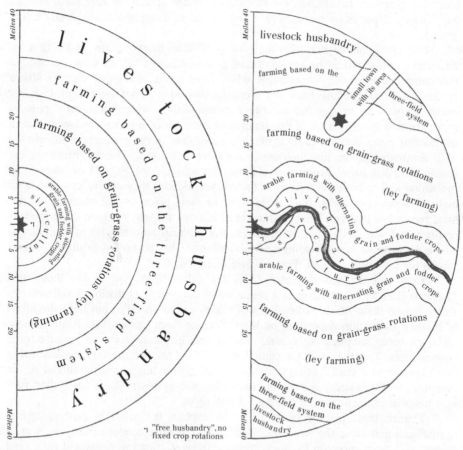

*Fig.1: Johann Heinrich von Thünen: Agricultural zones of the Isolated State*
Source: J. H. v. Thünen 1875/1966, pp. 390–391

In the adjoining zone the less intensive field-grass system or regulated ley farming has to be chosen, (von Thünen used the term "Koppelwirtschaft" common in Schleswig-Holstein and Mecklenburg) in which in the individual enclosed plots (or sections of common strip fields, as practiced in the peasant villages of Schleswig-Holstein on the traditional medieval field pattern after the introduction of "Koppelwirtschaft" in the 16th century) a fixed number of years under grain and rape are followed by a fixed number of years of ley which is used for intensive grazing and hay. The droppings of the grazing animals improve the soil.

In the most distant market-grain zone the least intensive three-field system ("Dreifelderwirtschaft") has to be applied, in which the arable covers only one third of the farmland, the remaining two thirds consist of unimproved rough pasture for grazing farm animals. Hence droppings during grazing time are lost as dung, which is only collected in the stables during the night, a very labour extensive practice. In addition the stubble fields are grazed. Winter feed, besides hay, largely depends on straw, with a low output of dung and consequently a low grain harvest.

For the zone beyond the limit of profitable market-grain production an alternative option would be production of commodities of higher value per weight unit compared to grain, for example wine, fibre crops or cheese; meat is also a profitable option because beef cattle can be moved on hoof instead of in horse drawn wagons as in case of the other commodities, which drastically reduces transportation costs. Wages in this outermost zone are comparatively lower than in the interior zones because grain is much cheaper than in the grain exporting circle.

Surprisingly, in his theoretical considerations von Thünen came to the conclusion, that forest production ("Forst-Wirtschaft", see Fig. 1) of timber and fuel wood had to be located close to the market because they are bulky goods with high transportation costs, high in relation to their price on the market. But we must remember that von Thünen thought of *planted* forests, while in early modern reality wood was still exploited from existing *natural* forests with no capital and labour investment for replanting and silviculture. This permitted to utilize natural forests for timber at a larger distance from the market.

Thus von Thünen deducted from his calculations that in his theoretical "Isolated State" there must exist a sequence of concentric agricultural and forestry zones around the central market, each of which has its specific, but limited options of market production depending on their distance from the market. Each of these different zones supplies the central market and in return receives manufactured commodities. Together with the central market (and the commercial manufacturing core) they form a spatial economic system. Of course this is not the model of the complete world-system, but of an important sub-system. There are other sub-systems of manufacturing, of mining, of fishing etc.

Thünen built his theory in order to explain the actually existing spatial agricultural pattern of the world he lived in: of the early modern European world-system with transportation by horse drawn carriages, floating boats and rafts, and sailing-ships. Therefore, the real zonation of agricultural market production of that era must in general lines have conformed to the theoretical von Thünen pattern, though with many variations caused by the diversities of natural conditions of soil, climate etc.

A very important variation of the theoretical pattern which brings it closer to reality is caused by navigable waterways on seas and rivers (Fig. 1). Von Thünen's theoretical "Isolated State" constitutes a land-locked spatial system. But transportation on rivers by ships reduces transportation costs quite considerably to about 1/10 compared to overland haulage. This results in a ribbon-like extension of the concentric zones along the seas and rivers. The present author is convinced and will produce evidence in the second part of this article that the von Thünen-model especially in this second version can be used as a tool to explain part of the early modern world-system in the systemic way which R.Dodgshon demands.[5]

Von Thünen's theory of the spatial order of the early modern agricultural (European) world has attracted keen interest of economic historians like the German W.Abel.[6] As von Thünen based his theory on the assumption of the rational *homo oeconomicus*, he did (and could) not include the social repercussions of the capitalist market economy which e.g. led to the re-introduction of serfdom and coerced labour under feudal lordship. It may be due to these "omissions" that I.Wallerstein does not at all refer to von Thünen. F.Braudel includes von Thünen's circles as one step in the explanation of the hierachical differentiations of zones of the world economy, but he critizes von Thünen for not at all considering the obvious economic and social inequalities between the zones.[7] In a human-geographical approach we have certainly to consider these repercussion on the rural society.

After this general introduction, we will try (first) to demonstrate and explain the spatial pattern of the real agricultural zones, in von Thünen terms, of the early modern European world-system arranged around the core, with their collection points and transportation routes to the central market. In principle, each zone was individually linked to the core market, with Amsterdam as the main entrepôt especially for grain, with no functional interconnections between the specialized agricultural zones. But it can be shown - and this will be our second point - that there did exist co-operating zones which may be seen as sub-systems - as a second tier - within the general von Thünen system. A first case: The beef cattle breeding zone in the periphery of the system and the field-grass zone as the beef cattle fattening zone with an intermediate location between the former and the market.[8] A second case: Sub-systems which consist of a main von Thünen-zone and of its hinterland as its individual periphery which supplements it with special resources.

The systemic geographical approach has to include a third spatial tier: The individual von Thünen-zone which consists of several local regions which had joined or had been drawn into the respective zone of the early modern world-system. According to the von Thünen theory growing demand from the market would result in a rising price level which again would lead to the spatial expansion of the whole system which actually happened during the boom periods of the "long sixteenth century" (Wallerstein) and again during the eighteenth century. Hence, regions closer to the market were earlier drawn into the world-system than more distant ones, and with the widening of the whole system a region could switch from e.g. the three-field zone to the field-grass zone what actually happened to the Mecklenburg region since the late 18th century. These regions were in earlier, medieval times more or less intensively bound to local or regional market systems, with a strong emphasis on subsistence economy, too. It should not be forgotten, however, that the Hanseatic League did interlink the Baltic regions and the west European urban regions as early as the late Middle Ages, but these connections to external markets were of course strongly intensified and spatially expanded since the sixteenth century.

The new role of these regions drawn into the early modern European world-system demanded many adaptive changes in their internal structure in reaction to external forces. This is the approach of G.A.Hoekveld[9]: "World-system theory may help explain the development of geographical external relations, nodality, and the spatial context of areas or regional societies. Via these external relations, world-system theory may elucidate the areal and spatial structuring of these societies."[10] We just list some categories of changing regional elements: Innovative changes in agricultural landuse; reclamation of additional agricultural land; changes in social ranking, formation of new strata in the regional and local societies; changes in labour and capital relations; changes in property relations; demographic changes; changes in settlement structures and field patterns, introduction of new types of farmhouses; decline of old and evolution of new market networks and of market and port settlements with possible changes in the composition of merchants, with foreign traders and entrepreneurs, mainly from the core, displacing indigenous people; even natural conditions could be changed in response to market incentives, either in a positive direction by controlling natural hazards and converting nature into an economic resource, or in a negative sence through overexploitation with final destruction of the ecological balance. As most of these elements of the regional structure constitute a closely interrelated context, we have to study their changes in a systemic approach, with a strong emphasis on the common external causes of these changes which indirectly originate from the impacts of the world-system.

Hence the systemic approach applied here will analyse three spatial tiers or levels of the world-system: (I) The international von Thünen-system as a sub-system (or partial system) of the world-system; (II) cooperative sub-systems of the von Thünen-system with one zone supplementing the other; and (III) the individual von Thünen-zone as a regional system.

## 2. The Zones of the von Thünen-System in Early Modern Northern Central Europe

The first tier will be presented in an overview of the von Thünen-system as it did exist since the 16th century, but limited to a cross-section: the eastern continental sector along the North Sea and the Baltic Sea with the navigable rivers of the continent flowing into them (Fig. 2). This spatial pattern is explained by the second version of the von Thünen-model (see Fig. 1). The sequence of zones expands along the coasts and the navigable rivers. Thus land-locked inland parts of

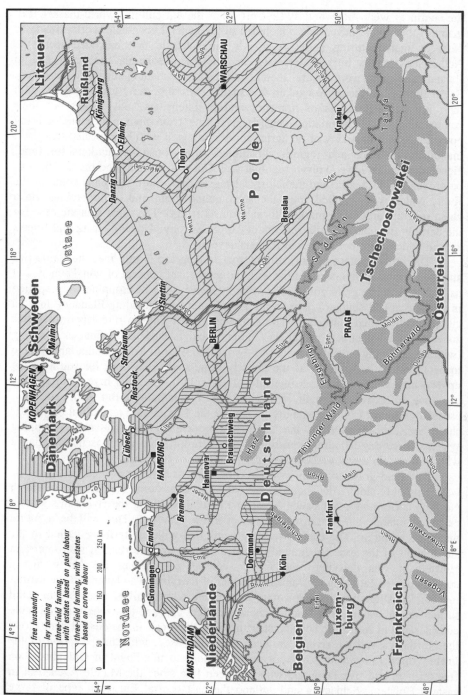

*Fig.2: Von Thünen zones in early modern northern Central Europe*

the continent were more or less shut off from the agricultural world-market. But many of them were nevertheless included in the general economic world-system as parts of manufacturing belts: A mainly rural textile belt extended from the Netherlands across interior Germany and northern Bohemia to Silesia and to Poland, and there were several metal producing and manufacturing regions, too. So far there is no theory to explain this spatial pattern of rural textile regions with their internal differences of branches and of intensities. But because of their comparatively high population densities resulting from numerous manufacturing households with only tiny or no agricultural holdings it is to be expected that these manufacturing belts and regions constituted large consumer markets for food. Hence there must have evolved regional von Thünen-zones around them. This can be proved for the textile and metallurgical region of western Germany south of the Ruhr, which was supplied from the surrounding grain and cattle regions via a chain of grain and cattle markets located on its fringe. The same was the case for the densely populated textile regions of Upper Saxony, Upper Lausitz and Silesia around which, too, belts of grain regions developed.[11]

The broad pattern of agricultural zones along the North Sea, the Baltic Sea and their tributary rivers conforms with the theoretical concept of von Thünen. Let us first look at the grain zones. The most intensive types of cropping were applied in the central and southern parts of the Low Countries, including Flanders, located closest to the urban markets.[12] The Rhine as a "von Thünen-river" permitted this zone to extend upstream as far as Cologne, and islands of continuous cropping (without fallowing) mainly devoted to dyeing crops (sold to the Netherlands) were to be found in the infields of the densely populated villages as far south as the upper Rhine valley. The high intensity of agriculture practiced here permitted the evolution of extremely small farms through fragmentation. But also demesnes, especially around Cologne, mainly owned by the Church and leased out on a half-share basis, changed from the three-field system to more intensive cropping systems. Free day-labourers constituted the working force which is in accordance with Wallerstein's wage-labour system in the core of the world-system.

The field-grass economy as the second landuse system, of medium intensity, was emploid along the coast from the north-eastern Netherlands through Northwest Germany, Schleswig-Holstein to Jutland. Along the rivers this system extended inland, but so far we don't yet know how far. This will be a future theme of research. The cattle economy combined with ley-farming varied in the sub-regions of the zone. In Jutland the main branch was breeding of lean-steers, in Holstein it was dairying, and in the coastal zone along the North Sea a combination of fattening of Danish oxen and of dairying. The latter and the former together constituted a complementary sub-system which will be discussed in detail further down. Apart from landuse characteristics, parts of this zone differed quite considerably in respect to societal structures. In the west, in the coastal zone along the North Sea and the rivers, the economy was based, with the exception of the lower Weser valley with some corvee obligations to the Count of Oldenburg, on free farmers and free labourers who had been free since the Middle Ages. But in eastern Holstein, Schleswig and Jutland the east central European type of large demesnes of the nobility and the Crown with corvee labour (second serfdom) prevailed everywhere. Demesnes were enlarged at the expense of peasant farms by eviction and their owners reduced to the status of unfree cottars, with up to four families in one cottage. A small number of peasant farms per estate were left which were enlarged to double and triple size (some to more than 50 hectares) in order to supply, as part of their corvee obligations, the horses for the demesne - up to 16 per farm along with four wagons and five men.[13] Hence the settlement pattern was characterized by splendid Renaissance and Baroque castles of the nobility and rather poor peasant and cottar villages.[14]

The third grain zone in which the least intensive three-field system had to be applied, covered large parts of east Central Europe. A. Dunin has drawn a detailed economic map for 16th century Poland which shows the spatial distribution of grain exporting regions.[15] They have been included in our map (Fig. 2). In complete accordance with the von Thünen rule, they extend along the coast of the Baltic Sea in the north and further inland along the Oder and Wistula rivers and their tributaries as far as they were navigable for flat boats. Further north the smaller rivers like Memel and Duena showed the same pattern of hinterlands. From the most remote grain exporting inland region in southern Poland the distance to Amsterdam as the grain entrepôt of the European world-economy was more than 2500 kilometers. According to von Thünen, overland haulage on a (theoretical) completely flat surface would have permitted a maximum distance of 230 kms only (see Fig. 1), but on waterways ten times more. Thus, early modern reality conformed to theory quite convincingly.

But again the societal and especially the labour system has to be taken into consideration. Since the 15th/16th centuries the regions east of the Elbe river saw the emergence of large estates which grew through recultivation of deserted peasant lands (after the late medieval phase of desertion), through clearing of forests and commons, and through eviction of peasants from their farms ("Bauernlegen").[16] Due to their feudal political power - as territorial lords of their estates - they were able to forcibly introduce second serfdom with ever growing demand of corvee labour (up to six days per week in the plowing and harvesting seasons). This quite considerably reduced labour costs for the demesnes of the nobility (and of the Crown and the Church as well). It is for this reason that estates of even very distant regions could participate in grain production for the world-market.

But at the same time and as the result of the growth of the demesnes depending on excessive corvee obligations, the peasants' economy fell into a severe crisis because their fields could no longer be cultivated as intensively as before. The area under cultivation had to be reduced, and the yields decreased. There were additional damaging effects through wars, but it is obvious that the peasant economy suffered comparatively much more than that of the demesnes which can only be explained by the negative impacts of excessive detraction of labour from the peasant farms. The Polish historian J. Topolski has calculated that the decrease in production of the demesnes for the period from the end of the sixteenth to the end of the 18th century was only 6%, "while that of the peasant farms reached approximately 37%".[17] As Topolski has shown, the flourishing of the demesnes and the prosperity of the nobility and of the big magnates who were able to construct large palaces on their estates and in the regional and national capitals as well, was accompanied by a severe deterioration of the dependant peasant villages with almost delapidated farmsteads and the arable land considerably reduced, "poorly sown and attacked by weeds, a hazard often mentioned in the sources".[18] The grain economy for export of the large feudal farms led to an overall reduction of the national Polish grain production.

Furthermore the free rural market towns severely suffered from the crisis of the peasants who were no longer able to sell as much grain to the market as before and buy from merchants and artisans. What was even more destructive for the small towns: The nobility sent their grain immediately to Gdansk via the grain ports along the rivers, and bought luxury goods from there. For the rural market towns all this led to a severe economic crisis which resulted in shrinking population, reduction in the number of houses and growing dependence of the townsmen on additional farming just for survival. But there were exceptions: Magnates founded and favoured patrimonial new towns on their large estates which should serve as their headquarters, with a large palace, and as well controlled markets for the surplus produce of their peasants which could be exported by the lord. On the other hand, the port towns and especially Gdansk as the main grain port for Poland saw a splendid growth, with numerous new Renaissance houses after the Dutch mode.

Even these destructive changes in the Polish countryside can be related to the von Thünen-system: In those agricultural regions which were located closer to the market, the nobility could -comparably- save transportation costs. Hence they had not to depend to such a degree on coerced labour as those in distant parts of the grain zone. Topolski points to the cases of the Gdansk region and to Silesia: Around Gdansk the large farms mainly worked with day-labourers, and the peasant farms with only slight corvee obligations, if at all, could keep their economic standard and participate in the market economy. Large parts of rural upland Silesia and the towns as well had a flourishing linen industry with a large and ever growing workers population which formed a market for the demesnes of the nobility and the peasantry of lowland Silesia as well. The demesnes mainly engaged paid cottars ("Gärtner") and "the Silesian peasants worked on their lords' lands for a mere 50-70 days a year".[19]

The western section of the three-field grain zone, e.g. the region along the upper Weser, also saw the development of large demesnes of the nobility who reacted to the new chances offered by the world-market. But due to their lesser socio-political power and to protective legislation of the princes for the peasants they were not able to increase corvee obligations of their subjects to more than two or three weeks per year, and therefore had to rely mainly on paid labour of cottars. As around Gdansk this does not seem to have posed problems to the demesnes because they were located much closer to the market of Amsterdam than those in interior Poland. The number of day-labourers in the upper Weser region increased quite considerably since the 16th century and it was through this socio-economic development that former hamlets grew into large compact villages ("Haufendörfer").

The main outlet for grain export from the Baltic three-field zone was Gdansk/Danzig located on the mouth of the Wistula. Secondary in rank because of their smaller river hinterlands were Bremen (Weser), Hamburg (Elbe), Stettin (Oder), Königsberg (Pregel), Memel (Memel river) and Riga (Duena). The grain export trade from the Baltic ports was almost completely in the hands of Dutch merchants and it was due to their influence that Danzig became a city of a strongly Dutch character. Since the mid 16th-century annually 1000-2000 Dutch ships arrived at Danzig.[20] Along the rivers indigenous grain merchants as intermediaries had established numerous granaries at the port towns to which the grain was brought during winter on sleds and in spring on boats.[21] Many of the large grain stores were owned by the nobility as the main suppliers most of whom immediately sold to the big grain merchants of the river ports or even of Danzig.[22] Together with the main Baltic ports like Danzig they constituted a grain-trading network.[23]

The interior parts of Poland beyond the limits of profitable grain export also practiced the three-field economy. But grain could be sold on local and regional markets only. A chance for export was offered by converting grain into alcohol, a low weight-high value commodity, in Poland and Lithuania into Wodka.

Another option for regions beyond the limits of profitable grain export was to grow flax and hemp and sell the fibres and high quality seed to the world market. There was a good demand for these commodities in rural linen weaving regions of Central Europe to supplement the local production of fibres, regions such as Silesia, the Upper Weser and even Galicia in northern Spain which received its supply via Amsterdam (Fig.3). This place itself with its shipyards was an important consumer market for hemp which was used for rope and sail making. Again it were Dutch merchants who organized the trade from Riga and other Baltic ports, with local tradesmen serving as their intermediaries who had to advance large sums - which they again received as credit from the Dutch - to the nobility of the hinterlands as the main suppliers. In reaction to growing demand the supply regions could expand along the rivers as far as interior Lithuania and White Russia, even to the region around Smolensk in the Dnjepr river basin from where transportation was carried out during winter on sleds. Big magnates extracted flax and hemp as feudal dues from their serfs. But closer to the ports also peasants were able to send these primary commodities to the market. Drawing the nobility and the peasants into indebtedness was the strategy of the port merchants and their Dutch partners to secure a continuous and cheap supply which again made it possible to cover the costs of transport over such large distances as from Smolensk to Amsterdam and even to Galicia in northern Spain (Fig.3). And no doubt was it the feudal system of forced extraction of a considerable portion of the peasants' produce that brought the prize down, a good illustration for unequal exchange.[24]

The beef cattle zone, too, extended - as it should according to von Thünen - beyond the grain circle, from Denmark[25] to the interior parts of eastern East

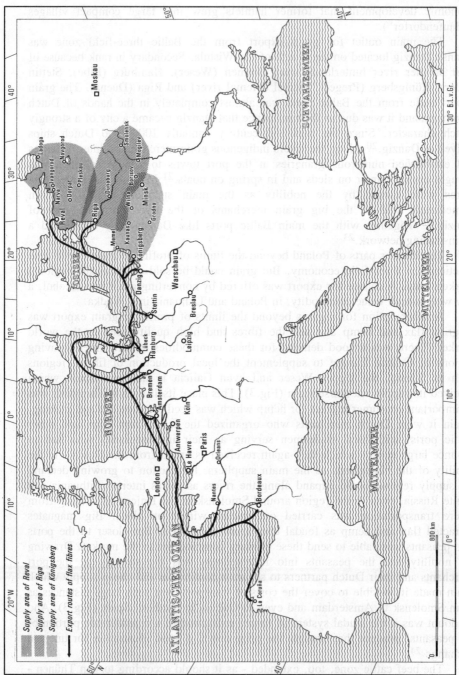

*Fig.3:* Export of flax fibres from the Baltic territories and Russia in the 18th century
Sources for the supply areas: −Jensch, Georg: *Der Handel Rigas in 17. Jahrhundert* Riga 1930
−Eckert Walter: *Kurland unter dem Einfluß des Merkantilismus (1561–1682)* Marburg 1926
*(draft: M.-I. Rommel)*

Prussia, Wolhynia, Podolia, Ruthenia, Ukraina and even Moldavia and Walachia to the east of the Karpatian Mountains which are now parts of Romania and the former USSR. From there the cattle were moved in large herds to Poland and further west to the markets of Danzig and Germany, as far as Nuremberg, Frankfurt and the towns of the upper Rhine valley, with a market limit north of Berlin, Brunswig, Kassel and Marburg.[26] The Polish economic historian J.Baszanowski has prepared a very detailed map of the numerous routes on which the oxen were driven west. As the animals had to move more than 2000 km, there arises the question of intermediate feeding grounds. How were these problems solved? As can be read from the map of Baszanowski, the seven or eight drow routes from the eastern pasture regions converged to a corridor of about 160 km width near Przemysl in eastern Poland and then diverged again in central and western Poland to double that distance. Here they passed through the three-field-zone where it may well be that enough natural pastures were available for them and that the practice of cattle pens on fallow fields might even have supplied additional dung. H.Szulc[27] in her discussion of the demesnes of the Silesian nobility refers to those located in the Oder valley which were mainly devoted to a cattle pasture economy. It may well be that the extensive natural pasture grounds of the valley also served the herds of oxen on their drow to the western markets. As a matter of fact, the Oder valley is located half way from the eastern convergence of drows to the west German markets. The importance of studying the cattle routes and their spatial organisation is underlined by V.Zimanyi's study of the respective routes from Hungary to Venice[28] and by our own discussion of the intermediate pasture regions for Danish oxen (further down).

If we look beyond our cross-section, the southeastern part of the cattle belt was Hungary which sent her oxen to Central Europe as well as to Venice.[29] On the Balkans and in the Mediterranean as the southern section of the zone of export-based animal husbandry, due to climatic and natural pasture conditions cattle were for the most part replaced by sheep, with transhumant grazing. Its main primary commodity supplied to the world-market was of course wool, which came from the Balkans, then under Turkish rule, from southern Italy - the kingdome of Naples, and Apulia - , and from Spain. But the densely populated and urbanized region of northern Italy, especially Venice and Florence, had a high demand for meat and cheese as dietary supplements and formed strong export markets for the south Italian and Alpine sheep regions. Castrated rams and lambs were moved on hoof from the Foggia region to the north.[30] But also in interior parts of northern Central Europe beyond the export grain limit we find regions of cattle raising and wool production. An example will be given further down. To close the von Thünen circle of cattle and sheep grazing regions we have to point to Scotland, Ireland and the western highlands of Britain.[31]

According to von Thünen's theory, the zone of forests supplying bulky fuel wood should have been located as close to the core market as possible. Since the Middle Ages most of the natural forests of the Netherlands had been cleared or through overexploitation for fuel and timber reduced to heathlands. Forest plantations which according to von Thünen's model should have been created, were started in the 19th century only. For fuel, there was a cheaper alternative: The Flemings and the Dutch exploited the peat bogs for turf which after thorough drying serves as excellent fuel. Thus peat areas close to the urban and rural consumption areas in the Low Countries were, even as early as the Middle Ages, transformed into fuel producing zones which were connected to the markets by a dense network of canals which considerably reduced the transportation costs for the very bulky turf.[32] In the "veens" of the eastern Netherlands south of Groningen where due to exhaustion of older supply regions and to growing demand from the commercial centres a new fuel region was opened in the 16th century, business was organized by companies which ran the exploitation as large scale enterprises, each of which engaged hundreds of saisonal migrant labourers from the adjoining inland regions outside the grain zone.[33] Beyond the Dutch/German border the natural peat belt continues, and here, too, entrepreneurs intended to compete on the Dutch fuel market. But protective legislation by the Dutch government impeded export to Amsterdam. Thus, the market was limited to the German "intermediate zone", with the main outlet to the comparatively rich belt of grain farms along the coast. Large scale entrepreneural peat turf production proved unprofitable, and after having constructed the canals, the companies leased numerous small strip plots of peat land, 3 to 4 hectares each, to cottars who with their families produced the turf and even carried out transportation to the markets by boats.[34]

Highly valuable commodities from the forest like timber, ship masts, boards, potassium and naval stores which include pitch and tar, were produced in the zone of vast natural forests beyond the grain zone in the northeast and east, in the coastal zones of Norway, Sweden including Finland, along the rivers of the Baltic countries, of the Dukedom of (East) Prussia and of eastern Poland. The interior forests of the Grand-Dukedom of Lithuania east of Poland floated their forest produce via the Bugo and Wistula rivers to Danzig and from western Lithuania via the Njemen and Memel rivers to port Memel and via the Duena river to Riga. Lithuania supplied much more forest produce than Poland, which for her larger part belonged to the three-field zone of grain export.

The remote location of these export-oriented forest regions with high transportation costs was possible 1. because *natural* forests could be exploited, and 2. because forest commodities like sawn timber, pitch and potassium had a much higher value than fuel wood. Extensive forests of East Prussia and of Lithuania were claimed as state monopoly by the Princes. They contracted with big private entrepreneurs, including merchants from Danzig and Konigsberg and members of the nobility who again engaged subcontrators who worked with free wage-labourers.[35] Only where the Duke or the nobility undertook the woodwork on their own account, they employed their corvee labour force, mainly for moving logs and timber to the rivers.[36]

Due to excessive transportation costs, exploitation for export was possible only under the condition of floatable watercourses. In his case study of Sweden, I.Layton (in this volume) has discussed the conditions of this zone and its links to the world market in detail. For an interpretation of the forest zone after von

Thünen's location theory a short reference will be made to its internal spatial zonation which evolved from different transportation costs for the various forest commodities depending on their specific labour intensity and value per weight unit. Von Thünen himself did not discuss this internal spatial structure but it can easily be deduced from his general rules. Such a von Thünen pattern could be reconstructed by the German historical geographer Friedrich Mager from 17th and 18th century sources for the forests of East Prussia.[37] It is a remarkable proof for the general von Thünen rule that growing distance from the river as the transportation basis for export to the market meant rising costs for overland haulage to the river, so that commodities of higher value had to replace those of lesser value per weigh unit: Thus fire wood was produced in the zone nearest to the river, beams, masts and planks in a second zone, and labour-intensive high-value products like tar, pitch and potassium in an even more distant zone. Glass, for the production of which potassium as well as charcoal was needed, iron, also based on smelting with charcoal, and finished wooden products (e.g. dishes) were produced in the remotest zone, but it seems that they went to regional markets only.

A characteristic feature of these various activities of forest exploitation was the short lived labourers' camp which after exhaustion of nearby recources had to be shifted to another place. Thus these semi-permanent settlements for about 50 to 80 people consisted of simple boothes only, and this is why all types of wood exploitation mentioned above were termed "Budenwerk" (literally "boothwork").[38] The mobile labourers, especially those who prepared potassium and pitch, were called "Budniks".

When through reckless exploitation the easily accessible forest recources of the Baltic region, especially the oak forests of East Prussia, had almost been exhausted by the beginning of the 18th century[39], the range of exploitation was extended further north[40] and upstream the Rhine, Weser and Elbe and their tributaries. The Black Forest as well as the forests along the Main were tapped for "Holländerstämme" ("Dutch trunks"). As early as the 16th century the forests along the tributaries of the lower Weser river were exploited including the southern Lüneburger Heide where extensive forests were through ruthless timber cutting and subsequent pasturing converted into bare heathland. Demand shifted the frontier of timber cutting to the hill forests along the upper Weser and finally even to the upper Werra from where the Thüringer Wald (Thuringian Forest) could be tapped.[41]

## 3.   Interregional Supplementary Sub-Systems

The second functional category in our systemic von Thünen approach to the world-system are interregional supplementary sub-systems. The first case: The remote zone of beef cattle breeding included Denmark as its northern wing.[42] For oxen the trail to the core market with Cologne and Amsterdam as the main places of consumption covered a distance of about one thousand kilometers. As they were moved on hoof for several weeks they naturally lost much of their weight and would have arrived at their destinations as mere skeletons. The economic solution of this problem was to use intermediate fattening zones. The most important one was the field-grass zone along the coast of the North Sea including the river marshes of the lower Elbe and Weser. The practice of regulated ley-farming included several years under grass which could be used for dairy as well as for beef cattle. Fattening Danish oxen was a well-paying option. In spring Danish cattle traders (since the mid 17th century also German and Dutch traders) exported some tens of thousands of oxen from Denmark. They had been brought up as lean steers on peasant farms and semifattened on the estates of the nobility and the Crown. A large part was sold to German and Dutch traders on the intermediary cattle markets ("Magermarkt", literally "lean market") near Hamburg (the most important was at Wedel). To get them fattened the cattle traders contracted with farmers of the ley-farming zone for pasturing them during the summer season. By the end of the summer the fat-stock continued the trail to urban markets in the west, via Münster, a centre of cattle traders, to Cologne, and across the Ems river to the Netherlands. The field-grass regions could easily adapt themselves either to grain booms or beef booms by changing the field-grass ratio.

The second case of dual sub-systems is the formation of subsidiary regional supply-hinterlands for main von Thünen zones. A section of the coastal field-grass zone and its hinterland, located between the Weser and the Ems rivers may serve as an example.[43] The field-grass zone exported grain to Amsterdam, fattened beef cattle to Amsterdam and Cologne, and dairy products to the regional urban markets, including the large relay-port towns of Bremen and Emden. During the peak season of grain harvest the regional labour market could not cover the demand. Migrant labourers were attracted from the adjoining inland zone beyond the limit of market-grain production who stayed for several weeks and returned to their home villages in autumn. There they invested part of their wages to establish tiny cottage farms on the heathland commons of their home villages and in new "colonies" on state owned peat lands. Frequent poor harvests on their subsistence holdings compelled them to buy grain from the bigger farms. Thus this hinterland served as a labour reservoir for the grain exporting zone. In addition to grain, the farms of the field-grass zone had a labour intensive dairy branch, which included fattening pigs on skimmed milk and barley. To reduce labour costs, the farmers left the labour intensive raising of piglets to the peasants of the hinterland and bought them on special stock-markets which developed on the border to the hinterland. Here they also bought young draught-horses, again raised on hinterland farms which of course also supplied lean steers for fattening. Thus this hinterland zone as a functional periphery of the field-grass zone supplemented it with important labour-intensive resources. This was possible because of the lower grain price beyond the transport limit for export.

## 4. Adaption of a Regional Society and its Cultural Landscape to the Role of a von Thünen Zone in the World-System

The last tier in this systemic approach to the world-system is the individual von Thünen zone. Its analysis, too, has to include more than purely economic aspects. Von Thünen based his model on the assumption of the *homo oeconomicus*. The actors on the stage of his theoretical economic world were rational, well informed, entrepreneurial people, not bound to any societal restrictions. In reality, societies were different, with many inherited social, economic and political power relations, with weak and strong societal strata. How did these societies adapt to the new conditions of the world-system in its various zones? This can be studied only on the lowest geographical tier, on that of Hoekvelds "area"[44] - the individual regions and localities with their regional and local societies. As has been said in the introductory remarks of this paper, various elements beyond the economic characteristics have to be studied as parts of the regional socio-economic structure which can be viewed as a system that reacted to the external impacts of the world-economy.

To demonstrate this approach, a regional example has been selected from the same field-grass zone of the marsh lands which we just referred to as the one spatial "partner" in the dual sub-system of the beef cattle economy. As all the elements of the regional structure are linked together, the sequence in which they will be presented may be arbitrary. We have tried to follow some lines of "hierarchical" reactions.[45]

The first element of the areal structure which members of the regional society started to change in reaction to the growing demand of the world-market for grain, was the area of arable which was considerably extended. Expansion was possible (first) by reclamation of new lands from the sea by diking. Step by step former ingression bays were transformed into polderlands. This labour and capital intensive business was undertaken mainly by the territorial princes who claimed the property of all virgin land. If they lacked the capital they sold or leased the area to be diked to contractors, in the Dollart Bay even to Dutch entrepreneurs. Diking demanded hundreds of workers who were engaged as migrant labourers from the aforementioned supplementary inland zone. The new lands were subdivided into large grain farms and either leased or sold. The old farm areas which prior had mainly been devoted to cattle grazing on permanent grass were now converted into arable under the ley-system. In response to booms on the international grain market with a rise of the price level, the arable areas were - in accordance with the von Thünen law -extended into inferior moist pasture land of the interior marshes. In slump periods they were turned to pasture again. In boom periods it was even profitable to improve inferior marshy soils by "mining" the calcareous subsoil and spreading it as fertilizer on the arable surface. All this labour-intensive work could only be carried out with the help of cheap migrant labour from the inland zone. Thus we can see that when drawn into the international market economy the peasants changed into efficient farmers and the princes into entrepreneurs.

Another adaption: Planting a new type of settlement on the grain polders and changing old settlement and field patterns in response to the requirements of a successful and rational convertible husbandry. In an optimal way this aim could be arrived at in the new polders which were subdivided into rows of large isolated farms as compact holdings of 30 to 70 hectares, large in comparison with the traditional holdings. The emerging local society of the polders consisted of big wealthy farmers and of permanent farm-hands who were drawn from the migrant labourers. The farmhouses were placed amidst their fields in the polders. For the cottages of the labourers, the old "dormant" dikes (in German "Schlafdeiche") between the polders offered cheap ground. This resulted in a very strict spatial separation of the two social classes.

In the old farm areas the medieval settlement pattern had consisted of villages built on artificial mounds, so-called "Wurten" ("terpen" in Dutch). Since the invention of dikes in the Middle Ages many farmers had moved out of the villages into isolated farms. But the former field pattern of intermingled plots had - up to the 16th century - no yet been fully reshuffled into consolidated holdings. Now this process of consolidation by individual exchange of the intermingled plots was accelerated under the demands of rational farming to reduce the timewasting distance to remote property. The growing number of cottars as wage-labourers of the farms concentrated in the *wurten* villages.

In the river marshes along the Weser the farmsteads remained in the villages, but the traditional open fields which consisted of numerous narrow strips were consolidated into larger plots, either step by step through mutual exchange or in one single action carried out at the request of the respective village community by an officer of the Count. In case of the villages east of the Weser this was achieved by laying out a well planned chequerboard field pattern, designed and carried out by the individual village councils as early as the 16th century.[46] The main purpose was to create larger pastures for fattening Danish oxen, a business which in the field-grass economy of the river marsh region proved to be even more paying than grain. These processes of deliberate restructuring of the field patterns have to be interpreted as adaptive reactions to the economic demands of the European market economy.

Another adaption to the new grain economy of the coastal marshes was the innovation of a new type of farmhouse (Fig.4). It was first invented on the Dutch wing of the coastal field-grass zone and from there diffused to the east up to the Jade bay. Dutch entrepreneurs who were engaged in making polders in Eiderstedt in Schleswig-Holstein introduced the Dutch "stelp" version of this farmhouse type (known as "Haubarg"). Where grazing of oxen and dairying were the main branches, with less area under grain, as was the case in the river marshes along the Weser, the new farmhouse type was not introduced or only very late because the traditional cattle-house still served the purpose. Thus it becomes obvious that the new type was, for its main purpose, a grain farmhouse with a huge barn for storing the harvest until threshing in winter.

Engagement in the grain market economy with its booms and slumps proved to be risky. Slump periods meant severe crises especially for small farmers who

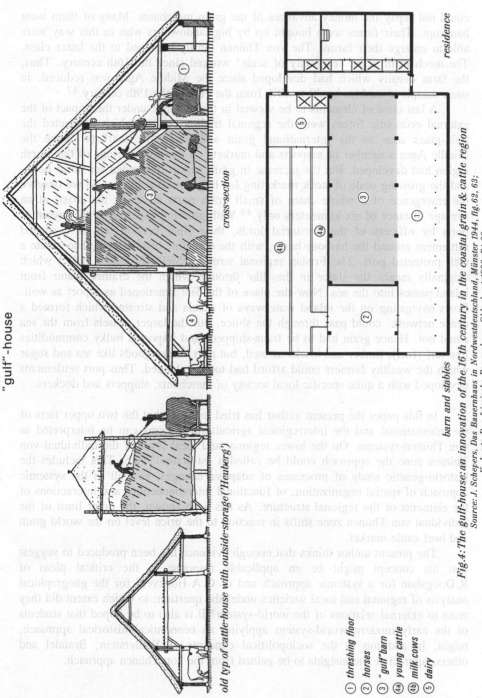

"gulf"-house

residence

cross-section

barn and stables

old typ of cattle-house with outside-storage ('Heuberg')

① threshing floor
② horses
③ "gulf" barn
④a young cattle
④b milk cows
⑤ dairy

Fig.4:The gulf-house:an innovation of the 16 th century in the coastal grain & cattle region
Source: J. Schepers, Das Bauernhaus in Nordwestdeutschland, Münster 1944, fig. 62, 63;
K. Junge, Das friesische Bauernhaus, Oldenburg 1936, fig. 23

could not repay the money advances of the grain merchants. Many of them went bankrupt. Their farms were bought up by big landowners who in this way were able to enlarge their farms. The von Thünen family belonged to the latter class. The mechanism of an "economy of scale" worked since the 16th century. Thus, the farm density which had developed since the Middle Ages was reduced, in some sample townships by 30 to 40% from the 16th to the 19th century.[47]

A last class of elements to be viewed in their change under the impact of the external economic forces were the regional trading centres which connected the field-grass zone to the international grain and beef-cattle markets. Since the Middle Ages a number of seaports and markets on the inland fringe of the marsh region had developed. But the increase in grain export demanded for more ports, and the growing scale of stock marketing for larger market places. So we observe the emergence of a whole chain of small grain ports all along the coast, at an average distance of six kilometers only.[48] With growing expertise in the layout of ports by officers of the territorial lords, their final version was a U-shaped settlement around the harbour basin, with the dike line bended inward to create a well protected port. The Frisian regional term is "Siel" (in Dutch "zijl") which originally means the sluice in the dike through which the drained water from inland passes into the sea. Now the place of the *Siel* functioned as a port as well. Boats navigating on the inland waterways of canals and streams which formed a dense network, could pass through the sluice, but the larger vessels from the sea could not. Hence grain had to be trans-shipped, and imported bulky commodities like turf (fuel), timber and bricks as well, but also luxury goods like tea and sugar which the wealthy farmers could afford had to be unloaded. Thus port settlements developed with a quite specific local society of merchants, shippers and dockers.

In this paper the present author has tried to show that the two upper tiers of the international and the interregional agricultural systems can be interpreted as von Thünen-systems. On the lower regional and local level of the individual von Thünen zone the approach could be called a "structural" one. This includes the historio-genetic study of processes of adaptive changes, as well as the systemic approach of spatial organization, of functional interconnections and interactions of the elements of the regional structure. As has been shown, the outer limit of the individual von Thünen zone shifts in reaction to the price level on the world grain and beef cattle market.

The present author thinks that enough evidence has been produced to suggest that his concept might be an applicable response to the critical pleas of R.Dodgshon for a systemic approach and of G.A.Hoekveld for the geographical analysis of regional and local societies under the question: to which extent did they react to external relations of the world-system? It is also to be hoped that students of the early modern world-system applying an economical historical approach, might, in addition to the sociopolitical concepts of Wallerstein, Braudel and others, appreciate the insights to be gained from the von Thünen approach.

**Notes and References**

1.  I.Wallerstein, The Modern World-System. Vol.I (1974), II (1980), III (1989) (London).

2.  F.Braudel, Civilisation matérielle, économie et capitalisme, XVe-XVIIIe siécle, Tom.1-3 (Paris 1979).

3.  F.Braudel, op.cit., tome 3 (Le temps du monde), quoted after the German edition (Aufbruch zur Weltwirtschaft), (Munich 1986) p.36-38.

4.  J.H. von Thünen, Der isolierte Staat in Beziehung auf Landwirtschaft und Nationalökonomie. Of main importance for economic geography is part 1 "Untersuchungen über den Einfluß, den die Getreidepreise, der Reichtum des Bodens und die Abgaben auf den Ackerbau ausüben", which was first published in 1826, with a revised edition in 1842. The third edition of 1875 (Berlin) is the first of the complete work of three parts. It was brought out as a reprographic reprint in 1966 by Wissenschaftliche Buchgesellschaft (Darmstadt). Quotations refer to this edition.

5.  See R.Dodgshon, The early modern world-system: A critique of its inner dynamics, in this volume.

6.  W. Abel, Agrarkrisen und Agrarkonjunktur. Eine Geschichte der Land- und Ernährungswirtschaft Mitteleuropas seit dem hohen Mittelalter (2nd ed. Hamburg 1966), pp.105-113.

7.  F.Braudel, op.cit. (note 2), vol.3, pp.35-36.

8.  The Danish breeding zone as one part of this dual subsystem is discussed in the article of K.E.Frandsen, production of beef cattle in the European periphery: A case-study from the island of Zealand, Denmark, in this volume.

9.  See the contribution of G.A.Hoekveld, World-system theory: Implications for historical and regional geography, in this volume.

10.  ibid., p.57

11.  See the contribution of H.Szulc, The impact of the evolution of large estates on the rural settlements of Silesia and Pomerania, in this volume, pp.192-194 with reference to the evolution of a bread grain supply belt in lower Silesia which favoured the growth of large feudal grain farms.

12.  For details see P.Saey and A.Verhoeve, The southern Netherlands - part of the core or reduced to a semi-peripheral status? in this volume, pp.99-103.

13.  W.Prange, Das Adlige Gut in Schleswig-Holstein im 18. Jahrhundert. In: C. Degn and D. Lohmeier (eds.), Staatsdienst und Menschlichkeit. Studien zur Adelskultur des späten 18.Jahrhunderts in Schleswig-Holstein und Dänemark (Neumünster 1980), pp. 57-75.

14.  I.Leister, Rittersitz und adliges Gut in Holstein und Schleswig. Forschungen zur deutschen Landeskunde 64 (Remagen 1952).

15.  A.Dunin, Economic life in 16th century Poland, Historical Atlas of Poland, English edition (Wroclaw 1986), p.19.

16.  For details see H.Szulc, op.cit. note 10, in this volume. For an interpretation from the view of an historical economist see W. Kula, An Economic Theory of the Feudal System. Towards a model of the Polish Economy 1500-1800, Foundations of History Library, New Left Review Editions (London 1976), first Polish edition 1962. A different position is held by the economic historian M.Malowist, Croissance et regression en Europe, 14-17 siécles (Paris 1972).

17.  J.Topolski, Economic Decline in Poland from the Sixteenth to the Eighteenth Centuries. In: P. Earle (ed.), Essays in European Economic History 1500-1800 (Oxford 1974), pp. 127-142, quotation p.131.

18.  ibid.,p.134

19.  ibid., p.139

20.    M.Bogucka, Danzig an der Wende zur Neuzeit: Von der aktiven Handelsstadt zum Stapel
       und Produktionszentrum, Hansische Geschichtsblatter 101 (Cologne 1983), pp. 71-103.

21.    See the map of A.Dunin, op.cit. note 14.

22.    D.Krannhals, Danzig und der Weichselhandel in seiner Blütezeit vom 16. zum 17.
       Jahrhundert (Leipzig 1942), pp.100-110.

23.    A.Dunin-Wasowicowa, Spatial Changes in Poland under the Impact of the Economic
       Dynamics of the 16th and 17th Centuries, in this volume, pp.172. An excellent review of
       early modern urban development in Poland is given by M.Bogucka, Die Städte Polens an
       der Schwelle zur Neuzeit. Abriß der soziotopographischen Entwicklung, W.Rausch (ed.),
       Die Stadt an der Schwelle zur Neuzeit, Beiträge zur Geschichte der Städte Mitteleuropas IV
       (Linz 1980), pp.275-91.

24.    G.Jensch, Der Handel Rigas im 17.Jahrhundert. Ein Beitrag zur livländischen Wirtschafts-
       geschichte in schwedischer Zeit, Mitteilungen aus der livländischen Geschichte 24/2 (Riga
       1930).

25.    See K.E.Frandsen, op.cit. note 8, in this volume.

26.    J.Baszanowski, Z dziejów handlu polskiego. Handel wolami (History of the Polish trade.
       Trade of oxen) (Gdansk 1977), especially pp.58-90. The present author is grateful to
       Dr.A.Dunin-Wasowiczowa for a review of the chapter relating to cattle trade.

27.    H.Szulc, op.cit. note 11, p.193.

28.    V.Zimanyi, The Hungarian economy within the early modern world-system, in this volume,
       p.234.

29.    A detailed discussion for Hungary is given by V.Zimanyi, in this volume, pp.239-249.

30.    J.Marino, Pastoral Economics in the Kingdom of Naples. John Hopkins University Press
       (Baltimore and London 1988), pp.207-208; U.Sprengel, Die Wanderherdenwirtschaft im
       mittel- und südostitalienischen Raum, Marburger Geogr. Schriften 51 (Marburg 1971).

31.    For Ireland see K.Wheelan, Ireland in the early modern world-system, in this volume,
       p.207.

32.    See the article of T.Stol, The northern Netherlands: Centre and periphery in the European
       core, in this volume, pp.87.

33.    H.J.Keuning, De Groninger Veenkolonien, een sociaalgeografische Studie (Utrecht 1933),
       J.Lucassen, Migrant Labour in Europe 1600-1900, Beckenham, Croom Helm, 1987,
       F.Bolsker-Schlicht, Die Hollandgängerei im Osnabrücker Land und im Emsland. Ein
       Beitrag zur Geschichte der Arbeiterwanderung vom 17. bis zum 19.Jahrhundert,
       Emsland/Bentheim. Beiträge zur neueren Geschichte 3 (Sögel, Verlag der Emsländischen
       Landschaft 1987).

34.    J.Bünstorf, Die ostfriesische Fehnsiedlung als regionaler Siedlungsform-Typus und Träger
       sozial-funktionaler Berufstradition, Göttinger Geographische Abhandlungen 37 (Göttingen
       1966).

35.    F.Mager, Der Wald in Altpreußen als Wirtschaftsraum. Ostmitteleuropa in Vergangenheit
       und Gegenwart 7 (2 vols.) (Cologne 1960), vol.II,pp.2-5.

36.    ibid., vol.II, p.34-35. Wallerstein's claim that coerced labour was a general phenomenon for
       the periphery, seems to be valid for feudal agricultural estates only. Besides the forest
       economy, the Hungarian oxen economy too was completely based on wage labour (see
       V.Zimanyi, op.cit.,in this volume p.238).

37.    F.Mager, op.cit. note 31, vol.I, p.255.

38.    ibid.,vol.I, p.263.

39.    ibid., vol.I, p.270, vol.II, p.165.

40.    See the sequence of maps in I.Layton, The timber and naval stores supply regions of
       Northern Europe in the early modern European world-system, in this volume, Fig.4 to 10.

41.    D.Ebeling, Rohstofferschließung im europäischen Handelssystem der Frühen Neuzeit am
       Beispiel des rheinisch-niederländischen Holzhandels im 17./18.Jahrhundert. In: Rheinische

Vierteljahrsblätter 52 (1988), pp. 150-169. For timber floating on the tributaries of the lower Weser see A.Peters, Die Schiffahrt auf der Aller, Leine und Oker bis 1618. Forschungen zur Geschichte Niedersachsens 4, Heft 6 (Hannover 1913); for the upper Weser and her tributaries see J.Delfs, Die Flößerei im Stromgebiet der Weser. Veröffentlichungen des Niedersächsischen Amtes für Landesplanung und Statistik, Reihe A I, 34 (Bremen 1952).

42.  A detailed discussion of the spatial and functional pattern of the Danish lean-cattle economy is presented by K.-E.Frandsen, op.cit. note 8, in this volume. For the history of the organization of oxen trade and drove trails see H.Wiese, Der Rinderhandel im nordwesteuropäischen Küstengebiet vom 15. bis zum Beginn des 19.Jahrhunderts. In: H.Wiese und J.Bölts, Rinderhandel und Rinderhaltung im nordwesteuropäischen Küstengebiet vom 15. bis zum 19.Jahrhundert. Quellen und Forschungen zur Agrargeschichte XIV (Stuttgart 1966), pp.1-132.

43.  For a more detailed interpretation see H.-J.Nitz, Transformation of old and formation of new structures in the rural landscape of northern Central Europe during the 16th to 18th centuries under the impact of the early modern commercial economy. In: Tijdschrift van de Belgische Vereniging voor Aardrijkskundige Studies 58 (Leuven 1989), pp.267-290, esp. pp.273-286.

44.  G.A.Hoekveld, op.cit. note 9, in this volume, p.50.

45.  For a more detailed discussion see H.-J.Nitz, op.cit. note 43, pp.273-282.

46.  H.A.Pieken, Die Osterstader Marsch. Werden und Wandel einer Kulturlandschaft. Bremer Beiträge zur Geographie und Raumplanung 23 (Bremen 1991), pp.36-45.

47.  F.Swart, Zur friesischen Agrargeschichte. Staats- und sozialwissenschaftliche Forschungen 145 (Leipzig 1910), pp.368-376.

48.  A.Schultze, Die Sielhafenorte. Göttinger Geographische Abhandlungen 27 (Göttingen 1962).

# Part II   The European Core

## 4

# The Northern Netherlands
## Centre and Periphery within the European Core

Taeke Stol

In the 16th century the King of Spain was the sovereign ruler of the Low Countries. In the 1580s, the revolt against the Spanish King Philip II resulted in a division into the Northern Netherlands - the Dutch Republic - and the Southern or Spanish Netherlands, today Belgium. The Dutch Republic was a confederation of seven provinces (Fig. 1), each of which jealously guarded its own rights and liberties. Nevertheless, the period between the 1580s and 1750 was a very prosperous one for the young Republic. The zenith of this so-called Dutch Golden Age was in the first half of the 17th century. The origins of this wealth have never been fully explained (Slicher van Bath, 1979; Van Zanden, 1988). It was a combination of internal and external factors, together with a number of coincidences. The geographical location of the Republic played a role: it was situated between the Baltic in the North and the Mediterranean in the South, and between the European continent in the East and the British Isles in the West. Nevertheless political, military, religious and economic reasons were also important. The Dutch took advantage of the weakening of the Old World empires Spain and Portugal, and of the unrest in other European countries. The wars on the continent disrupted the traditional trade channels, stimulating the development of new production centres. In the year 1585, troops loyal to the King of Spain took Antwerp, and this immediately triggered a blockade of the river Scheldt by the Dutch rebels. The leading role of Antwerp came to an end, and after a while Amsterdam took over.

A closer look at the location of Amsterdam (Fig. 1) shows that it was all but suitably situated. Sea-going ships could not easily reach the town. The route through the Zuiderzee was especially problematic, due to the great number of sandbanks. Approaching Amsterdam, ships had to pass by the shallow Pampus. When larger ships were built, Pampus became a real bottle-neck. The problem was solved by a technical innovation, called a 'ship's camel'. Two large caissons, similar to the halves of a ship, were fastened on to either side of a vessel. The water was then pumped out of these caissons and the vessel raised itself sufficiently enough to enable it to cross the sandbank. This is just one example illustrating that the Golden Age of Amsterdam and the Northern Netherlands was certainly not the result of an excellent natural site. The explanation has to be found in more structural features, and above all, the role human agents played.

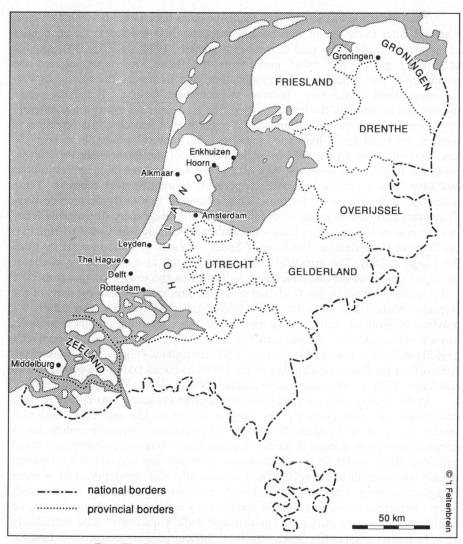

*Fig.1: The Dutch Republic of seven provinces around 1650*

*Drenthe was part of the Republic, but was not represented in the States General because of its relative unimportance.
In the south some areas were directly ruled by the States General (nowadays, from east to west, the provinces
of Limburg, North Brabant and part of Zeeland).*

The foundation of the Dutch Golden Age dates back to the late Middle Ages, when activities in the countryside became increasingly connected to the economies of the towns, leading to a kind of symbiosis (Jansen, 1974, pp. 255-256). Imports of primary goods, like bread grain from the Baltic, and meat, offered the possibility of specialization and intensification, thus increasing the role of capital. Commercialized agriculture became a common feature. Relatively huge investments to maintain the productivity of the land were necessary from an early stage, especially in the humid areas with soils of peat and clay. The creation of polders, the building of dikes and, since the 15th century, windmills to keep the land drained, were all carried out at the expense of the landowners themselves. From the 13th century onwards, they established a number of drainage institutions that gave them a vehicle to secure cooperation in improvement schemes. These water control boards were, for those days, organized in a fairly democratic way. They came into existence due to an initiative by the landowners. It was only later, that the rule of the sovereign ruler - the Bishop of Utrecht, the Count of Holland, the Duke of Gelderland or the King of Spain - became of some importance (Van der Linden, 1988). The principle of private enterprise also stood at the cradle of the development of the manufacturing industries in the towns. In this respect, I feel it worth mentioning that feudalism had been almost absent in most parts of the Northern Netherlands since the late Middle Ages. Around the year 1500, the province of Holland, and we may assume the other maritime provinces as well, had a proto-capitalistic, proto-industrial structure. In 1525, 44% of the population in Holland lived in towns of more than 2,500 inhabitants (Fig.2). The basis of the 'take-off' of the Dutch Golden Age in the 1580s had been laid in the period before (De Vries, 1976, p. 94; Atlas van Nederland, Vol. 2, 1984; Van Zanden, 1988).

In the 16th century, a capitalistic world-economy came into being. A good description has been given by the Belgian historians Lis and Soly: Sixteenth century capitalism was essentially commercial capitalism, extremely mobile due to minimal investment in durable means of production. Merchant entrepreneurs could withdraw their capital from one business or region and transfer it to another, quickly and with little financial loss. Consequently, the development of western European industry was characterized by a continuous 'redrawing of the map'. Since commercial capitalism brought about so few technological improvements, no single town or region during its 'golden age' took a qualitative lead sufficiently great to remain competitive when structural difficulties were confronted. Instead, new and better situated centres took over. At the national or international level, such transformations only signified replacement and compensation. At the local or regional level, on the other hand, the results were disastrous (Lis and Soly, 1979, pp. 67-68).

How did the introduction and diffusion of the capitalistic way of production take place in the Northern Netherlands during the 16th and 17th century? The available data of a number of products enable us to have a closer look. In fact I will start with the conclusion: the type of diffusion in this period was hierarchical diffusion (Baker,1975, pp. 27-28); the spread of innovations was undertaken by individuals, mostly merchant entrepreneurs.

A highly relevant example of this kind of diffusion can be seen in the introduction and development of tobacco growing (Roessingh, 1979). Commercial production of this cash crop in the Netherlands started between 1610 and 1620. Tobacco growing was an innovation in Dutch agriculture. Initially, cultivation was stimulated by Amsterdam tobacco merchants who saw in inland production a source for the reserve they needed in order to influence the price of overseas tobacco on the Amsterdam market. Others - such as the ruling urban upper class, rich citizens and the gentry who had ties with trade as well as agriculture - initiated tobacco growing around several small towns in the middle and east of the Netherlands (the provinces Utrecht, Gelderland and Overijssel). Prominent farmers with some education and commercial initiative were quick to adopt tobacco as a crop and small farmers followed their lead. In other West European countries, we also see the spread of tobacco growing from merchants and other urban people or gentry to farmers and peasants. The commercial production of other cash crops, like hop (for the production of beer) and hemp (to make ropes for the shipping industry), was similarly distributed (Bieleman, 1981).

So we see a continuing commercialization of the agricultural sector and increasing linkages between towns and countryside. One of the most striking examples is the relation between the growth of towns and industry on the one hand, and the development of the peat-digging industry in the Netherlands on the other. In a country with no coal and few forests that can be exploited for fuel, people used peat instead. This had already been mentioned in the 1st century by the Roman Pliny, and later in the 10th century by an Arab diplomat. Commercial peat-digging started in the Southern Netherlands in the late Middle Ages; the most important and largest towns were located there, i.e. Bruges, Ghent and Antwerp, and peat was available immediately adjacent to these towns. When these supplies were exhausted, one turned to the vast peat areas in the Northern Netherlands. From the middle of the 16th century onwards, commercial peat-digging took place in the border area between the provinces of Utrecht and Gelderland, and soon after it was also started in Friesland, Groningen and Drenthe. Large-scale peat-digging requires high investments - the digging of canals, the building of sluices and bridges - before profits start to flow. Only wealthy people and institutions could afford that. Usually they spread their risks by cooperating in an organization called a peat company. Most of them had shares in different peat companies, as well as in other sectors of the economy. We can follow the diffusion of these commercial peat-digging companies particularly well. It started in Flanders and Brabant, and the introduction into the Northern Netherlands was undertaken in 1550 by a company of merchants based in Antwerp at the time. Diffusing it further north was carried out by important citizens of Utrecht and Holland. Once again we discover spreading by means of hierarchical diffusion (Stol, 1990).

However, I should like to advise some caution on this subject. Firstly, the shift of commercial peat-digging from the Southern to the Northern Netherlands has nothing to do with the shift, at approximately the same time, from Antwerp to Amsterdam as world centre. It was caused only by exhaustion of peat-supplies in the South. Secondly, we should reject the idea that the Dutch Golden Age was based on the easy accessibility of a huge amount of peat to supply their need for energy, just as the British Empire was based on coal, and as the dominance of the United States was founded on oil and natural gas (De Zeeuw, 1978). The explanation of prosperous periods in certain areas is, moreover, highly complex, and most of the credit has to be given to human agency.

All this being said, it should not be forgotten that the Golden Age of the Dutch Republic was essentially that of a maritime empire. The foundation had been laid in the late Middle Ages with the shipping of grain from the Baltic. Amsterdam became the entrepôt of grain for the whole of Europe. This entrepôt served many regions in occasional years of harvest failure, but it served the Dutch Republic every year. By far the largest part of Baltic grain exports was consumed in the Low Countries (De Vries, 1976, pp. 71-72). The ability to import grain was the key factor behind the process of specialization in the Northern Netherlands. The Baltic grain-producing regions can be regarded as the first 'colonies' of the Netherlands. Characteristic of this trade was the transportation of bulk goods. In the 17th century, trade with the colonies in Africa, Asia and the Caribbean brought the transportation of luxury goods. The Dutch trade hegemony was the result of these two different kinds of commodity: bulk goods and luxury goods (Israel, 1989). This was a new element in the economy of a world centre. In Venice, Genoa and Antwerp the activities had mainly been restricted to luxury goods.

The shipowners created companies in order to spread the risks and profits. Shares of 1/256 of one ship were not uncommon (Lambert, 1985, p. 199-200; Boxer, 1988, pp. 6-7). This system was already common practice in the Baltic trade in the 15th century, and was not exclusively restricted to the Netherlands (Ketner, 1946, p. 146; De Vries, 1976, p. 118; Braudel, 1982, Vol. II, pp. 433-455). It was also used in the fishing industry, especially herring.

During the 16th and 17th century labour wages and purchasing power in the Dutch Republic were high, compared with other European countries (De Vries, 1976, pp. 93-94; Noordegraaf, 1985, p. 187). This obvious prosperity acted as a lodestar to the unemployed and under-employed of neighbouring countries. Not only Flemings and Walloons, but also Scandinavians and Germans swarmed into the Republic in the belief that the streets of Amsterdam were paved with gold (Boxer, 1988, pp. 64-65). The political and religious turmoil set in motion groups of refugees whose skills were particularly advanced. After 1585, for instance, thousands of mostly protestant inhabitants of Antwerp fled to the Dutch Republic, bringing with them their skills, money and commercial contacts. They took a great part in the hierarchical diffusion of innovations.

Much research still remains to be done on the spatial patterns of all these commercial products. The key factor is obviously the distance to the urban markets. A location near to waterways, enabling cheap transport of bulk goods, was a favourable one. In the 17th century, a complete network of canals, so-called 'trekvaarten', was created. Although initiated by private enterprise, this transportation network stimulated the consolidation of an urban system of virtually autonomous cities (De Vries, 1978).

In historical studies, the Northern Netherlands are usually divided into the maritime (peat and clay) provinces in the west and north (Holland, Zeeland, Utrecht, Friesland and Groningen), and the inland (sand) provinces in the east and south (De Vries, 1974; Van Zanden, 1985). The former were fully integrated in the Dutch trade economy. The main countryside products for the market were: dairy products, cash crops, horticultural products and peat. The inland provinces were dependent on subsistence, being a peasant economy and rarely producing an exportable surplus. The farmers did not see their farms as a means for making maximum profits. Their initial aim was to assure their own survival. Risk minimizing was one of the main principles that ruled farming in these provinces, compelling the farmers to run a very diversified type of farming.

However, recent research shows that, to a certain extent, the inland provinces were also connected with the commercial economy of the maritime provinces. I have already mentioned the development of hop and tobacco growing. In the 17th and 18th century, the eastern part of the province of Gelderland exported grain to the towns in Holland (Wildenbeest, 1983). In the open-field villages in the province of Drenthe, farmers kept a great number of cattle. This cattle-breeding for the Dutch and Flemish towns accords with the zoning theory of von Thünen. Large parts of Drenthe belonged to the peripheral part of this zonation, which also included Denmark and parts of Northern Germany (Bieleman, 1987, pp. 666-667). Since the 17th century, however, a huge amount of peat had been dug out in Drenthe and exported to the towns in Holland as well. The peat-producing regions can be regarded as the second zone of von Thünen's model, the zone of forests/fuel. Thus, the province of Drenthe is in two different zones. Probably the explanation might be found in the well-developed canal traffic system in the Netherlands.

Many researchers consider that, because of the small scale and the open economy of the Netherlands, it is imperative to describe its regional development in terms of the international context (De Smidt, 1987, p. 135). To them, the theories of Wallerstein are a useful conceptual framework, though some researchers regard them as a simplification. And indeed, the economy of the Netherlands does show great variation in time and in space. We usually see expansion in one region or one sector of the economy and contraction in another region or sector (Klein, 1979, pp. 81-82; Zijp a.o. ,1989, pp. 10-12). This can be seen by examining the demographic development (Fig. 2). The 16th and 17th centuries saw a tremendous growth of urban population, especially in Holland. In the late 17th and 18th century, a process of de-urbanization characterized this province (Table 1). With the exception of Amsterdam, Rotterdam and The Hague, the towns all saw a sharp decline in their number of inhabitants. This was particularly the case in Enkhuizen and Leyden. One cause of this decline can be found in the protectionist measures taken first by England and France and later by other countries as well. But above all, the competition with merchants and manufacturers from other countries became increasingly tough. The export-oriented industries, mostly located in the towns of Holland, suffered heavily. Only the three large towns, in particular Amsterdam, could cope with such a period of

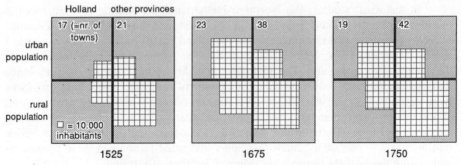

Fig.2: Urbanization of the Netherlands in 1525, 1675 and 1750
(Source: Atlas von van Nederland, Vol.2, p.17, map 20)

|             | 1600  | 1650  | 1700  | 1750  | 1800  | 1850  |
|-------------|-------|-------|-------|-------|-------|-------|
| Amsterdam   | 65    | 175   | 200   | 210   | 217   | 224   |
| Rotterdam   | 13    | 30    | 48    | 44    | 57    | 90    |
| The Hague   | 10    | 18    | 33    | 38    | 38    | 72    |
| Alkmaar     | 12    | 13    | 12    | 11    | 8     | 10    |
| Hoorn       | 12    | 16    | 13    | 10    | 10    | 9     |
| Enkhuizen   | 17    | 22    | 14    | 7     | 7     | 5     |
| Leyden      | 25    | 67    | 55    | 38    | 31    | 36    |
| Delft       | 20    | 24    | 18    | 15    | 17    | 18    |
| Middelburg  | 20    | 30    | 25    | 24    | 20    | 16    |
| Groningen   | 19    | 20    | 20    | 21    | 24    | 34    |
| Netherlands | 1.500 | 1.850 | 1.900 | 1.950 | 2.000 | 3.000 |

Tab.1: Number of inhabitants (x 1000) in ten towns in the Netherlands
(Source: Schmal, 1988, p.291)

contraction with some degree of success (Schmal, 1988). Summarizing, we see a concentration of formerly wide-spread activities in a small number of towns.

On the other hand, we see an increase of the rural population in the other provinces in the 18th century. This is partly the result of some industries moving to the countryside; for example, the labour-intensive parts of the Leyden textile industry. This happened on account of the labour wages being lower in the country. Most of the growth in these provinces, however, can be explained by developments in the agricultural sector.

Can these developments be fully explained by changes in the world economy, in which the Dutch Republic lost its leading position? Certainly some developments can be explained in such a way, but not all of them, and only to a certain extent. Dynamism and dynamic individuals had not all suddenly disappeared. The agricultural sector was, and remained, highly commercialized, export-oriented and industrialized. In contrast to the description of Lis and Soly I quoted before, this sector took a qualitative lead which was sufficiently great to remain competitive when structural difficulties were met with. In the towns, some of the industries remained competitive as well, particularly, if they were highly specialized and capital intensive. An example of this is the diamond industry in Amsterdam that has remained there to the first half of the 20th century (Van Tijn, 1989). Finally, we should keep in mind that commercial capitalism dominated Dutch economy up till the 19th century. This is at least one explanation of why it took so long before industrialization started in the Netherlands.

To conclude, it can be seen that for a better understanding of the role that the different parts of the Northern Netherlands played within the world economy, two kinds of research are necessary. First, regional studies concentrating on one economic sector: the distinction Braudel makes between a basic, a market, and a world economy will be of help to understand the functioning of a certain region on different levels. Hoekveld (in this volume) has already mentioned the necessity of distinguishing several levels of analysis. Secondly, we need more systematic investigations into specific products, such as cash crops or industrial commodities. How and where did the introduction and diffusion take place, and where did the necessary capital come from? Both kinds of research should not merely be restricted to the period of expansion, but should also include the period of contraction.In this way, we will be able to find out which regions were, at a given time, parts of the centre, and which were parts of the periphery of the Northern Netherlands, which at that time was the European core of the world system.

## References

Atlas van Nederland. Vol. 2. (1984): Bewoningsgeschiedenis. 's-Gravenhage

Baker, A.R.H. (1975): Historical Geography and Geographical Change. London

Bieleman, J. (1981): De noord-drentse hoptcelt. In: Nieuwe Drentse Volksalmanak, pp. 61-78

Bieleman, J. (1987): Boeren op het Drentse zand 1600-1910. Een nieuwe visie op de 'oude' landbouw. Wageningen

Boxer, C.R. (1988): The Dutch Seaborne Empire 1600-1800. London

Braudel, F. (1982): Civilization and Capitalism 15th-18th Century. 3 Vols. New York

Israel, J.I. (1989): Dutch Primacy in World Trade, 1585-1740. Oxford

Jansen, H.P.H. (1974): Middeleeuwse geschiedenis der Nederlanden. Utrecht/Antwerpen

Ketner, F. (1946 ): Handel en scheepvaart van Amsterdam in de vijftiende eeuw. Leiden

Klein, P.W. (1979): De zeventiende eeuw (1585-1700). In: J.H. van Stuijvenberg (ed.), De economische geschiedenis van Nederland.Groningen , pp. 79-117

Lambert, A.M. (1985): The Making of the Dutch Landscape. An Historical Geography of the Netherlands. London [etc.]

Linden, H. van der. (1988): De Nederlandse waterhuishouding en waterstaatsorganisatie tot aan de moderne tijd. In: Bijdrageh en mededelingen betreffende de geschiedenis der Nederlanden, 103, pp. 534-553

Lis, C., and H. Soly (1979): Poverty and capitalism in pre-industrial Europe. Bristol

Noordegraaf, L. (1985): Hollands welvaren? Levensstandaard in Holland 1450-1650. Bergen

Roessingh, H.K. (1979): Tobacco Growing in Holland in the Seventeenth and Eighteenth Centuries: A Case Study of the Innovative Spirit of Dutch Peasants. In: The Low Countries History Yearbook; Acta Historiae Neerlandicae XI (1978). The Hague , pp. 18-54

Schmal, H. (1988): Patterns of de-urbanization in the Netherlands between 1650 and 1850. In: H. van der Wee (ed.), The Rise and Decline of Urban Industries in Italy and in the Low Countries (Late Middle Ages - Early Modern Times). Leuven, pp. 287-306

Slicher van Bath, B.H. (1978): De economische toestand van de Republiek in de 17e eeuw. In: B.H. Slicher van Bath, Geschiedenis: theorie en praktijk. Utrecht/Antwerpen, pp. 360-374

Smidt, M. de (1987): In pursuit of deconcentration: the evolution of the Dutch urban system from an organizational perspective. In: Geografiska Annaler, 69B, pp. 133-143

Stol, T. (1992): De veenkolonie Veenendaal. Turfwinning en waterstaat in het zuiden van de Gelderse Vallei, 1546-1653. Stichtse Historische Reek 17

Tijn, Th. van (1989): De Amsterdamse diamanthandel en -nijverheid. In: Holland, regionaal-historisch tijdschrift 21, pp.248-262

Vries, J. de (1974): The Dutch Rural Economy in the Golden Age, 1500-1700. New Haven/London

Vries, J. de (1976): The Economy of Europe in an Age of Crisis, 1600-1750. Cambridge

Vries, J. de (1978): Barges and capitalism. Passenger transportation in the Dutch economy, 1632-1839. In: A.A.G. Bijdragen 21, pp. 33-398

Wildenbeest, G.W.F. (1983): De Winterswijkse scholten: opkomst, bloeien, neergang. Een antropologische speurtocht naar het fatum van een agrarische elite. Amsterdam

Zanden, J.L. van (1985): De economische ontwikkeling van de nederlandse landbouw in de negentiende eeuw, 1800-1914. Utrecht

Zanden, J.L. van (1988): Op zoek naar de 'missing link'. Hypothesen over de opkomst van Holland in de late Middeleeuwen en de vroeg-moderne tijd. In: Tijdschrift voor sociale geschiedenis 14, pp. 359-386

Zeeuw, J.W. de (1978): Peat and the Dutch Golden Age. The historical meaning of energy-attainability. In: A.A.G. Bijdragen 21,pp. 3-31

Zijp, R.P. a.o. (eds.) (1989): Barre tijden. Crisis en sociale politiek rondom de Zuiderzee, 1650-1850. Zutphen

# 5

# The Southern Netherlands
## Part of the Core or Reduced to a Semi-Peripheral Status?

Pieter Saey and Antoon Verhoeve

## 1. Introduction

In this paper an attempt is made to answer the two following questions:
1. Are there changes in the geographical development of the Southern Netherlands that support the thesis of Wallerstein -the inception of a world-system in the 16th century with the reduction of the Southern Netherlands from a candidate core-area to a semi-peripheral area?
2. Does the geographical analysis of the internal differentiation suggest anything about the degree of integration of the supposed system?

The main purpose of this paper is to elucidate evidence for, or against, Wallerstein's theory, which has been produced independently of Wallerstein's work.

The two questions will be answered by means of investigating the explanatory power of three theories:
1. An attempt will be made to apply von Thünen's theory to agricultural landuse in the Southern Netherlands.
2. An explanation will be made of Van der Wee's theory of the urban development of Flanders and Brabant.
3. Comments will be presented relating to Mendels' model of proto-industrialization as applied to Flanders.

The choice of these three theories will be clear at the end of the paper.

## 2. Agricultural Production Systems in the Southern Netherlands and von Thünen's Theory

### 2.1. The von Thünen Model and Theory

In 1826 von Thünen, farmer and landed proprietor, published the first part of 'The Isolated State' concerning 'the effect of grain prices, fertility and taxation on agriculture' as the subtitle reads. In this book von Thünen elaborated an econometric model *avant la lettre* of agricultural landuse. It relates to a completely homogeneous plain with a very large town at the centre. The town obtains all its provisions from the rural areas in the plain in exchange for all the salt, metals and

industrial products the countryside needs. The landuse in the plain is characterized
by a succession of concentric rings: free cash cropping, forestry, crop alternation
system, 'improved' (field-grass-) system, three-field system, stock-farming.[1] Most
of the industrial crops are grown in the outermost ring (see Fig. 1 in the article of
Nitz).

The explanation for the succession of concentric rings is threefold:

1. The existence of an innermost ring of free cash cropping is possible
because the fertility of the soil is raised by manure brought from the town.

2. The rings of the crop alternation system, the 'improved' (field-grass-)
system and the three-field system are grain producing zones. The crop alternation
system is the most intensive system, the three-field system the most extensive one,
with the term 'intensive' being used to refer to the number of cost inputs that are
applied to the economic margin (von Thünen 1966, p.XXXI). The succession is
brought about by the increasing costs of transportation that compel the farmer,
living at greater distances from the town, to save on production costs. An intensive
system of cultivation with larger costs of production yields also a larger product,
but the revenue thereof is progressively reduced by the transport costs as the
distance to the town increases. Consequently, from a certain distance a less
intensive system starts to provide a larger net revenue than a more intensive one
because of its lower production costs. At a still greater distance grain production
(for the town) will stop altogether and stock-farming becomes the sole activity that
pays.

3. The location of forestry (in the second ring) and of industrial crops (most
of them in the farthest ring) is determined by the costs of transportation in relation
to their value.

Forestry and industrial crops figure in the model only if their price is high
enough in relation to the grain price. The existence of the grain producing belts
themselves depends on the grain price. An intensive system of cultivation is
remunerative only in the case of high grain prices. When the grain price rises (in
comparison with other prices), the area under intensive cultivation will increase,
the distinct rings of grain production will expand and the three-field system will
occupy a part of the former zone of stock-farming. When the grain price falls (in
comparison with other prices), the grain production rings will contract.

In applying the model to reality, two considerations should be taken into
account:

1. The interaction of prices, transport costs and savings on production costs,
as depicted by von Thünen, does not necessarily lead to a succession of concentric
rings; the different systems of cultivation that result from this interaction (which
henceforth we call the von Thünen mechanism) may constitute a more complex
and irregular spatial pattern.

2. A pattern of concentric rings can be created from methods other than the
von Thünen mechanism.

In other words, the von Thünen mechanism is neither a sufficient nor a
necessary condition for a concentric-belted pattern of farming systems.

Ad 1.

The succession of concentric rings is generated only because a) the plain around the central town is isotropic, i.e. transport can take place in any direction in the same manner, and b) all soil, with the exception of the innermost ring, has the same fertility. As river transport is so much cheaper than transport overland, the pattern of concentric rings will be disturbed if the central town is located on a navigable river (von Thünen 1966, pp.172, 215-216). 'Fertility' does not mean the natural suitability of the soil, but the level of fertility a soil has attained under cultivation. Soil fertility operates in the same way as the price of grain (ibid., p.74). Higher production costs are remunerative when soil fertility is high as they do not pay in the case of infertile soils. As a consequence, agricultural landuse will show an irregular spatial pattern when the fertility of the soil varies in an equally irregular fashion (ibid., p.174). In addition, industrial crops that greatly exhaust the soil may be cultivated near the town, instead of far away from it, if the soil fertility has been raised to the required level e.g. through the application of a crop alternation system (ibid., p.186). It should be noted that the von Thünen model does not imply a uniform distribution of the population or a uniform farm size. On the contrary, the crop alternation system, for example, is feasible on small or at most medium-sized holdings only (ibid., p.83). More generally, the population density will vary with the intensity of the system of cultivation. Because of the larger gross product, the crop alternation system can support a larger population than the 'improved' (field-grass-) system. Von Thünen cites the population figures for the western part of Belgium and the Département du Nord in France to prove this (ibid., p.87-88). And while the belt of stock-farming will, generally speaking, support but a very thin population, the population density will be larger when flax production takes place in this ring (ibid., p.185). When flax is produced in the belt of crop alternation, a very high density of population will be attained. Von Thünen cites the example of East Flanders (ibid., p.185).

Ad 2.

In von Thünen's theory population density, farm size, and farm structure are dependent variables. The essence of the theory is maximization of profit in the shape of land rent. Land rent is "that part of the total (gross) product of land that remains as a surplus after deduction of all costs, including interest on invested capital" (ibid., p.LI). A piece of land is subjected to the landuse that yields the largest land rent, and all components of agricultural activity are adapted to it. However, variations in population density, and in size and structure of the holdings, may also be the outcome of historical developments in which the specific market relations of the von Thünen mechanism do not play the same decisive role in generating the model (either because they do not exist, or because they take place on a scale other than the scale of the said variations). Once the market relations are established, these variations may prevent the creation of the pattern of concentric rings that, in their absence, should result from the functioning of the rent maximization process. If, for example, a dense population of small peasant farmers occupies the ring which according to the theory should be subjected to forestry, this type of production will not take place, because the peasants simply want a living for all members of the family. They do not try to maximize profits; their labour is not reckoned as a cost (Eysberg 1979). On the other hand, even in the absence of a rent maximization process, the variations in farming systems can constitute a pattern of concentric belts when the factors that are at the origin of these variations (for example soil suitability) vary in space in the required manner. It does not require much imagination to construct e.g. a connection between the nature of the soil and landuse (Figs. 1 and 2) in the Southern Netherlands, independent of any market relations.

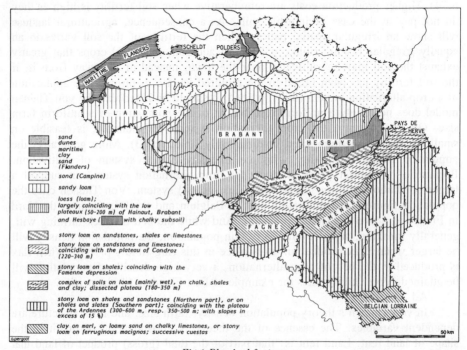

*Fig.1:Physical factors*
Source: Ameryckx, J., Verheye, W. & Vermeire, R. (1985), Bodemkunde. Ghent: Prof. Ameryckx, fig. 84. Tavernier, R.+Maréchal, R. (1959), De bodemassociakaart van België. Natuurwetenschappelijk Tijdschrift, 41, p. 161–204.

Von Thünen equates the maximization of land rent with rational farm management. Real world deviations from the theory are explained as the outcome of irrational behaviour either by ignorance or by inertia. In this respect von Thünen undeniably is a faithful adherent of Adam Smith's theory of the invisible hand. In our opinion, one gets a better insight into reality by envisaging social relations that make the individual partners in these relations pursue interests, that, upon analysis, appear to be social class and other group interests. In the context of agricultural activity, social relations are defined, in the first place, by proprietary relations (such as large landowners versus tenants versus landless labourers) and the size of the holdings (such as large tenants versus small tenants and peasants). The degree to which a rent maximization process might govern landuse depends on the promotion of the said interests because the way in which this is done determines how much room there is for depth investments and by whom these depth investments are made.

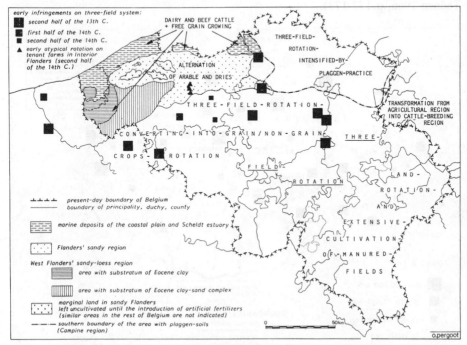

*Fig.2:Rotation systems in the Southern Netherlands in early modern times*
*(cf. Appendix)*

## 2.2.  The Systems of Cultivation in the Southern Netherlands

### 2.2.1. Descriptive Comparison of the Southern Netherlands with the Model of von Thünen

The map of the systems of rotation in the Southern Netherlands in early modern times (Fig. 2) reveals the existence of a zonal pattern of differing intensities that resembles the succession of grain producing belts of increasing intensity in von Thünen's model. Of course, the Southern Netherlands are not built up around one single central town as is the Isolated State (Fig. 3), but that is no real problem. Von Thünen himself explains what happens in a zone with many towns. If a large town in this zone gets its grain from remote areas, the grain price will be high in that large town because of the transport costs. But a high grain price in the large town means a high grain price in the small towns too, otherwise the farmers living in areas adjacent to the small towns will send their grain to the large town. As a result, agriculture will be intensive everywhere in this zone (von Thünen 1966, p.172-173).

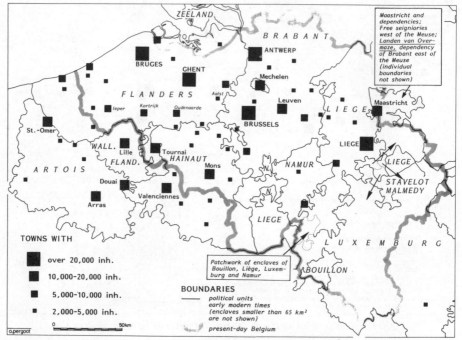

*Fig.3: Towns with more than 2,000 inhabitants in the late 15 th century*
(according to Bolckmans e. al. 1980, p. 48)

In the von Thünen model, the grain growing systems are arranged along the axis crop alternation/convertible husbandry/three-field system. A specific form of convertible husbandry, the 'improved' (field-grass-) system, figures as the point of reference. In chapter 18 of the Isolated State von Thünen presents a more complete series of six farming systems to be applied according to soil fertility and grain price. The series ranges from the pure crop alternation system to the pure three-field system. It also includes, besides of course the pure 'improved' field-gras-system system, three transitional systems that are chosen from the set of innumerable graduations by which mixed systems may approach, or deviate from, the pure forms. In addition, von Thünen specifies the circumstances under which the infield-outfield variation of the 'improved' (field-grass-) system is recommended (ibid., p.102-103).

In the Southern Netherlands the farming systems can be arranged along the axis crop alternation/three-field system/infield-outfield. The point of reference is the three-field system. The stage of crop alternation in the sense of von Thünen (a system without fallow) is not yet attained. Therefore we use the term 'grain/non-grain crops rotation' to refer to infringements on three-field rotation and to atypical rotations with fallow. On the other side of the point of reference, we find at the extreme of the axis a very extensive form of infield-outfield: the farming system of the Ardennes. A variation of the infield-outfield is practised in the Campine.[2] On the map (Fig. 2) it is marked as intensified three-field rotation. This obviously refers to the manured fields only. Heathland was used for sheep breeding and the consequential production of manure.[3] The alternation of arable and *dries* (impoverished arable used as pasture) falls under the category of temporary cultivation. In Flanders it was either the characteristic way of cultivating the outfield of large holdings, or a farming system in its own right, particularly on insufficiently drained soils. In what is now Central Belgium (with the exception of its northernmost and western parts) temporary cultivation, together with the two-field system, preceded the three-field system as the prevailing method of cultivation. After the holdings turned to the three-field rotation, the *trieu* (the Walloon equivalent of *dries*) continued to occupy a significant part of the land. In the Ardennes the outfield land rested for such a long period that one might speak just as well of land rotation.[4]

Free grain growing does not fit in with the general pattern. In the Isolated State the concept of free cash cropping contains two elements: commercialized gardening (including milk production and the cultivation of potatoes) and the absence of a fixed rotation of crops (as manure was brought from the town). However, in Maritime Flanders and the adjacent part of the Flemish sandy loam region, the staple product was grain and the absence of a fixed rotation was due to significant numbers of livestock, fed, to a rather high degree, with fodder plants. On the other hand, the first infringements on the three-field system in Brabant and Flanders bore the mark of an extension of certain practices of gardening to field cultivation and possibly the use of manure from the town.

## 2.2.2. The Belt of Intensive Cultivation.

In spite of the rather striking parallelism, described above, between the spatial pattern of the systems of cultivation in the Southern Netherlands and the von Thünen model, any attempt to explain this pattern exclusively in terms of the von Thünen mechanism would be a futile exercise. Some of the basic features of the distribution of farming systems in the Southern Netherlands date back from periods in which market relations cannot have played the role such an explanation would ascribe to them. However, from the moment the process of urbanization began and the towns became an integral part of the economic fabric, the von Thünen mechanism might have contributed to the important changes of the agricultural production system in some regions and to its conservatism in other regions.

The debate around the dating of the intensification of Flemish and Brabantine agriculture is of particular interest here. In a paper published in 1956, Verhulst tried to refute the thesis of some noted historians that the three-field system was infringed upon as early as the 13th century. A text from 1252 concerning a dispute about the tithing on rape and similar produce in a village near Oudenaarde was interpreted mistakenly as evidence of the cultivation of crops on the fallow (rape and pulses were cultivated on separated small, mostly enclosed, fields near the farmstead, outside the courses of the three-field rotation). As this text was the sole piece of direct evidence of infringements on the fallow in 13th century Flanders put forward by the adherents of the said thesis, Verhulst was led to the conclusion that the intensification of Flemish agriculture started only in the 14th century and in a very incidental fashion: the number of cases of growing crops on the stubble fallow, or on the fallow in the 14th and 15th centuries, described by Lindemans, sank into insignificance beside the number of records, gathered by Verhulst himself, that pointed to the strict application of the three-field rotation in the same period. Verhulst thought it probable that famine and the sudden increase of the population after an epidemic (frequent phenomena in the 14th and 15th centuries) were the origin of short and disputed infringements on the fallow in order to meet the temporarily increased needs for agricultural products. If this were true, an explanation in terms of the von Thünen mechanism would be beside the point.

However, since 1956 a number of texts have been discovered (in some cases rediscovered), that attest to the intensification of agriculture in Brabant, French Flanders and Artois in the 13th century. This intensification came from two sides: The cultivation of pulses and turnips; and the cultivation of dye plants (woad and madder). Van Uytven explains:

> "The large proportions that the textile industry assumed in the Brabantine towns in the 13th century meant an increase in the demand for dye plants and gave a significant impetus to agriculture. The urbanization that attained this industrialization forced up the demand for food, e.g. turnips, a basic ingredient of the daily *potagie* (hotpot), and also for rape, a fodder plant for the cattle breeding that had to supply the urban population with meat and dairy-produce. As a result, infringements on the fallow and the stubble fallow were already occurring in the Brabantine towns and their surroundings in the 13th century. Conversely, the concentration of people in the towns offered the opportunity of utilizing manure from the town in order to maintain the fertility of the arable land, while, at any rate in later centuries, the waste products of woad dyeing were likewise a much desired manure ... A similar interaction between industrialization on the basis of the cloth industry, urbanization and intensive agriculture took place in French Flanders and Artois in the second half of the 13th century" (1984, p.69; author's translation).

In the meantime Verhulst, who himself contributed to the discovery of the necessary records, had reconsidered his view (Verhulst 1988, p.16-17). In this manner, the problem of dating the first infringements on the three-field system in Flanders and Brabant has been solved in a way that paves the way for a von Thünen interpretation of the intensification of the agriculture in these regions.

A second problem that may throw light upon the relevance of the von Thünen theory is found in the work of Thoen on the agriculture in the Castellanies of Oudenaarde and Aalst. This author has established that in this part of Flanders fallow diminished by 60 % on small and medium-sized holdings from 1397 to 1429, but increased by a factor of 2.5 from 1430 to 1567. In the latter period the area under *dries* quadrupled. On the large holdings the area under *dries* was on the increase too, whereas the percentage of the land that lay fallow remained rather constantly at a high level. Thoen interprets this evolution as the result of, on the one side, intensified cultivation of fertile soils, and, on the other side, the extension of cultivation to poor soils (Thoen 1988, p.748-760). His calculations of the average yields provided the main argument for this interpretation. It appeared that the highest yields were obtained in the first half of the 15th century (ibid., p.813). The most plausible explanation for the decrease in average yields is indeed the extension of cultivation to marginal land and the reduction of fallow on fertile soils.

Thoen's interpretation sounds like a straightforward application of the theory of von Thünen, but it is not. Apparently inspired by Boserup (1965), who explains the intensification of agriculture in pre-industrial societies as the result of population growth, Thoen relates the evolution of the fallow and *dries* to the growth of the population and not to an increasing demand from the town. It goes without saying that the growth of the population can include an increasing demand from the town, and in this case, it does. Nevertheless an unqualified explanation using von Thünen's theory is out of the question. The smaller holdings led the way in the intensification of agriculture, but they were far less orientated to the market than the large holdings (with respect to food production). Moreover, it seems not unlikely that they were expelled from the most fertile parts of the open fields and furlongs with their common rotation system (ibid., p.841). Thus the larger holdings, i.e. the holdings that were the most active on the market, occupied the land on which the three-field system held out. On the other hand, in the vicinity of the towns infringements on the three-field rotation were easily allowed on these same fields (ibid., p.741-742). Summing up, we get the following idea of the farming systems in this part of Flanders: the agriculture of the larger holdings is determined to a certain extent by the von Thünen mechanism; near the towns grain/non-grain crops rotation, elsewhere three-field rotation, whether or not in combination with arable/*dries* rotation on poor or peripheral soils (that constitute a kind of outfield). The farming system of the smaller holdings is determined by the evolution of the population, operating as an independent variable.

As stated above, breaches of the three-field system started with the cultivation of dye plants. By 1500 flax had become the most important industrial crop in Flanders and its production on the smaller holdings formed an integral part of the intensification of Flemish agriculture. However, unlike the infringements due to the cultivation of dye plants, the intensification caused by the production of flax cannot be subsumed under the von Thünen mechanism. There are two reasons for denying the possibility of a von Thünen interpretation:

1. The production of flax and its attendant activities (processing, spinning, weaving) developed into an export industry in Flanders during the last decade of the 14th century in a period of agricultural crisis. It assumed constantly increasing proportions (except for the period of crisis in the second half of the 15th century)

even after 1550, when the general agrarian economy fared badly again. This rise and further development in times of crises prove that the flax industry operated as a means of absorbing hidden unemployment (cf. Van der Wee 1978, p.22).

2. The demand from the towns of the Southern Netherlands constituted but a minor part of the total demand for linen (and thus for flax). Even the local trade of linen and flax was not monopolized by the towns.

A sample of 67 holdings in the surroundings of Oudenaarde has enabled Thoen to calculate the area under flax according to the size of the holding for the period 1541-1550. It appears that, except for the dwarf-holdings of less than 0.5 ha, the farmers aimed at a flax area of 0.3 to 0.4 ha. In the case of the small holdings (0.5-4 ha) this meant 1/10 to 1/5 of the seeded area. Flax (and linen) were the main link which these holdings had with the market, for the corn they grew was chiefly destined for consumption by the grower. Winter corn occupied on average 60 % of the seeded area, summer crops barely 20 to 30 %. Only a very labour-intensive tillage of the land (e.g. by the use of the spade) made this kind of exhaustive cultivation possible. On the medium-sized holdings (4-10 ha) 6 or 7 % of the area was under flax. According to Thoen, these holdings approached the ideal model of a subsistence farm. 4 ha seemed to be the threshold for a remunerative use of a horse. Winter corn and summer crops were in balance. The possession of a loom for weaving linen was less common than on a small holding. On the large holdings the production of flax and linen was of minor importance (less than 5 % of the acreage was under flax), at least in comparison with the growing of corn for the market (winter corn occupied almost 70 % of the acreage; only the largest holdings grew more summer grain, presumably because of the cultivation of poor soils). The dwarf-holdings were operated by crofters and subsidiary farmers, who cultivated mainly winter corn (80 % of the acreage against less than 5 % for flax). A significant number of them bought flax to weave linen commissioned by the more thriving farmers. However, as only a minority appears to have possessed looms, an additional income must have been earned from wage labour of a purely agricultural nature.

Clearly, the frequency distribution of the holdings according to size is a key factor in tracing the impact of the von Thünen mechanism on agriculture in the belt of intensive systems of cultivation in the Southern Netherlands. The evolution in the size of the holdings from the 14th to the 16th century was a faithful image of the evolution of the population in rural areas (Thoen 1988, p.845-877; Van Cauwenberghe and Van der Wee 1978, p.151-158). Diminution of the average size was the general trend. In the 16th century this evolution took place at the expense of the medium-sized holdings. They were the least market-oriented holdings in an agricultural economy that was increasingly commercialized. The small holdings derived their viability from the production of flax and linen, the large holdings from the sale of corn to the towns. The latter benefited also from the exploitation of dwarf-holdings by agricultural labourers, which exercised a downward pressure on wages. By guaranteeing the productiveness of the large holdings, the market relations between the town and its surrounding countryside contributed to the process of expulsion of the medium-sized holding, turning the frequency distribution of farm size into a dependent variable (as the von Thünen theory implies).

## 2.2.3. Interzonal Trade.

Thus far we have dealt with the von Thünen mechanism

a) on the level of the individual towns and their respective surroundings within the belt of intensive cultivation (the way in which market relations between the town and the surrounding countryside determine the system of cultivation in the latter),

b) on the level of holdings (the way in which distance to the farmstead and variation of soil suitability give rise to a division into infield and outfield).

In this section we turn to market relations on the scale of the arrangement of the belts with different farming systems around, in the case of the Southern Netherlands, a highly urbanized zone; in other words, on a scale comparable with that of the Isolated State.

With regard to what he calls "the extensive grain growing exploitation" in Central Belgium, Lindemans writes:

> "This exploitation originated on the large holdings of the Walloon country (and also in France) in consequence of a more active production of grain [by which he means: production for the market]. In its original form, three-field farming was pure grain growing that produced grains solely for consumption. In the manorial system these were mainly destined to be consumed on the spot. After the disappearance of this system, these cereals were the only products that the three-field farm sold on the market. The increasing demand for bread-corn stimulated farmers to market ever-increasing quantities of grain. They did this, not by forcing up the yield of the existing acreage, but by enlarging the seeded acreage. The dries grasslands were tillaged ... The large tenant farms in Hainaut, Walloon Brabant, Hesbaye, Condroz were transformed from dries-farms into grain exploitation. This occurred probably at the end of the 16th century and in the case of many farms, not before the 17th century" (1952, p.39-40; author's translation).

When we read this quotation in combination with the work of other authors (as summarized by Hoebanx 1975, Périssino-Billen 1975, Thoen 1988,p.761-762, Verhulst 1990, pp.67-68 and 125-126), we are led to the following conclusion: It seems definitely established that in the central part of Hainaut, nowadays in France, the three-field system in its most stringent form (*Flurzwang* on three common fields into which the land of a village unit was divided) developed in the 13th-14th centuries. Less stringent forms of the three-field system were found in the neighbouring regions up to Hesbaye, Pays de Herve and Condroz. The system became predominant as a result of the demand for grain on the market, thereby expelling the two-field system and all types of temporary cultivation (from land rotation to the more advanced forms).

There are two problems. The first problem is one of dating the period of domination. Lindemans states that the transformation of *dries*-farms into grain exploitation did not take place until the end of the 16th century. It is not quite clear how this statement can be reconciled with what he says about the connection between grain production for sale and the three-field system. The difficulty arises from the fact that changes in the trading pattern of grain exerted a profound influence on economic development and agriculture earlier in the 16th century. The literature offers three divergent pieces of evidence of this influence:

1. The simultaneity of the import (via Antwerp) of grain from the Baltic (15th-16th centuries), the economic decline of Brabantine Hesbaye and the flourishing of the brewery industry in Leuven-Hoegaarden. The supposed link between these phenomena is the following. Before the imports from the Baltic became a permanent feature, Hesbaye was one of the suppliers of grain to the Flemish and Brabantine towns. Then the imports from the Baltic countries undermined the economic base of Hesbaye and its grain producers started to supply the brewery industry (Van Cauwenberghe and Van der Wee 1978, p.146; Van der Wee 1978, p.10-11).

2. At the end of the 15th and in the early 16th century the country of Namur experienced an economic decline, whereas the region of Maastricht was prospering. The only explanation that could be found for this contrast seems to be the differential price evolution of spelt (exported by Namur) and oats (exported by Maastricht) (Jansen 1978, p.83; 1979, p.67-74).

3. In the 16th century the Pays de Herve was transformed in a very short time (a matter of decades) from a grain producing area into an area of dairy production. The transformation was the result of the prohibition of the export of grain by Charles V (1520) , who wanted to prevent food shortages in the Spanish Netherlands (one should know that the Pays de Herve was separated from the more westerly located regions of the Spanish Netherlands by the independent episcopal principality of Liège). The export of dairy produce, however, was not forbidden. Moreover the farmers hoped to escape from tithing, as Charles V had declared that tithes should not be levied on new products (Ruwet 1943).

Obviously, if we are to equate grain growing for the market with the prevalence of the three-field system, the dating by Lindemans of the transformation of *dries*-farming is late by at least a century, unless he meant by *dries*-farming (in this part of the Southern Netherlands) a combination of three-field rotation and temporary cultivation, and by the transformation simply the final stage of a process of gradual disappearance of the *trieu* through the expansion of the division of the land in three courses. This brings us to the second problem.

The prevalence of extensive grain growing exploitation (meaning monoculture of grain) implies that extension of the cultivated area was preferred to intensification of the existing cultivation. This preference is criticized by Lindemans who argues that, as a result of the reduction of the *dries*-grassland, holdings had to rely on marl and straw to procure manure: "From the standpoint of agronomics this extensification of grain growing was a mistake, and also a retrogression as compared with farming that procured manure from grassland" (Lindemans 1952, p.40). This standpoint of Lindemans makes us think of the divergence of opinion between von Thünen and Thaer. Thaer wanted farmers to intensify the system of cultivation, because high levels of investments bring higher returns. Von Thünen questioned this thesis and proved, in his Isolated State, that the superiority of a more intensive system is indeed not absolute. Using the argumentation of von Thünen, one can put forward the hypothesis that, whatever the drawbacks of monoculture of grain may be, grain growing on the basis of the three-field system fits in with the spatial pattern of farming systems that is adapted to the way in which the demand of the urbanized regions in the Low Countries for food could be adequately met. Yet the demand from the towns failed to increase sufficiently to intensify farming in Central Belgium and possibly to incorporate also the Ardennes into the system of interzonal trade. The imports from the Baltic reduced the sales potential for other suppliers and political events (the food-policy of Charles V, the Revolt against Spain, the separation of the Northern Netherlands) broke up the developing network of centripetal trade. Brabantine Hesbaye - north of the watershed between the basin of the Scheldt and the basin of the Meuse - remained linked to the belt of intensive cultivation and found a way out in the production of grain for the beer industry, in conformity with the von Thünen mechanism. The Pays de Herve switched to cattle-breeding, mainly to supply Liège with dairy produce. In the long run this switch led to the growth of grain production in the area north of the Pays de Herve (nowadays in the Netherlands) through the intensification of cultivation on large holdings (late 17th century - early 18th century) (Jansen 1979, p.102-103). In the central and southern parts of the Southern Netherlands, intensification had to wait until the population pressure and the demand from the towns within the regions in question attained the required level.[5]

## 3.    Urban Development in Flanders/Brabant as Explained by Van der Wee

In order to assess the explanatory power of von Thünen's theory we were obliged to search for the pieces of the puzzle ourselves. With regard to the theses of Van der Wee and Mendels our task is much easier. We have merely to summarize those elements of their work that are relevant to the questions we put in the introduction to this paper.

In his explanation of the three cycles of urban growth/decline (Fig. 4).[6] the Belgian economical historian Van der Wee tries to answer the question of whether these three cycles are to be related to the development of a European mono, or multinuclear network of cities based on commerce (the Braudel - De Vries hypothesis as Van der Wee calls it), or whether the internal dynamics of the industrial activities of the towns account for the cyclic development. According to Van der Wee,

> "urban development [in the Low Countries] cannot be reduced to a process wherein the commercial factors act as the only crucial variable. This is because the export industry was central to the emergence of towns in the Low Countries. The subsequent development of these towns was also governed by industrial factors. The urban export industry in the Low Countries had its own dynamics, ruled by a specific life-cycle pattern" (1988, p.316).

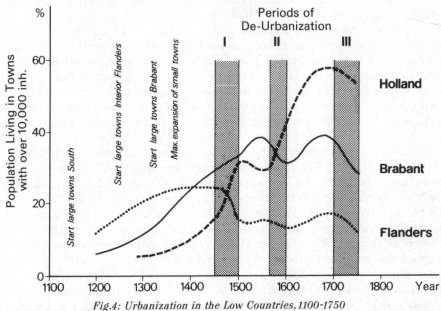

*Fig.4: Urbanization in the Low Countries, 1100-1750*

In the introductory outline of his viewpoint Van der Wee does not discriminate between the three cycles as regards the weighing of the importance of the internal dynamics of industrial activities against the importance of the network of cities based on commerce. Yet, in the course of the detailed analysis of the cycles differences in degree emerge. Precisely these differences, especially those between the first and the second cycle, are of interest to us.

According to Van der Wee, the first cycle of urban growth/decline (12th - 15th century) can be explained by the internal dynamics of the drapery industry. The drapery industry is a standardized system of production under the control of the merchant-entrepreneurs who apply the putting-out system. There are, of course, many intervening factors: the difference between old, new and light drapery, the English tax policy, and so on. But the fundamental point is that such a standardized form of production is characterized by an easy diffusion and easy spatial shifts. As a putting-out system does not require technological progress, no production centre can build up a technological superiority. A new centre can easily take over the production and push old centres from the market. The decisive role is played by the labour costs. At the moment labour costs rise too much, the merchant-entrepreneurs will move the production to new centres. This is what happened to the cloth industry. It shifted from the South in a north-easterly direction and from the large towns to the small towns and then to the countryside, including the English countryside, where its production eventually undermined urban production in Flanders and Brabant, hence the first period of decline (Fig. 5).

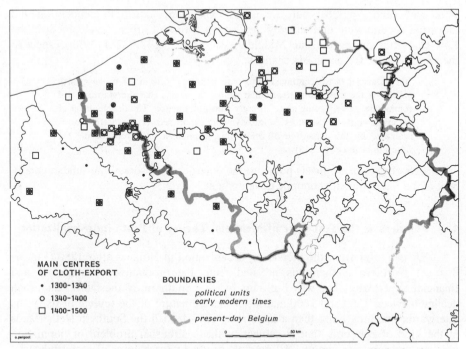

*Fig.5: The shift in the location of the cloth industry in the Southern Netherlands between 1300 an 1500 (main centres of cloth-export, according to Van Houtte & Van Uytven 1980)*

The second cycle was supported by the production of luxuries. It had already started in the 14th century in the large towns, especially in Brabant, which were able to react against the decline of their cloth industry in a productive way. The production of luxuries is in the hands of the craft guilds. In contrast to a standardized form of production, production of luxuries takes place in a limited number of centres. High labour costs are no problem. On the contrary, they are a sign of skilled labour and quality control. The growth and decline of luxuries production is determined by a multiplicity of factors. The growth in the 14th, 15th and 16th centuries was a matter of demand for non-essential goods on the European market. Moreover, luxuries can bear high transport costs, whereas standardized systems of production cannot, and this difference is important in times of decline in international trade. The decline of luxuries production at the end of the 16th century was due to emigration of the skilled members of the guilds of Flanders and Brabant caused by religious persecution after 1560. The urban decline in the first half of the 18th century was, as far as the luxuries production was involved, due to the mercantilistic policies of the great national states of Europe.

Van der Wee concludes from his survey of the first cycle that "of course, the expansion of commercial capitalism in Europe ... exercised an important influence" (ibid., p.328), but "industrial expansion, industrial competition and industrial decline were undoubtedly the driving forces of urban development in the Low Countries during the late Middle Ages" (ibid., p.327-328). With respect to the second cycle, the emphasis changes:

> "the industrial factor continued to play an important role in the urban growth pattern of the Low Countries, but it cannot be denied that the maritime and commercial factors came more strongly into prominence. The development of the urban crafts and services ... into dynamic export sectors, was indeed much promoted by the expansion of commercial capitalism in Europe during the sixteenth century" (p.345).

This change in emphasis points to the emergence of something fundamentally new in the international economic development.

## 4.    Flanders as the Test Case of Mendels' Theory of Proto-industrialization

We can infer from the increasing urbanization in Holland after 1580 (Fig. 4) that the Northern Netherlands profited from the productive, commercial and financial superiority that was built up by the Southern Netherlands. Its urban decline between 1560 and 1600 and the restricted nature of the revival of its towns afterwards - nothing more than a mere recovery - landed the Southern Netherlands in the kind of economy that is subject to the Malthusian problem of population pressure and food production. All possible productive responses of a pre-industrial society to population pressure were attempted in the Southern Netherlands: increase of the area in cultivation, reduction of the fallow, increase of crop yields, switch to higher yielding crops, regional specialization and increase of the non-farm income, for example by day labouring and domestic industry (Vandenbroeke and Vanderpijpen 1978, Vandenbroeke 1979b, Verhulst and Vandenbroeke 1979, Verhulst 1990).[7] The efforts were not in vain. In 1759-1791 the Southern Netherlands scored an export surplus of winter corn amounting to 5 % of production (the equivalent of one fourth to one third of the Baltic grain supplies between 1661 and 1783). The surplus rose to 7 % of production just before 1800 (Vandenbroeke 1979b).

One of the most striking examples of domestic industry was the processing of flax and the production of linen in Flanders. Yet the expansion of this rural industry is but another example of the development of so-called 'cottage industry' which became widespread in Europe in the 17th and 18th centuries. Mendels has coined the term proto-industrialization for this particular type of expansion of manufacturing activities. He suggested that it was a solution to the economic crises of the foregoing decades which paved the way for the real industrial revolution. Mendels is inclined to defend the thesis that proto-industrialization is a necessary (but not a sufficient) condition for modern industrialization (Mendels 1972, 1981, 1983).

The concept refers to a regional industry, located in the rural areas, that serves a market outside the region and that induces an increase in the market activities of the neighbouring towns. The majority of the labour force is composed of landed peasants and landless agricultural labourers, who devote their summers to agricultural activities. Proto-industrialization is accompanied by a change in the spatial organization of the economy, viz the bifurcation into commercial agriculture in some regions and agriculture-cum-cottage industry in other regions.[8] Mendels took Flanders as a test case and thought that the situation in Flanders confirmed his ideas. He was wrong.

Of course, there is no doubt about the existence of a dominating cottage industry in the 17th and 18th centuries. The most important export industry in Flanders, the linen industry, was a rural industry practised by farmers, the members of their familiy and agricultural labourers.[9] However, as stated in 2.2.2., the production of linen developed into an export industry in Flanders during the last decades of the 14th century. Flanders' linen industry became predominant in the 16th century as a result of the discovery of America. Flanders superseded Hainaut (which had become the first producer during the Hundred Years' War) and was able to do so because of lack of regulation. A non-regulated production was more adaptable to the demand for coarse linen in the Spanish colonies. The linen industry was located in other parts of the countryside of Flanders than the cloth industry. It was found in the rural areas around the towns which in the 14th century had prohibited the cloth industry in their surroundings, whereas diffusion of the cloth industry in the countryside remained restricted to the areas around Ieper (Ypres) and west of Kortrijk (Courtrai), where the interdiction either did not exist, or was not strictly enforced (Fig. 6) (Sabbe 1975).

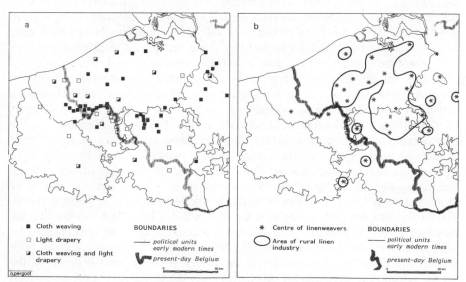

*Fig.6: The location of the drapery industry (a) and the linen industry (b) in the western part of the Southern Netherlands around 1550 (according to Soly & Thijs 1979)*

Clearly the early development of the cottage industry in itself does not contradict Mendels' theory. It does so only in combination with the early differentiation between Maritime Flanders and Interior Flanders. It is indeed this combination (on the one side rural industry of which the labour force is composed of subsidiary farmers and crofters, and, on the other side, areal differentiation) that makes proto-industrialization distinct from mere cottage industry (Mendels 1981, p.278-280). As the differentiation between Maritime and Interior Flanders reaches back to the Middle Ages, the validity of Mendels' view is questionable. There is no evidence that a bifurcation took place. The peculiar development of Maritime Flanders as a region with dairy and meat production and grain growing, reaching back to the 12th century and the rise of the linen industry as an export industry in the countryside of Interior Flanders since the late 14th century are just parallel occurrences. There is nothing more to them.

## 5.    Conclusion

We return to our two questions.

1. The explanation by Van der Wee of the changes in urban development provides independent support for Wallerstein's thesis. The fact that the first cycle of urban growth and decline can be explained by the internal dynamics of the drapery industry, whereas the explanation of the second cycle invokes a different set of factors, seems to point to the emergence of something fundamentally new in the international economic development. Moreover, the contrasting urban development in the Northern Netherlands and in the Southern Netherlands after 1560-80 and the fragmentation of the developing network of centripetal trade are evidence of the reduction of the latter region to a semi-peripheral status.

2. If the von Thünen theory can be incorporated in Wallerstein's theory, a Thünian explanation of the geographical pattern of systems of cultivation in a certain area would at once be evidence of a high degree of integration in that part of Wallerstein's world-system. Likewise, if the theory of Mendels can be incorporated in Wallerstein's theory, the territorial bifurcation of commercial agriculture and agriculture-cum-cottage industry would at once be proof of a similar high degree of integration. However, in the present state of research, no definite answer can be given to the question of to what degree the pattern of farming systems in the Southern Netherlands can be interpreted in Thünian terms. The theory does not apply to the growing of flax that is processed into linen on the farm itself for purposes of sale on foreign markets. But the theory may be applicable to the differentiation between a belt of intensive systems and a belt with the three-field system. The validity of the theory in this respect depends on the precise relationships between the urban economy, the growth of the population and social and proprietary relations - an intricate network of relations, still to be unravelled by thoughtful research of a theoretical and empirical nature. Social and proprietary relations determine to what degree a rent maximization process - the cornerstone of von Thünen's theory - governs landuse. We have not embarked upon this problem because of the lack of relevant research. However, there is one

single exception; the dissertation of Thoen, who analyses the evolution of population and of agricultural economy in terms of the endangered surplus-extraction by the ruling classes (the so-called crisis of feudalism). Thoen (1988, pp.595-604 and 929-940) presents evidence for the thesis that farmers, and not the (large) landowners, were responsible for the depth investments needed to intensify agricultural production. Yet, his research is limited in time (15th - 16th centuries) and space (two castellanies in S.E.Flanders) and any generalization with a view to the incorporation of von Thünen's theory into Wallerstein's theory would be premature. As regards Mendels' theory, it is of little help to us, because the internal differentiation of Flanders into two regions, of which one is a region with cottage industry, dates from a period before the era of proto-industrialization and can hardly be interpreted as a bifurcation.

This paper on the Southern Netherlands cannot be finished without a word on the role of the putting-out system. According to Wallerstein (1980, p.195), the putting-out system, controlled by foreign merchant-entrepreneurs, is one of the characteristic features of a semi-peripheral area. Undoubtedly the putting-out system has been operating in the Southern Netherlands. But:

a) The putting-out system was absent in the most important export industry of Flanders. The linen industry worked with the *Kaufsystem*.[10] Wallerstein (1980, p.195) mentions this system in a footnote, whereas he deals with the putting-out system in the main text. This threatens to create a false impression about the actual importance of both systems in the Southern Netherlands.

b) The putting-out system was not controlled by foreign merchants. The control by the merchants of Leiden of the cloth industry of Verviers, in the southern part of Pays de Herve, is the only real example we could find in the literature, but the control was already at an end by the 1630's (Posthumus 1939).

### Notes

1.    These terms are taken from the English translation of *Der Isolierte Staat* by Carla Wartenberg, edited by Peter Hall and henceforth referred to as 'von Thünen 1966'.
2.    It bears resemblances to the so-called *eeuwigdurende roggebouw* (perennial growing of rye, all-corn system) in Drenthe and the Veluwe, that is classified by Slicher van Bath (1976, p.67) in the same group of systems of cultivation as the infield-outfield system of Scotland and Ireland. Sods cut in the heathland were brought into the byres or folds and after enrichment with manure they were spread on the arable land that surrounded the villages. Because some sand was always picked up with the sods, the level of the fields gradually rose and a thick (more than 60 cm) humid topsoil was built up. Due to this *plaggen* practice an intensified three-field rotation could be applied. The usual rotation was winter corn (mostly rye), winter corn (idem), summer grain (oats or buckwheat). There was a short fallow between the harvest of the second winter corn and the sowing of the summer grain. In many cases it was a green fallow cropped by spurry. Another rotation consisted of a sequence of three times rye, alternating with green fallow of 2-3 months. Both types of rotation could be combined in a six-course system.
3.    Fig. 2 is a map of rotation systems, not of farming systems. As a consequence the concept of intensity acquires another meaning. With respect to rotation in itself, it relates to the length of the intervals in the tillage of the arable, and not to the cost input of the (marginal)

holding. The two meanings of the concept may lead to a different ranking: the intensity of the tillage of the manured fields in the Campine is offset by the small size of these fields in proportion to the vast area of the heath.

4.  Temporary cultivation is the literal translation of the Dutch term *tijdelijke verbouw* and the French term *culture temporaire*, which Slicher van Bath (1967, p.67) and Bloch (1961, p.35) used as a synonym for the German term *Feld-Gras-Wirtschaft*. The length of the *dries* ranged from 2 or 3 years in Flanders to two decades or more, if we consider the cultivation of the outfield in the Ardennes as a form of arable/*dries* alternation. In the Ardennes two methods of cultivation prevailed, the first one being reminiscent of land rotation. The arable land was obtained either by sod-burning, and in that case one harvest of rye and sometimes a second harvest of oats could be earned, or by scrub-burning, and this technique of clearing the land made one or two harvests of rye possible. Besides these *terres à sart* manured fields, the *terres à champ*, were found near the villages (the second method of cultivation). The *terres à champ* rested on the average for 8 years, the *terres à sart* for 20, 25 or 30 years. Some of the sarts were called *trixhe*, the local word for *dries*, or *trieu*, the more general Walloon term.

5.  The corroboration of this set of hypotheses would throw a new light on the so-called 'traditionalism' of Walloon farming (including Walloon Brabant, but not the north-western part of Hainaut). A lot of research needs to be done in this context, especially on the influence of farm size and of location relative to the towns. An assessment of the part played by the rural industry is also badly needed (rural industry, based on hired labour of subsidiary farmers and crofters, was widespread not only in Flanders, but also in Wallonia: see Van Uytven 1975).

6.  Ratios including smaller towns would not significantly alter the general trends.

7.  Of all the productive responses to population pressure the introduction of the potato was the most important. It reduced the per capita consumption of cereals by 30 % and guaranteed (in combination with the cottage industry) the viability of dwarfs-holdings.

8.  In addition, Mendels refers to the poverty that results from the interaction between economy and population, but this thesis of a vicious poverty trap is of minor importance for the questions we deal with in this paper.

9.  In the late 18th century, 1/5 or 1/4 of the population of Interior Flanders, and in some regions even half of the population, was engaged in the linen industry (Vandenbroeke 1979a).

10. Linen was woven in the farmstead itself or by village weavers who purchased the raw material from the farmers. It also occurred that operators of dwarf-holdings wove linen by order of large farmers, who supplied them with the raw material. Merchants who lived in the towns or in the villages purchased the linen at the weavers' homes and exported it. Other village merchants operated as middlemen who bought on the local markets. In the 16th century the towns tried to monopolize the trade, but the countryside managed to keep part of it. The towns played the role of intermediaries for Antwerp, where the trade had become centralized. The linen was exported by foreign merchants who traded in a variety of products, or by linen merchants from the centres of production, some of whom acted as middlemen for the foreigners. Some linen merchants employed *facteurs* in the centres of production who bought up, bleached and sent on the linen. In Ghent the linen traders were usually local partners of foreign firms. After the surrender of Antwerp (1585) the so-called *kutsers*, middlemen between weavers and merchants, took advantage of the disorder on the local markets and tried to get hold of the weavers who used to bring their products to the market. They also succeeded in involving the inhabitants of an ever-increasing number of villages in the flax industry. Many conflicts between the towns and these *kutsers* were the result. As in the former centuries, so after 1585 the organization of the linen industry in Flanders was different from the putting-out system. The weavers purchased some worked

flax with their own small capital. Spinning and weaving were done in the home. The village weavers brought their fabrics to the urban markets and sold them to the merchants or entrusted them to the *kutsers*. The merchants had the linen bleached at their own expense (Sabbe 1975, Soly and Thijs 1979, Thijs 1980).

# References

Ameryckx, J., W.Verheye and R.Vermeire (1985): Bodemkunde. Ghent
Bloch, M. (1961): Les caractères originaux de l'histoire rurale française. Tome 2. Paris
Bolckmans, W.P., G. Pieters, W.Prevenier and R.W.M.von Schaïk (1980): Tussen crisis en welvaart: sociale veranderingen 1300-1500. In: Algemene Geschiedenis der Nederlanden: 4.Middeleeuwen, Haarlem, pp.42-86
Boserup, E. (1965): The conditions of agricultural growth. London
Eysberg, C.D. (1979): Gebruik en misbruik van het model van von Thünen in de geografie. In: Geografisch Tijdschrift 13, pp.15-31
Hall, P. (ed.) (1966): Von Thünen's Isolated State. Oxford
Hasquin, H. (ed.) (1975): La Wallonie: I.Des origines à 1830. Brussels
Hoebanx, J.J. (1975): Seigneurs et paysans. In: Hasquin, H. (ed.), pp.161-211
Jansen, J.C.G.M. (1978): Tithes and the productivity of land in the south of Limburg (1348-1790). In: Van der Wee, H. and E.Van Cauwenberghe (eds.), pp.77-95
Jansen, J.C.G.M. (1979): Landbouw en economische golfbeweging in Zuid-Limburg, 1250-1800. Assen
Klep, P.M.M. (1988): Urban decline in Brabant: the traditionalization of investments and labour (1374-1806). In: Van der Wee, H. (ed.), pp.261-286
Lindemans, P. (1952): Geschiedenis van de landbouw in België. Antwerp
Mendels, F. (1972): Proto-industrialization: the first phase of the industrialization process. In: Journal of Economic History 32, pp.241-261
Mendels, F. (1981): Industrialization and population pressure in eighteenth century Flanders. New York
Mendels, F. (1983): Sur les rapports entre l'artisanat et la révolution industrielle en Flandre. In: Ungarische Akademie der Wissenschaften, Veszprémer Akademische Kommission: International Symposium of Handicraft History, Veszprém 21-26.8.1982, pp.19-48
Périssino-Billen, C. (1975): Des campagnes sous le joug des traditions. In: Hasquin, H. (ed.), pp.295-312
Posthumus, N.W. (1939): De geschiedenis van de Leidsche lakenindustrie. The Hague
Ruwet, J. (1943): L'agriculture et les classes rurales au Pays de Herve sous l'Ancien Régime. Liège
Sabbe, E. (1975): De Belgische vlasnijverheid. Kortrijk
Slicher van Bath, B. (1976): De agrarische geschiedenis van West-Europa, 500-1850. Utrecht/Antwerp
Soly, H. and A.K.L.Thijs (1979): Nijverheid in de Zuidelijke Nederlanden (1490-1580). In: Algemene Geschiedenis der Nederlanden: 6. Nieuwe Tijd, Haarlem
Tavernier, R. and R.Maréchal (1959): De bodemassociatiekaart van België. In: Natuurwetenschappelijk Tijdschrift 41, pp.161-204
Thijs, A.K.L. (1980): Nijverheid in de Zuidelijke Nederlanden, 1580-1650. In: Algemene Geschiedenis der Nederlanden: 7. Nieuwe Tijd, Haarlem
Thoen, E. (1988): Landbouwekonomie en bevolking in Vlaanderen gedurende de late Middeleeuwen en het begin van de Moderne Tijden: Testregio: de Kasselrijen van Oudenaarde en Aalst. Belgisch Centrum voor Landelijke Geschiedenis 90, Ghent

Van Cauwenberghe, E. and H.Van der Wee (1978): Productivity, evolution of rents and farm size in the Southern Netherlands agriculture from the fourteenth to the seventeenth century. In: Van der Wee, H. and E.Van Cauwenberghe (eds.), pp.125-161

Vandenbroeke, C. (1979a): Sociale en konjunkturele facetten van de linnennijverheid in Vlaanderen (late 14de - midden 19de eeuw). In: Handelingen van de Maatschappij voor Geschiedenis en Oudheidkunde Gent, Nieuwe Reeks 33, pp.117-174

Vandenbroeke, C. (1979b): Landbouw in de Zuidelijke Nederlanden, 1650-1815. In: Algemene Geschiedenis der Nederlanden: 8. Nieuwe Tijd, Haarlem

Vandenbroeke, C. and W.Vanderpijpen (1978): The problem of the 'agricultural revolution' in Flanders and in Belgium: myth or reality? In: Van der Wee, H. and E.Van Cauwenberghe (eds.), pp.163-170

Vandenbroeke, C. and P.Vandewalle (1980): Landbouw in de Zuidelijke Nederlanden, 1490-1650. In: Algemene Geschiedenis der Nederlanden: 7. Nieuwe Tijd, Haarlem

Van der Wee, H. (1978): The agricultural development of the Low Countries as revealed by the tithe and rent statistics, 1250-1800. In: Van der Wee, H. and E.Vancauwenberghe (eds.), pp.1-23

Van der Wee, H. (1988): Industrial dynamics and the process of urbanization and de-urbanization in the Low Countries from the Late Middle Ages to the Eighteenth Century: a synthesis. In: Van der Wee, H. (ed.), pp.307-381

Van der Wee, H. (ed.) (1988): The rise and decline of urban industries in Italy and in the Low Countries (Late Middle Ages - Early Modern Times). Leuven

Van der Wee, H. and E.Van Cauwenberghe (eds.) (1978): Productivity of land and agricultural innovation in the Low Countries (1250-1800). Leuven

Vandewalle, P. (1986): De geschiedenis van de landbouw in de Kasselrij Veurne (1550-1645). Gemeentekrediet, Historische Uitgaven, reeks in -8° 66, Brussels

Van Uytven, R. (1975): Die ländliche Industrie während des Spätmittelalters in den Südlichen Niederlanden. In: Kellenbenz, H. (ed.): Agrarische Nebengewerbe und Formen der Reagrarisierung im Spätmittelalter und 19./20.Jahrhundert. Forschungen zur Sozial- und Wirtschaftsgeschichte, Stuttgart, pp.57-77

Van Uytven, R. (1984): Vroege inbreuken op de braak in Brabant en de intensieve landbouw in de Zuidelijke Nederlanden tijdens de 13de eeuw. In: Tijdschrift van de Belgische Vereniging voor Aardrijkskundige Studies 53, pp.63-72

Verhulst, A. (1956): Bijdragen tot de studie van de agrarische struktuur in het Vlaamse Land: 2. Het probleem van de verdwijning van de braak in de Vlaamse landbouw (XIIIe-XIVe eeuw). In: Natuurwetenschappelijk Tijdschrift 38, pp.213-219

Verhulst, A. (1988): Agrarische revoluties: mythe of werkelijkheid. In: Mededelingen van de Faculteit Landbouwwetenschappen Rijksuniversiteit Gent 53, pp.1-26

Verhulst, A. (1990): Précis d'histoire rurale de la Belgique. Brussels

Verhulst, A. and G.Bublot (eds.) (1980): De Belgische Land- en Tuinbouw, verleden en heden. Antwerp

Verhulst, A. and C.Vandenbroeke (eds.) (1979): Landbouwproduktiviteit in Vlaanderen en Brabant, 14de-18de eeuw. Studia Historica Gandensia 223, Ghent

Wallerstein, I. (1974): The Modern World-System: Capitalist Agriculture and the Origins of the European World-Economy in the Sixteenth Century. New York

Wallerstein, I. (1980): The Modern World-System II: Mercantilism and the Consolidation of the European World-Economy 1600-1750. New York

# 6

## Northern Italy

## Secondary Core or Reduced to a Semi-Peripheral Role?

Wilhelm Matzat

In order to avoid misinterpretation, I wish to make clear, that I will use in this paper the regional distinction: Northern, Central and Southern Italy. Northern Italy consists of the plains and hills of the Po Valley, the so-called Padania, Central Italy embraces Tuscany, Umbria and Lazio, Southern Italy the former kingdom of Naples. The question above was raised by Prof. Nitz and he asked me to try to provide an answer which certainly is no easy task. The question involves two main problems: one of regionalization and one of ranking. In which way can different regions be compared, how can their ranking - if there is one - be evaluated, either in qualitative way or operationalized by quantitative means? Wallerstein distinguishes only three types of regions, because he closely follows the conception of Marx and others, who believe that the mode of production is so important in characterizing different social and economic structures. As there are only a few different modes of production, he can construct only three regions: core, semi-periphery and periphery. He would therefore consider the distinction suggested above: "secondary core or semi-periphery" as invalid and would argue that they are just different names for the same. We on the other hand do not hold the modes of production to be of any significance and consider them to be just as one attribute among many others. So if we want to stick to a difference between secondary core and semi-periphery, we have the liberty to construct more levels of ranking, maybe: primary, secondary, tertiary core, intermediate zone, semi-periphery, inner and outer periphery etc.

My task at this meeting it is not to discuss the fundaments and principles of Wallerstein's world-system theory. This has already been done by more competent authors and it may suffice to mention just two publications (Dodgshon 1977, Kearns 1988) and some of the papers given at this conference. I will only take a look at Wallerstein's macro-level statements on Northern and Central Italy and test them in the light of a more regional scale. I shall put the accent on the western part of the Padania, especially Lombardy, as Salvatore Ciriacono will deal with the eastern part, the Venetian state in, his paper.

Wallerstein's conception reads as follows: in high and late medieval times the Low Countries, Western and Southern Germany, Northern and Central Italy were the core area, the dorsal spine in the European economy. But when economic crisis and depression spread there after 1620/30 Northern and Central Italy became a semi-peripheral area. - This statement is too undifferentiated. I think it is advisable to make a distinction between the agrarian and the manufacturing sector. If we look at the agrarian economy, there have been in Lombardy since early

modern times until today at least five distinct agrarian zones north of the river Po:
the Bassa Pianura, the Fontanili zone, the Alta Pianura with the Altopiani, the Hill
zone and the Alps. These zones differ from each other in settlement and social
structure, soils, irrigated or dry farming, crops, size of holdings and, agrarian
contracts, among others. All these differences are subsumed by Wallerstein under
the label "Northern Italy" and "semi-periphery". Regarding the agricultural
production he says:

> "The periphery (eastern Europe and Hispanic America) used forced labour
> (slavery and coerced cash-crop labour). The core, as we shall see, increasingly
> used free labour. The semi-periphery (former core areas turning in the
> direction of peripheral structures) developed an inbetween form,
> sharecropping, as a widespread alternative.(...). Unable to move all the way to
> large estates based either on enclosure and tenancy as in England or coerced
> cash-crop labour as in eastern Europe, the landed classes of southern France
> and northern Italy chose the halfway house of sharecropping, as a partial
> response to the creation of a capitalist world-economy, in the form of
> semicapitalist enterprises, appropriate indeed to semi-peripheral areas"
> (Wallerstein, 1974, p.103 & pp.106-107).

These statements ignore the fact that sharecropping in Central and Northern
Italy had already developed in high and late medieval times (when this region was
a core area!). Sharecropping, therefore, cannot be "a partial response to the
creation of a capitalist world economy" (which, in my opinion, did not exist yet in
the 16th century). The Bassa Pianura of Northern Italy became, from the 15th to
the 18th century, a region of intensified agrarian production by expansion of
irrigation, by a more rational distribution of land ownership, by a change of
settlement structure and the size of holdings, by changes in land-use and rotation
and by a new mode of production that can be called "capitalistic large tenancy",
the latter starting in the 16th century (for a more detailed report on this see Matzat
1987). Wallerstein also assumes, incorrectly, that the capitalistic large-scale
tenancy is a more progressive mode of production. It in fact brought about a social
degradation of the former sharecroppers - they became landless labourers on the
estates (mostly with a one year contract) or day labourers. They were illiterates,
whereas the sharecroppers had a high degree of literacy and self-consciousness.
From an economic point of view *coltura promiscua* and sharecropping were a very
intensive form of agriculture. If one considers the agrarian economy alone,
Northern Italy and parts of Central Italy certainly were a core region or secondary
core in 17th and 18th century Europe.

Quite a different picture arises if one considers the cities of Venice and
Milan, where the production of manufactured goods, especially of textiles and
metallurgical wares, experienced a dramatic collapse after 1620. The reasons for
this in Milan were manifold (Sella, 1981, passim): pest epidemics reduced the
population, building activities declined, important markets were lost in Germany
in the aftermath of the Thirty Years' War, exports to France were hindered either
by wars or because of the latter's mercantilistic policy, which raised the tariff
walls. On the other hand the wages in Milan remained on a high level, rendering
the competition with cheaper foreign goods difficult. Italian historians such as

Cipolla (1959) and Romano (1968) paint a pessimistic picture of the "economic decline" of Italy in the 17th century. More recent research (Ciriacono 1988, Sella 1981) has stressed the point that the decline was a relative one and that manufacture shifted from large cities as Venice and Milan to smaller towns and the countryside. To demonstrate this a few passages from Sella's paper (with the programmatic title "The two faces of the Lombard economy in the 17th century") may be cited :

> "The large Lombard cities, their economy and notably their manufactures failed either to recover even a fraction of their past prosperity or to generate replacements. Rural Lombardy, on the other hand, presents a sharply different, although less well-known story: unlike that of the cities, it is one of remarkable endurance, adaptation and resilience. The basic structure of the rural economy, unlike its urban counterpart, survived intact despite losses and contractions in the crisis of the 1630's and 40's. In regard to industrial activities in the countryside, the record is mixed: while there was no noticeable progress in iron metallurgy and in the making of coarse woollens, after the mid-17th century the production of mixed cotton fabrics and of silk yarn very definitely expanded. By the middle of the 18th century the Gallarate district, producing over 100,000 bolts of fustian annually, had far surpassed Cremona at its zenith two centuries before; and around 1750 there were nearly 800 silk mills in the hill zone as against a paltry 34 in the city of Milan. The striking contrast between the dying urban and the more vital rural economies deserves greater attention than it has received. The dramatic collapse of the manufactures of Milan, Como, Lodi, Cremona and Brescia ca. 1600 in the face of foreign inroads by English worsteds, French silks, German linen and fustians and Dutch firearms clearly indicates that their craftsmen and merchant-manufacturers, earlier virtually unchallenged, were ill-prepared to meet the competition. Therefore the bourgeoisie abandoned the sinking ship of the urban economy and channelled capital and enterprise toward the countryside, where post-war reconstruction, works of irrigation and reclamation and the diffusion of cottage industries and power-driven silk mills presented fresh and in the long run far more promising opportunities. Conditions in the countryside contrasted sharply with those prevailing in the cities. With production geared to foodstuffs, raw materials and simple, low-grade manufactured goods, the countryside had never enjoyed the protective shield afforded the cities by the relative superiority of their technology and skills. The absence of guilds and, more importantly, the well documented scope and intensity of seasonal migration of labour, meant that labour mobility was not impeded. Part-time employment of peasant labour during slack periods in the agricultural cycle by rural Lombard handicraft industries meant relatively low labour costs which the city, with a fully specialized labour force totally dependent on wages, could not match. These and other circumstances gave the rural economy a greater toughness and resilience than its urban counterpart." (Sella 1975)

We come back to our original question: Northern Italy - secondary core or semi-periphery? As a résumé of what was said above we can state that the agriculture of Lombardy (and parts of Piedmont) in the 17th and 18th century was very much advanced, so that it could export rice, grain, wine, cheese, raw and thrown silk to neighbouring states. Production and export of manufactures seized to a large extent and the latter had to be imported. Therefore it appears justifiable to conclude for that period: in agriculture it was a European core region, in industry a "semi-periphery", using this term in a rather broad sense and not in the manner Wallerstein conceptualizes it. His conception of the relations between core, semi-periphery and periphery is confused and unrealistic, as Kearns (1988, p.288) has shown, and can be discarded.

As the State of Milan (and also Piedmont) was landlocked, it appears to be difficult to find out, from where and what kind of foreign goods came in. One is amazed to find in the voluminous books on the Duchy of Milan and its history in the 17th and 18th century (Sella 1981, Sella u. Capra 1984) no mention whatsoever of imports and exports. Apparently there are no relevant data to be found in the archives. For harbour cities like Venice or Genoa there are better sources available, as it was easier here for the customs to check and register the goods of incoming and outgoing ships. Of course these "official" statistics have to be interpreted carefully, contraband being "almost a separate industry".

I agree completely with Dodgshon's (1987, pp.349-50) view, that an economic world-system had not yet come into existence by the 16th and 17th century and that we have to "avoid the mistake of confusing the achievement of interconnectedness with the more advanced state of interdependence. The linkages which we see being threaded through the different parts of Europe from the 16th century onwards did not produce the kind of complementarity which lies at the heart of Wallerstein's 'world system' until much later". In other words, what existed in early modern times were national or regional markets. Looking at 17th and 18th century Italy in a regional scale, it is not difficult to detect cores and hinterlands there with some form of von Thünen arrangement for the agrarian production. Tuscany seems to be a fairly good example. Around Florence and the other cities one finds the intensive *coltura promiscua* with sharecropping, and in this first zone, vegetables, fruits, wine and olive and mulberry trees. The Tuscany hills and the Apennine Mountains formed, in the vertical direction, the second, the forest zone, which produced timber and charcoal for cooking and heating. In higher parts of the Apennines (e.g. Pratomagno) beef cattle could be grazed. The third zone, the grain belt, was in Sicily, which also supplied sugar. Sicily, not Amsterdam and its granaries, was the chief supplier of surplus grain for Italy. As the government in Naples was always short of funds and had to take up loans, it had leased much of Sicily's grain export to Genoese and Florentine bankers. The ranching zone is represented by the Maremma and Puglia, where the transhumant sheep flocks provided wool and mutton.

For the world-based economy of Europe in the 16th and 17th century the comparative approach seems to be the appropriate method, whereas a synoptic view should only be applied when dealing with the real world economy, which emerged in the 19th century as an entity.

# References

Cipolla, Carlo M. (1959): Il declino economico dell'Italia. In: Storia dell'economica italiana. Ed. by C. Cipolla. Torino

Ciriacono, Salvatore (1988): Mass consumption goods and luxury goods: the de-industrialization of the republic of Venice from the 16th to the 17th century. In: H. Van der Wee (ed.): The Rise and Decline of Urban Industries in Italy and in the Low Countries. Leuven Univ. Press

Dodgshon, Robert A. (1977): The Modern World-system. A Spatial Perspective. In: Peasant Studies, vol. VI, No. 1, pp.8-19

Dodgshon, Robert A. (1987): The European Past. Social Evolution and Spatial Order. London

Kearns, Gerry (1988): History, geography and world-systems theory. In: Journal of Historical Geography, vol. 14, No. 3, pp. 281-92

Matzat, Wilhelm (1987): Early mercantilistic and agrarian capitalistic economy and its impact on the agrarian landscape in the Stato di Milano (West Lombardy) in the 15th to 18th century. In: H.-J. Nitz (ed.): The Medieval and Early Modern Rural Landscape of Europe under the Impact of the Commercial Economy. Göttingen, pp. 129-36

Romano, Ruggiero (1968): L'Italia nella crisi del secolo XVII. In: Studi Storici, vol.IX, pp.723-41

Sella, Domenico (1975): The two faces of the Lombard economy in the 17th century. In: F. Krantz & P.M. Hohenberg (eds.): Failed Transitions to Modern Industrial Society: Renaissance Italy and 17th century Holland. Montréal, pp. 11-15

Sella, Domenico (1981): L'economia lombarda durante la dominazione spagnola. Bologna

Sella, Domenico and Carlo Capra (1984): Il Ducato di Milano dal 1535 al 1796. Torino

Wallerstein, Immanuel (1974): The Modern World-System I. Capitalist Agriculture and the Origins of the European World-Economy in the 16th Century. New York

# 7

# The Venetian Economy and its Place in the World Economy of the 17th and 18th Centuries. A Comparison with the Low Countries

Salvatore Ciriacono

Economic models serve a purpose: Namely they force us to look at what is essential, at the basic truths of a particular historical era, without getting led astray by fine distinctions and secondary features. However, in their determination to create a comprehensive picture, that at all costs presents a basic interpretation of the period, historians must not alter the significance of these secondary features, when attempting to place them into a pre-conceived scheme. Starting from this premiss I would try to use the concept of world-economy as elaborated by Braudel and Wallerstein to analyse the development of the Venetian state during the 17th and 18th centuries.

Given that the northern Low Countries became the core of the world-economy in the 17th century - after the heyday of the Venetian economy at the turn of the 16th century and the brief economic supremacy of Genoa and Antwerp at the end of the 16th century - a comparison obviously has to be made between the situation found in Venice and that found in the Netherlands.[1] Whilst there can be no doubt that the city's market share had begun to fall behind that of North-Western Europe (in terms of both the value and quantity of goods exported), problems arise when one tries to quantitatively analyse the decline of Venice. This is something which happens every time the rhythm of acceleration, or deceleration, of a given economy is determined. This is especially so when a comparison is attempted with the economies of those other nations that were late, and often unwelcome, newcomers on the economic scene. From this point of view it can be seen that Venice and Holland are also comparable in that they are both the subject of a historical debate to establish the nature, extent and date of their decline: Whilst most historians would date Venice's decline in the 16th century and that of Holland two centuries later (when, due to the so called "Industrial Revolution", the country was overtaken by England), there is actually no uniform interpretation of the extent of the crisis. This has led to the introduction of the concepts of "relative" and "absolute" decline (Rapp applies them to 17th century Venice, and Johann de Vries to 18th century Holland.[2] De Vries, in particular, was followed by a series of scholars (Spooner, Riley, Jan de Vries, Van Zanden) who further elaborated his work. A different timetable of decline was proposed, greater emphasis was laid on one or the other sector of the economy. For example agriculture was associated with the fall in grain prices, or the absence of a Dutch Industrial Revolution was stressed, or, finally, the increasingly tough competition Dutch traders had to face was highlighted. As a result a new methodological and statistical approach was tried, especially in van Zanden's work. However, as Israel has commented recently, "despite a growing mass of statistical data on the Dutch economy of the eighteenth century, scholars continue to arrive at strikingly different conclusions".[3]

Similarly for Venice, the idea of a downward curve that starts at the end of the 16th century and continues uninterrupted until the fall of the Republic in 1797 is no longer accepted without debate. Even in the political sphere [4] the concept of a Venetian aristocracy totally opposed to any type of reform has, through necessity, been modified. This modification accounts for a situation in which there was an attempt to renovate, or at least remove hindrances to the development of the regional economy.

In effect, Wallerstein's model can turn out to be insufficient for an analysis of the role played by an individual market place, if different interpretative parameters are not identified. Hoekveld's call for "different intermediary theories" is, therefore, very timely.[5] These theories would not only consider a city in relation to the region around it, but would also examine the internal mechanisms of the region itself. However, it was the same proto-industrial model that led many historians, both Italian and foreign, to take a new look at what was happening outside cities. This "new look" involved viewing the "decline" of cities such as Venice, Florence, Genoa and Milan in relation to the changes that were taking place in their respective hinterlands.[6]

If one is to follow this line of enquiry, it is worth recalling the various theories of regional organization and the suggestions put forward by von Thünen - all of which are still very useful when analysing the Venice model. In fact, if the process by which Venice, Amsterdam, Genoa or Antwerp each became centres of a world-economy is described, it is not enough to look at the cities themselves. The complex relation between each city and the territory around it must also be considered. In addition, any analysis must include those neighbouring cities that were, more or less, under central political dominion. This author agrees with Hoekveld on one point: Namely, that the *organized capitalism* of a world-economy can become *disorganized capitalism* when new equilibria become necessary. These equilibria can be imposed by market forces, geographical features, political considerations, or simply by the need to survive, and it is the historian's job to precisely describe them without falling into the trap of false generalization. The career of each city is a specific one, and is characterized by the special political and economic relationship between the capital and the neighbouring cities of lesser economic weight. In the case under consideration these cities were totally dominated by Venice.

There is no doubt that the 17th century in Europe was marked by important changes and economic reorganization. Thus, for a city that had developed its production of medium-high quality goods during the Renaissance period, a new economic equilibrium might depend on the expansion of luxury industries. This can be seen in several Italian cities which will be reiterated below, but also in numerous trading centres in the southern Low Countries, in particular Flanders. However, a similar reconversion of production has been observed in some German cities, such as Augsburg, Pforzheim and Cologne. Other cities derived new

economic stimulus from the fact that they were the seats of regional governments:
The decline in the number of inhabitants being balanced by the quality of
administrative bureaucratic services.[7] These considerations must be born in mind
when looking at Venice and Amsterdam, especially during their period of
international "decline". In fact, if population or manufacturing capacity alone are
considered, a limited view of the fortunes of a large urban centre will be obtained.

It is no accident that one historian of urban development, Hohenberg, has
recently reconsidered a factor that generally plays only a marginal part in the
urban history: agriculture and its modernization.[8] In fact, this is a crucial point for
a comparison of the Venice and Low Countries economies. Not that Wallerstein's
study ignores it, but the analysis requires geographical distinctions as well as
chronological and socio-political ones. One should distinguish between the western
(above all, Lombardy) and eastern parts of the Po Valley without forgetting that,
at that time, key provinces such as the areas around Bergamo and Brescia (which
now form part of Lombardy) were part of the Venetian Republic. This means that
an analysis which describes the Po Valley as becoming part of the semi-periphery
during the course of the 17th century, and thus as being characterized by share-
cropping that is typical of such zones, can only cause erroneous or misleading
conclusions.

However, it has to be admitted that the Po Valley was falling behind the
Low Countries core, in particular the northern Low Countries while in the south
there were the same problems of adaptation and delicate reorganization that
affected the Italian regions. It should not be forgotten that the decline of a
geographical area, or the core of a regional economic system, must be interpreted
not only in terms of the internal weakness that become evident with the passage of
time, but also with regard to the international situation as a whole. Sometimes,
though, history seems to emphasize internal mechanisms (basing its interpretations
on social and political factors), and sometimes it seems to ignore internal weakness
altogether (concentrating its attention on the international context). It is no
coincidence that, as far as Italy is concerned, a lot has been written about the
sixteenth-century "trahison de la bourgeoisie" and about the conservatisms of the
aristocracy and urban élites, as well as about the inability of the guilds to adapt to
new circumstances. All of these factors are of undoubted importance, and will be
returned to in a moment. However, the following old, though still valid,
observation should not be forgotten: Namely, that the movement of trade towards
the Atlantic coasts was an objective factor that made it more difficult for Italy's
centres of trade and government to maintain their status. This shift meant, of
course, that the geographical position of Antwerp (and, later, of Amsterdam)
certainly helped those ports to become centres for the exchange of information,
key points in international strategy and important trading centres for new goods
from distant lands. As a consequence Venice became divorced from the import
trade based on new colonial products (coffee, cocoa and sugar) that replaced or
supplemented the old spice trade, and missed out on the trading in precious metals
from the New World. Nevertheless it would be wrong to deduce from this that
there was a terrible collapse in living standards and consumption during the course

of the 17th and 18th centuries. In fact the very opposite is true: The capital maintained appearances for a long time, continuing to act as the administrative and political centre of an important Italian state.

Two factors that help to explain how Venice maintained its standard of living are its colonial empire (even if this gradually shrank in size) and, more importantly, the terrafirma (with its numerous, well-populated cities). Unlike the Dutch colonies of the 17th and 18th centuries, the Venetian colonies were not intercontinental. However it must be remembered that up to 1571 Venice could rely on Cyprus for its supplies of cotton, sugar, vines, oils and raisins. Subsequently it relied on Crete (up to 1669) and the Ionian Islands (up to the Fall of the Republic).[9] These colonial products attracted the attention of other trading powers especially France, but also England and Holland (whose presence in the Mediterranean was far from negligible).[10] It is nevertheless true to say that by the second half of the 17th century the Republic's naval and trading power was unmistakably weaker than it had been. A balance was re-established by drawing on the resources of the terrafirma. Such resources enabled Venice to resist the economic pressures exerted by the more dynamic nations of north-west Europe and to create an intermediate position for itself in the international division of trade. There is a lot of truth, therefore, in Wallerstein's conclusion that most of the Italian regions became semi-periphery ; this is particularly clear if one looks at the new core of the world-economy, the Low Countries. Venice had, in fact, been a different type of core. Dodgshon is certainly correct when he suggests that a distinction can be made between the specific characteristics of the various cores that follow on from one and other.[11] Holland extended its trading to countries that were not only new, but were also further away than the nations it had previously been possible to reach (for instance America and Asia). What is more, its trade relied on large-scale financial and economic instruments which prefigured modern capitalism, if not "capitalism" (which had already been practised by medieval cities, particularly in Italy.[12]

However, not only must one avoid a general, and in some ways erroneous, theory that equates semi-periphery and share-cropping, a distinction must be made between the various Italian regions and sub-regions that can be described as semi-peripherical. Attention will now be concentrated on the two sectors of economic activity that underwent the greatest change during the period under consideration: agriculture and manufacturing industries. As far as the latter are concerned, it is a well-known fact that in the cities of central and northern Italy the production of luxury and cheaper goods maintained a sustained pace throughout the 16th century. In Venice, for example, there was a marked increase in the number of bolts of woollen cloth, which served to balance out certain losses made by the city's silk industry. The technical expertise required for glass-blowing and the making of mirrors was still a Venetian preserve (as was the pearl industry). Wax, soap, paper and arms were other areas that were vital to the Venetian economy. Reasons of space prevent a consideration of the characteristics of the other main Italian manufacturing centres (for example Milan, Genoa and Florence) which, together with Venice, make up the quadrilateral referred to by Braudel. Suffice to say that to foreign travellers in the late 16th century these cities seemed to be amongst the richest in Europe.[13] Recent research has also suggested that their careers are very similar to that of the Venetian economy. On the one hand, they continued to concentrate on the production of luxury goods, trying to conserve their market-share. On the other, much of the city's production capacity shifted out to the surrounding rural areas and the smaller urban centres. More than the notion of a world-system, or the theories of regional organization put forward by Christaller and Perroux, or even von Thünen's concentric circles, it is the research into proto-industrial models that can be of assistance in the analysis of this situation. Tuscany alone can serve as an example of what happened in the 17th century. In this region there was an irreversible decline in the Florentine wool industry (an essential part of the city's economy during the 16th century) and the expansion of a typical "luxury" industry: silk-weaving. Contemporary with this industrial "conversion" in the region's capital, was the expansion of Pisa to become Tuscany's second most important silk-weaving centre (working inferior-quality fabrics[14]. A similar process was taking place in the Venetian state around the same time. Venice continued to produce its brocades and damasks (a tradition that had deeper roots than Florentine silk-weaving), whilst there was, parallel to the development of Pisa, a growth in Vicenza's silk industry and the other cities of the Venetian state (whose products were of a lower quality than those produced in Venice itself). Similarly there was a decline in the Venetian wool industry, which had grown in size quite notably during the 16th century.[15] In Liguria, the weaving of silk cloth was already a rural industry by the 16th century, so there was no need to wait for the "re-structuring" that took place in the 17th century. This fact also disqualifies the thesis that all proto-industrial activity was concerned with common consumer goods.[16] The first effect of this was that the number of looms in Genoa dropped (in 1675, only 480 of the 2064 looms in operation on the *Riviera di Levante* were located in the city). Subsequently, the city also ceased to produce damasks and black velvet, limiting itself to the production of satin.[17] If the competition in foreign markets from the Lyons silk industry is then considered, it can be seen how the Italian industry was being forced to struggle to maintain a significant market share both for its various finished and unfinished products.

Two important phases in the production of silk are "spinning" and, above all, "milling", and it is no coincidence that the Po Valley (from Piedmont[18] to Lombardy[19] and Friuli) contained numerous silk mills of the Bolognese type. These ran on water power and, according to Poni, were an example of capitalist production facilities *avant la lettre*.[20] In the Venetian state, the mills tended to be concentrated in the areas around Bergamo and Bassano, where there were abundant supplies of water power (the two towns are in the Alpine foothills).[21] It was the plentiful supply of water, along with the absence of city guilds and the abundance of cheap labour, that gave the mainland cities a decisive advantage over Venice itself. Whilst the capital maintained its monopoly on the production of intensively-worked luxury goods, it had to leave the production of what might be considered as relatively "common" consumer goods (lower quality silks, wool, linen and hemp cloth, arms and ironmongery) to its provincial cities and towns.

The resulting situation in the region was inevitably complex, so much so that one gets the impression that the central authorities had some difficulties in organizing it efficiently. One reason for this was the fundamental opposition between a capital that was trying to protect its own industries and commercial interests, and a whole series of cities and towns taking an increasingly independent view of the opportunities offered by the international markets of the day. This is the context within which the "re-structuring" and "adaptation" have to be viewed. These have already been identified in other Italian regions.[22] In Lombardy, for example, great emphasis has been laid on the role played by the rural areas, in particular, the Alpine foothills, in the process of industrialization. It has been suggested that there is a certain continuity between the ruralisation of industry in the 17th century and the "industrial revolution" of the 19th.[23] This author would argue that this approach has proved very fruitful in the analysis of the Venetian state as well; that is provided it is not forgotten that throughout the 17th and 18th centuries the region was becoming increasingly "polycentric". At that time, the state did not have a metropolis such as Milan (with all its attendant economic effects) but rather a whole series of medium-small towns and cities, each with its own specialist industries. It is particularly clear that the conclusions drawn a few years ago by Cipolla have turned out to be misleading. When he observes that "the cities lost their vitality" (which is true) "and Italy not only failed to industrialize, but she actually moved away from any possible kind of industrialization" (which is an exaggeration), his conclusions must be treated with caution.[24] In fact, research must concentrate on the way the Italian regions settled down and adapted to an average-to-poor level of economic activity, if we are to understand the timing and course of the industrial advances of the 19th century.

However, this author would also reject the hypothesis that argues for an interpretation that is almost diametrically opposed to Cipolla's. Analysing the changes in the Italian economy over the course of the 17th century, Malanima paid particular attention to the question of whether there was a drop in the amount of wealth being produced (especially in terms of *pro capita* income). He concluded that

> "the arrest in economic development did not cause a decline in the production
> of wealth - the change in the country's income was more one of quality than
> quantity".[25]

This is maybe an excessively optimistic interpretation, in this author's view. Notwithstanding the nuances that have already been discussed, there can be no doubt that the "crisis of the 17th century" marked an important turning point in Italian history, and that from that moment on the country was both poorer and of less political consequence.

However, no one could deny that throughout the country, and particularly in Venice, there was a clear desire to maintain, to learn about and, if possible, adopt the technical know-how that had been imported from north-west Europe since the 16th century (see Hondschoote's "light drapery" for example). Nevertheless, Venice used its position as a capital to put substantial barriers in the way of foreign production processes and products that threatened its own economy.

Imported wool cloth, for example, clearly posed a threat to the local wool industry, and so in 1640, the city decreed that

"all foreign dark, black or coloured woollen cloth should be banned, and that only cloth of limited value should be accepted".[26]

However, Venice was not alone in resorting to such protectionism (which is always a clear sign of weakness): In the same decades Holland was taking similar measures against Flemish cloth, which was of good quality and

"continued to be regarded as the main rival down at least to the 1640s".[27]

In fact, up until 1647, Dutch cloth exports to Italy had only consisted of "says" from Leiden. This was a sure sign of the limited Dutch penetration into European markets. Then came what has been described as the economic zenith of the Low Countries (1647 -1672), when, as was to be expected, a more liberal and uninhibited trade policy was followed.[28]

Whilst Venice remained an almost impregnable fortress, Dutch craftsmen (such as Jan Bukerij and Roy & Pieter Comans) had more success in the Alpine foothills of the state, from Treviso to Bergamo; weaving "in the Dutch style" offered the chance to keep in touch with the new core by imitating its methods.[29]

However, Dutch improvements were not only a question of new production methods. French commentators have already remarked upon the fact that by using the new Dutch looms and production processes cloth could be made that, in France, would have required a higher labour intensity and up to a third more wool. The Low Countries also had a further commercial advantage in that they controlled the imports of Spanish wool. It was said that control of this trade guaranteed one a position of strength in the industry. As a consequence, when they subsequently managed to become the number one importer of mohair thread from Turkey, it is no surprise that "the Republic emerged as the main European producer of camlets".[30] A further point has been brought out in the work of Posthumus: In the second half of the 17th century, the value of the trade in Leiden "lakens" (a fine black cloth that was the symbol of the city) increased markedly (in 1630 the market was worth 110,000 guilders and in 1701, 4,200,000).[31]

It should be noted that Dutch technical superiority was not confined to the textile industry; it was still visible in a whole range of activities, such as paper production, Delft-ware pottery, shipbuilding, glass-making, "copper stills", "presses", diamond-cutting, wood saws etc. The result being that Venice was not the only state in which certain key sectors of industry were left behind. The same fate was suffered by countries of far greater political and economic importance (such as France). In the case of paper-making, for example, Holland had already established its technical supremacy over France by 1685;[32] and in Venice, the most substantial innovation in the industry - the "Dutch cylinder" - did not arrive until 1769! Throughout the 18th century, the supply of raw materials (rags) was a big problem for the Venetian paper industry.[33] This is further proof that a connection has to be established between trading strategies and the modernization of manufacturing industries.

The development of the Dutch glass industry was not as dramatic. Like the French glass industry it was still dependent on Venetian expertise in the second half of the 17th century, at least for high quality production.[34] It seems that the basic problems with the Dutch industry were shortage of raw materials and high wages, so that the finished product was not very competitive.[35] In fact, any analysis of the Seven Provinces in the 18th century must consider this problem of high wages, together with the connected problems of the concentration of the working population and the limited diffusion of rural industry in only a few areas (Overijssel, North Brabant and the border areas that were characterized by their relatively cheap labour costs.[36]

With regard to the silk industry, competition with Venice and the rest of Italy is still to be discussed exhaustively. It is, however, a fact that though the silk industry was limited to Amsterdam and Haarlem and was of little importance up to 1647, by the 1660's it was already competing with the French and Italian industries for Spanish markets. By 1700, the Dutch industry was one of the strongest in Europe, due to both the arrival of numerous French weavers after the revocation of the Edict of Nantes in 1685, and to new, guaranteed sources of raw materials. By that time Holland could count on a wide range of suppliers (for example Persia and India); so it was no longer dependent on the raw silk and thread imported from the East via Italy.[37] However, the Italian silk industry, i.e. trading in both raw silk and woven cloth, still retained its presence on international markets. There is no doubt, though, that it had already lost ground with respect to the Lyons silk industry, and, by the end of the 18th century, was beginning to find that the Dutch industry was putting it under some pressure.

Nevertheless, whilst it might be right to say that the establishment of the Low Countries as a core of the world-economy meant the disappearance of the "late-medieval polynuclear" urban system in Europe and the emergence of a "mono-nuclear" one, with Amsterdam as its centre[38], it would be wrong to extend this conclusion to cover the production of luxury goods. The silk industry in the 17th and 18th centuries seems to have had several lively centres, rather than being the preserve of one particular nation - be it France, Italy or the Low Countries (though, obviously, no one can doubt the prestige enjoyed by Lyons silk). In fact, one should not underestimate the German silk industry, which was gaining ground in the second half of the 18th century[39], nor the equally underestimated English silk industry.

With the market devided in this manner, the Venetian, and Italian, silk industry may have faced growing difficulties, but it still had good export opportunities for its products (to which, in fact, it added some new types of fabric).[40] As far as Venice was concerned, the main markets were in Central and Eastern Europe, as well as the Levant.[41] Not that this author would challenge the basic thesis that sees Northern Italy, and the Venetian state in particular, turning more and more to the exportation of raw materials (raw silk) and unfinished products (thread) rather than of finished cloth. This state of affairs is even clearer in 19th century Venetian state. An indication of it is the staggering increase in the amount of land used for the cultivation of mulberry trees during the course of the 18th century.[42]

It is the development of agriculture, therefore, which enables the summation of the relationship between the supposed *semi-periphery* of Venice and the *core* in Holland. In fact, this author would have no reservations in saying that agriculture must be analysed if there is to be understanding of the region's economic trends. In effect, there could have been no real development in manufacturing industries, nor any sort of expansion of trade, if the pre-industrial region had not guaranteed that its food supplies were sufficient to support its population. It is well-known, for example, that one of the keys to Dutch success was the control of the Baltic grain supplies. Not only did this grain figure largely in Dutch trade, but it also made it possible for domestic agriculture to concentrate on more remunerative crops: for example, linen and hemp (for industry) and woad and madder (for the dyeing industry, which was obviously directly linked with the textile industry) and tobacco. It is also well-known how much avant-garde agronomical research and experimentation was carried out in 17th century Holland. This research was drawn upon by the country's neighbours, in particular England which attracted an increasing number of Dutch technicians. It would seem, in fact, that the bases for the "agriculture revolution" in England were laid in Holland. The term "agricultural revolution" continues to be debated, however. Some argue correctly, in this author's opinion, that some of the agronomic techniques and knowledge referred to, date from before the supposed revolution of the 18th century.[43] Consider, for example, crop alternation: This replaced the fallow with the use of forage crops, and had already been proposed in 16th century Venice (by Agostino Gallo and Camillo Tarello). The most important point is that innovations require a dynamic social context if they are to be introduced successfully, and 17th century Holland provided such as context whereas late 16th century Venice did not. The stagnation in the Venetian state not only led to the Republic becoming of marginal geographical and colonial importance, it also affected production and the bargaining power of the Venetian property-owners.

For some time now it has been clear that the change which took place in the 17th century reinforced some of the privileges of the aristocracy to the detriment of an increasingly-weak rural labouring class. In the same way, research has elucidated the extent to which Venetian aristocrats increased the size of their land holdings on the mainland during the 16th century economic expansion. This does not mean that the small land-owner was completely eradicated. Rather this class was unable to resist the further advance of the aristocracy that took place in the first half of the 17th century. The contractual arrangements on large estates could only get worse. On the whole, estates were either rented out in small lots or else owners resorted to the traditional method of share-cropping. The availability of relatively cheap labour, therefore, meant that Venetian aristocrats and the mainland nobility did not apply those agronomical innovations that had been developed during the course of the second half of the 16th century.

The situation was very different in the Low Countries. Even in the late Middle Ages, settlers of agricultural land were better able to resist the large landowners (the reason seems to have been that it was difficult to cultivate recently-reclaimed land). Not only did small farmers have their own representative assemblies, they were also directly involved in the work of reclaiming and defending the land from the sea.

Water, in fact, occupies a key place in the history of the two countries. It became the crux of a number of changes in agriculture which marked the ascent, or decline, of the two nations in the world-system. The problem facing both states during the course of 16th century can be put in similar terms: Both states had to increase the area of agricultural land if they were to meet the needs of the population.

The increase in grain prices encouraged the speculative investment that was necessary to pay for costly drainage work. This explains why the Venetian aristocracy was so eager to participate directly in land reclamation consortia formed to drain new terrain that could be used for the cultivation of cereal crops: Terrain which they intended to add to their private estates. Throughout the 16th century, the two states kept in step with each other; so much so that the figures at our disposal suggest that the land reclaimed from sea and swamp by the end of the century was the same in both nations - about 70,000 to 80,000 hectares. Naturally, Venice had a further problem to deal with - the diversion of the rivers which threatened to silt up the entrances to the city's port, a problem which, in some ways, conflicted with the need to drain the land around the lagoon. Though a compromise solution was reached, there can be no doubt that rivers did tend to create swamping in the areas through which they were diverted. Of course Holland had similar problems - such as the diversion of the Rhine at Leiden or the cleaning of the Amsterdam canals. However, the conflict between the requirements of the port and the need to drain land was never, this author would say, as clear-cut as in Venice. During the first half of the 17th century, of course, the conflict became more muted: The stagnation in population growth, the drop in grain prices, and the sharp decline of Venice as a world-system core, all led to the stagnation in land-reclamation projects.

Agricultural historians, such as Slicher van Bath and Abel, have already elucidated the direct relationship between population growth and land drainage.[44] The situation in the Low Countries in the first half of the 17th century was the very opposite. On the one hand, the population continued to increase, but on the other, exchange rates and, in particular, corn prices continued to work in favour of the United Provinces. Finance was never lacking for new reclamation projects, and new terrain was drained (mainly the large inland lakes). The figures supplied by historians illustrate the large increases in the area of agricultural land (which were partly due to the use of up-to-the-minute technical equipment and methods).[45] It was in these decades that the Dutch started to use drainage windmills to reclaim swampland that would otherwise have remained useless. An expert hydraulic system was used to set up perfect networks of farms, which were designed to meet the needs of the market. Amsterdam's canals were kept clear and efficient by modern dredges, in which many other European nations, including Venice, showed an interest. In fact, the key to the relationship between the

*Serenissima* and the core was the city-state's interest in learning about equipment that could be used for her own preservation. This can be seen, for example, in the "Dutch shelters" that were installed on the littorals of the Venetian lagoon. Or again, in 1670 and 1675 respectively, with the interest shown in the dredges proposed by Gerhard Reighemberg, and the more famous Cornelius Meijer, for keeping the city's port open. And yet, Dutch land-reclamation technology does not seem to have had much impact in Venice. It is notable, too, there were less Dutch technicians here than in other North European countries.

However, Venetian agriculture seems to have been re-establishing itself during the course of the second half of the 17th century, but admittedly slowly. There are reasonably significant signs of this: A slight increase in irrigation and drainage projects, in the number of rice fields, in prices and in the number of inhabitants ( NB after the 1630 plague).[46] A clear example, therefore, of the different economic trends of North and South Europe is highlighted. The Low Countries go into economic depression in the later decades of the 17th century, when Italian economies are already showing some signs of coming out of it. Van der Woude has no doubts on this point. From 1650 onwards, in a key region of the Netherlands, the Noorderkwartier, the drop in population and grain prices discouraged further drainage projects similar to those undertaken in the first decades of the century ( a period which is referred to as the heyday of Dutch land reclamation). Of course, this is a long way from saying that all sectors of the economy were going through a crisis. The zenith of Dutch economic power has been dated as 1647-1672, and was followed by a "beyond the zenith" from 1672-1700. The question is still asked whether it was Dutch trade or industry that weakened most during the course of the 18th century?[47]

What is certain is that the agricultural sector seemed to turn in upon itself, both economically and socially. It is emblematic that Faber has used the term "oligarchiesering" to describe the situation in Friesland in the second half of the 17th century.[48] Anyway, a sense of realism should not be lost when discussing the legendary entrepreneurial spirit of Dutch farmers. It is true that they considered the needs of the market when deciding what crops to raise, or when financing costly drainage projects, but it is also true that not all these investors-speculators worked their own land. Some preferred to live in Amsterdam or neighbouring cities, entrusting the management of their agricultural affairs to a tenant.[49] This behaviour has always been taken as crucial to any definition of the characteristics that distinguish core and semi-periphery. So this fact about Dutch land-owners (which has also been observed with regard to English property-owners)[50] raises a doubt as to how the concepts of *entrepreneur* and *free labour* are to be considered as characteristic of the core. And can one really say that "capitalistic large tenancy" was really lacking in Northern Italy, even during the 17th century depression? Such clear-cut conclusions would seem rather risky when one considers the case of Lombardy: Recent studies have shown that investment had already picked up again by the second half of the 17th century, and that from the 1730's onwards continued at an even greater rate.[51] As Matzat has observed,

"when the state of Milan became a part of the semi-periphery after 1620, the
southern areas of the state (Bassa Pianura and Fontanili Zones) did not develop
share-cropping, as Wallerstein's model wants to persuade us, but rather the
contrary: It developed agrarian capitalism ( or maybe had already introduced
it)".[52]

Now even if the situation in the 17th century was not as rosy as Matzat
makes it out, there can be no doubt that this region had experienced forms of
capitalism since the late Middle Ages.

Nevertheless, the crop choices in 17th century Lombardy show that local
agriculture lagged behind its Dutch counterpart. There was a preponderance of
cereals over more remunerative crops. It seems that

"the situation in Lombardy was not like that de Vries describes in 17th century
Holland, where small farmers found it more profitable to produce butter rather
than grow rye because, with the butter they sold, they could buy more rye than
they could have produced themselves".[53]

However, within these limits, the Bassa Pianura and Fontanili Zones appear
to have been the most advanced agricultural areas in Northern Italy. They were
certainly ahead of the Venetian Po Valley area, where aristocratic land-owners
showed themselves to be incapable of laying the basis for modern agriculture. This
would only have been possible with direct management of estates according to
capitalistic principles which encouraged agronomic innovations and set up a single
circuit linking urban demand and crop choices. More forage crops and livestock,
which would have supplied milk, cheese and meat for the urban market should
have been chosen. However this did not happen in the Venetian state or else only
happened to a very limited extent. Most estates were still bound by the grain,
maize and vine trio. Livestock and irrigated fields are the absent component from
Venetian agriculture.[54] This helps to explain why, right up to the end, the regional
market was fragmented, turned in upon itself, and unable to keep up with that of
the more dynamic European regions. It is no coincidence that it has been observed

"where agriculture was healthy, as in Lombardy and Brabant, rural recovery
triggered the growth of individual cities as well as enlarged market ties".[55]

Holland had created this integrated market during its "Golden Age", and it
was this which enabled the country to offer some resistance to the commercial,
industrial and agricultural competition from England during the course of the 18th
century. In fact it seems, that after 1750 the doldrums into which the Dutch
polders had fallen (abandoned waterworks, halt in land reclamation) had a positive
effect. There was a new period of land drainage, encouraged by the increase in
agricultural prices which, once again, made investment in such projects
remunerative.[56]

The trend in 18th century Venice, however, continued to reveal links with
the North European core from which the region did not want to become totally
isolated. The most famous agronomists of the era (Duhamel du Monceau,
Beekman , Young) were well-known in Venice, and their ideas on crop rotation
were taken as a standard for experimentation. The enlightened patricians were,

finally, trying to apply the teaching of Camillo Tarello (who made some mark on Venetian agricultural after all). Given the interest that agricultural matters raised throughout Europe at the time, from 1762 onwards a whole series of agrarian academies were founded to debate both local and foreign discoveries. This trend was backed up by publishing ventures and university research. Several specialist journals and periodicals were set up (including Francesco Griselini's *Giornale d'Italia*) and the first Chair of Agriculture in Italy was established at Padua University in 1765. Irrigation and land reclamation work began again in the Republic of Venice. Of course, the irrigation cannot be compared with the system in Lombardy, which for centuries had had irrigation canals that could rival those in England.[57] Draining remained a real problem for Venice. Unlike the lower Lombardy area, the Venetian section of the Po Valley could not drain off its stagnant waters easily. Extensive zones (from the Verona *Valli Grandi* to the Polesine, and from the *Basso Padovano* to the Adriatic coast and the north-east of the lagoon) were still threatened by swamping and river flooding. This would continue to be so for a long time. From the data this author has gathered (which is, at the moment, being re-elaborated more precisely) it seems that there was a slow, continuous increase in the area of agricultural land from the 16th century onwards. Certainly the most dynamic period was at the end of that very century, when it is estimated that between 70,000 and 80,000 hectares of farming land were reclaimed from river, sea and swamp.[58] Romano has estimated that the total area of cultivated land between the 16th and 18th centuries was 150,000 hectares.[59] This author would conclude that this figure could be raised to, at least, 180,000-190,000 hectares (the land occupied by Venetian reclamation consortia).[60] In the same period, Holland seems to have reclaimed about 240,000 hectares[61] - so the acceleration of the 17th century caused the country to leave the Republic of Venice somewhat behind. However, Venice did not lose touch completely. Instead she showed a certain determination not to become isolated from the "Northern block" and, following Wallerstein's theories to their logical conclusion, one could say that the city was trying to avoid relegation to the periphery.

## Notes

1.   F.Braudel (1974): Civilisation matérielle, économie et capitalisme, XVe-XVIII[e] siècle, tome 3 (Le temps du monde), Paris, pp.12 and 24. I. Wallerstein(1974): The Modern World-System. Capitalist Agriculture and the Origins of the European World-Economy in the Sixteenth Century, New York.

2.   R.T. Rapp (1977): Industry and Economic Decline in Seventeenth-Century Venice, Harvard. Johan de Vries (1968): De economische achteruitgang der Republiek in der achttiende eeuw, Leiden.

3.   J.I. Israel (1989): Dutch Primacy in World Trade, 1585-1740, Oxford, p.377.

4.   P.Del Negro(1986): Proposte illuminate e conservazione nel dibattito sulla teoria e la prassi dello stato. In: G.Arnaldi-M.Pastore Stocchi (eds.): Storia della cultura veneta 5/II, Vicenza, pp.123-145.

5. See his article "World-System Theory: Implications for Historical and Reginal Geography", in this volume.

6. See the articles of P.Malanima, P.Massa Piergiovanni, A.Moioli and S.Ciriacono. In: H. van der Wee (ed.)(1988): The Rise and Decline of Urban Industries in Italy and in the Low Countries (Late Middle Ages-Early Modern Times), Leuven.

7. E. François(1986): Stagnation, régression, reconversion: remarques sur les "villes en déclin" de l'espace allemand, 1600-1800. In: J.de Vries-M.Gutmann (eds.): The Dynamics of Urban Decline in the Late Middle Ages and Early Modern Times: Economic Response and Social Effects. Ninth International Economic History Congress, Débates et controverses 2, Bern, p.129.

8. P.M. Hohenberg (1989): Urban Decline and Regional Economies: Brabant, Castile, and Lombardy, 1550-1750. Comparative Studies in Society and History 31, pp.441-442.

9. See M. Knapton(1986): Lo stato veneziano fra la battaglia di Lepanto e la guerra di Candia (1571-1644) and others essays. In: Venezia e la difesa del Levante. Da Lepanto a Candia, 1570-1670, Venezia, pp.233-241.

10. J.I. Israel (1986): The Phases of the Dutch *straatvaart* (1590-1713). A chapter in the Economic History of the Mediterranean. In: Tijdschrift voor Geschiedenis 99, pp.1-30.

11. See the article of R.A.Dodgshon: The Early Modern World-System: a Critique of its inner Dynamics, in this volume.

12. Braudel: Civilisation matérielle, op.cit. note 1, pp.73 sq.

13. P. Malanima (1989): L'economia italiana nel Seicento. Un'epoca di trasformazione economica. In: Storia della società italiana. La controriforma e il Seicento 11, Milano, p.149.

14. Idem: An Exemple of Industrial Reconversion: Tuscany in the Sixteenth and Seventeenth Centuries. In: The Rise and Decline of Urban Industries, op. cit. note 6, p.69.

15. Ciriacono, Mass Consumption Goods and Luxury Goods: the De-Industrialization of the Republic of Venice from the Sixteenth to the Eighteenth Century, ibid., pp.41-61.

16. See on this problem H.van der Wee, "A Synthesis", ibid., p.363.

17. P. Massa Piergiovanni: Social and Economic Consequences of Structural Changes in the Ligurian Silk-Weaving Industry from the Sixteenth to the Nineteenth Century, ibid., pp.17-22.

18. P. Chierici et al: Protoindustrialisation, marché et paysage rural: continuité et changements dans le processus de formation et de consolidation de l'état de Savoie. In: H.-J. Nitz (ed.): The Medieval and Early Modern Rural Landscape of Europe under the Impact of the Commercial Economy, Göttingen 1987, pp.247-274.

19. A. Moioli: De-Industrialization in Lombardy During the Seventeenth Century. In: The Rise and Decline of Urban Industries, op.cit., pp.86-87; L.A.M. Trezzi, A Case-Study of De-Industrialization of the City: the Silk Mills of the City and and Duchy of Milan from the Seventeenth to the Eighteenth Century, ibid., pp.139-151.

20. C. Poni (1976): All'origine del sistema di fabbrica: tecnologia e organizzazione produttiva dei mulini da seta nell'Italia settentrionale (secc. XVII-XVIII). In: Rivista Storica Italiana 88, pp.444 sq.

21. Ciriacono (1983): Protoindustria, lavoro a domicilio e sviluppo economico nelle campagne venete in epoca moderna. In: Quaderni Storici 18, pp.57-80.

22. Idem (1986): Mass Consumption Goods and Luxury Goods, op. cit.; idem, Venise et ses villes. Structuration et déstructuration d'un marché régional, XVIe-XVIIIe siècle. In: Revue Historique 276, pp.287-307.

23. D. Sella (1979): Crisis and Continuity. The Economy of Spanish Lombardy in 1ʰ Seventeenth Century, Cambridge Mass.; A.De Maddalena (1982): Dalla c'·· ⸵ al borgo. Avvio di una metamorfosi economica e sociale nella Lombardia spagnola, Milano. For a reminder of what was new about 19th and 20th century industrialization, see G.Mori, Il

processo di industrializzazione in sè e l'Italia. In: L.Segreto (ed.) (1984): La Rivoluzione industriale tra il Settecento e l'Ottocento, Introduction of V.Castronovo, Milano, pp.178 sq. On Italian industrialization, based on a close link between rural industry and the choice of crops to be cultivated - along with various sub-regional differerences - see A.Dewerpe (1985): L'industrie aux champs. Essai sur la protoindustrialisation en Italie du Nord (1800-1880), Rome. On these problems see also L.Segreto, La protoindustrializzazione nelle campagne dell'Italia settentrionale ottocentesca. In: Studi Storici 29 (1988), pp.253 sq.

24.    C.M. Cipolla(1975): The Italian Failure. In: F.Krantz & P.M. Hohenberg (eds.): Failed Transitions to Modern Industrial Society: Renaissance Italy and Seventeenth Century Holland, Montréal, p.10.

25.    Malanima, L'economia italiana nel Seicento, op. cit. note 13, p.188.

26.    State Archives, Venice, F.Cinque Savi alla Mercanzia, New Series, c. 87, 12 July 1647.

27.    Israel, Dutch Primacy in World Trade, op. cit. note 3, p.190.

28.    Ibid., pp.191-193 and 225.

29.    S. Ciriacono, Mass Consumption Goods, op. cit. note 15, p.50.

30.    Israel, Dutch Primacy in World Trade, op. cit. note 3, pp.225 and 262.

31.    N.W. Posthumus, De Geschiedenis van de Leidsche Lakenindustrie, quoted by Israel, op. cit., p.261.

32.    Israel, Dutch Primacy in World Trade, op.cit. note 3, pp.349 and 356.

33.    I. Mattozzi (1975): Produzione e comercio della carta nello stato veneziano settecentesco. Lineamenti e problemi, Bologna, pp.60 and 81.

34.    M. de Roever (1991): Venetiaans glas uit Amsterdam. In-trodutie van een luxe-industrie naar Italiaans model. In: M. de Roever (ed.): Amsterdam: Venetië van het Noorden, 's-Gravenhage, pp.156-173.

35.    Israel (1978): Dutch Primacy in World Trade, op.cit. note 3, p.353; Ciriacono, Per una storia dell'industria di lusso in Francia. La concorrenza italiana nei secoli XVI e XVII. In: Ricerche di storia sociale e religiosa 14, pp.181-202.

36.    For the problem of the high salaries, indicative of a high living standard, and not necessary a symptom of economic weakness, see especially the arguments of J.de Vries, An Inquiry into the Behavior of Wages in the Dutch Republic and the Southern Netherlands from 1580 to 1800. In: M.Aymard (ed.) (19882): Dutch Capitalism & World Capitalism, Cambridge-Paris, pp.37-61.

37.    Israel, Dutch Primacy in World Trade, op.cit. note 3, pp.227, 263-264, 352.

38.    Ibid., pp.4-6.

39.    U.-C.Pallach (1987): Materielle Kultur und Mentalitäten im 18. Jahrhundert. Wirtschaftliche Entwicklung und politisch-sozialer Funktionswandel des Luxus in Frankreich und im Alten Reich am Ende des Ancien Régime, München.

40.    S. Ciriacono (1981): Silk manufacturing in France and Italy in the XVIIth century: two models compared. In: The Journal of European Economic History 10 (1981), pp.194-199.

41.    R. Mazzei (1981): Traffici e uomini d'affari italiani in Polonia nel Seicento, Milano; A. Manikowski (1983): Il commercio italiano di tessuti di seta in Polonia nella seconda metà del XVII secolo, Warsaw; S. Ciriacono (1986): Esquisse d'une histoire tripolaire: les soieries franco-italiennes et le marché allemand à l'époque moderne. In: Etudes réunies en l'honneur de Georges Livet, Strasbourg, pp.317-326.

42.    See on these problems Poni, All'origine del sistema di fabbrica, op.cit. note 20, p.497.

43.    But for a reminder not to make too much of the crop yields in England at the beginning and the end of the 18th century, see M.Morineau (1987): Agriculture et démographie: l'évolution de la problématique, y compris un retour sur le cas anglais. In: A.Fauve-Chamoux (ed.): Evolution agraire et croissance démographique, Liège, pp.184-189.

44.    B.H. Slicher van Bath (1962): The Agrarian History of Western Europe, A.D. 500-1800 (transl. by Olive Ordish), London; W.Abel (1962): Agrarkrisen und Agrarkonjunktur, Hamburg-Berlin.

45.    R.H.A. Cools (1948): Strijd om den grond in het lage Nederland. Het proces van bedijking, inpoldering en droogmaking sinds de vroegste tijden, Rotterdam-'s-Gravenhage, p.131; P.Wagret (1959): Les polders, Paris, pp.89-96; C.T. Smith (1969): An Historical Geography of Western Europe before 1800, London, pp.507-509; A.Ventura (1970): Considerazioni sull'agricoltura veneta e sulla accumulazione originaria del capitale nei secoli XVI e XVII. In: Agricoltura e sviluppo del capitalismo, Rome, pp.534-537.

46.    S. Ciriacono (1981): Investimenti capitalistici e colture irrigue. La congiuntura agricola nella Terraferma veneta (secoli XVI-XVII). In: Proceedings of the Colloquium "Venezia e la Terraferma attraverso le relazioni dei rettori", Milano, pp.151-158.

47.    Israel, Dutch Primacy in World Trade, op.cit. note 3, pp.377 sq.

48.    J.A. Faber (1970): De oligarchisering van Friesland in de tweede helft van de zeventiende eeuw. In: AAG Bijdragen 15, pp.39-64.

49.    P. Burke (1974): Venice and Amsterdam, London, pp.106-107; A.M.Lambert (1971): The Making of the Dutch Landscape. An Historical Geography of the Netherlands, London-New York, pp.220-224.

50.    P.K. O'Brien (1987): Quelle a été exactement la contribution de l'aristocratie britannique au progrès de l'agriculture entre 1688 et 1789. In: Annales ESC 42, pp.1391-1409.

51.    Sella, Crisis and continuity, op.cit. note 23; L.Faccini, La Lombardia fra '600 e '700. Riconversione economica e mutamenti sociali, Milano 1988.

52.    W. Matzat, Early Mercantilistic and Agrarian Capitalistic Economy and its Impact on the Agrarian Landscape in the Stato di Milano (West Lombardy) in the 15th to 18th Century. In: The Medieval and Early-Modern Rural Landscape of Europe, op.cit., p.132; see also his article Northern Italy - Secondary Core or Reduced to a Semi-Peripheral Role?, in this volume.

53.    G. Chittolini (1988): La pianura irrigua lombarda fra Quattrocento e Cinquecento. In: R. Villari (ed.): Studi sul paesaggio agrario in Europa. Annali dell'Istituto A.Cervi 10, p.221.

54.    M. Berengo (1964): L'agricoltura veneta dalla caduta della Repubblica all'unità, Milano.

55.    Hohenberg, Urban Decline and Regional Economies, op.cit. note 8, p.458.

56.    A. Verhulst (1990): Precis d'histoire rurale de la Belgique, Bruxelles, pp.163-165; J.Materné: Modificazioni del paesaggio agrario nei Paesi Bassi (X-XV secolo). In: Studi sul paesaggio agrario, op.cit., p.100; M.Goossens, Modificazioni del paesaggio agrario nei Paesi Bassi (1750-1900), ibid., pp.103 sq.

57.    A. Young (1792): Travels during the Years 1787, 1788 and 1789 , quoted by Matzat, Early Mercantilistic and Agrarian Capitalistic Economy, op.cit. note 52, p.134.

58.    State Archives, Venice, F. Provveditori ai beni inculti; Ciriacono (1989): Venise et la Hollande, pays de l'eau (XV$^e$-XVIII$^e$ siècles). In: J.-L.Miège, M.Perney, Ch.Villain-Gandossi (eds.): L'eau et la culture populaire en Méditerranée, Aix-en-Provence, pp.99-114.

59.    R. Romano (1971): Tra due crisi: l'Italia del Rinascimento, Torino, pp.56-57.

60.    State Archives, Venice, F. Provveditori ai Beni Inculti.

61.    Cools, Strijd om den grond, op.cit. note 45, pp.131 and 151; Wagret, Les polders, op.cit. note 45, p.90; Smith, An Historical Geography, op.cit. note 45, p.507.

## Part III    Proto-Industrial Regions of the Semiperipherie

# 8

## Proto-Industrial Regions in England, with a Brief European Contextual Perspective

Robin A. Butlin

## 1.    Introduction

Of the existence in England of regionally-specific, rurally-located, industries in the period prior to that of intense, large-scale, capitally-intensive and generally urban-based industry there is no doubt. These appear to parallel the rise and development of 'proto-industrial' regions in many parts of Europe. The common characteristics of these industries include: the production of a wide variety of goods for markets beyond the immediate locale, sometimes for other regions of the same country or territory, sometimes for more distant, foreign, markets; the general tendency for proto-industries to be part of a marked dual-economy, in which farming and industrial production were combined, with a marked but not inevitable tendency for location in pastoral or woodland, that is marginal, environments; the creation of demand for foodstuffs by the development of this only partially-producing sector of the rural economy; the special relationship in the marketing process with towns in the region; the effect of such proto-industries on the processes of demographic change, notably by facilitating a lower age at marriage and therefore higher birth-rates and higher population density; and the expansion in due course to full industrialization.

## 2.    A World-Systems Approach

An important related question is that of the nature of the market forces which may or may not have tied production behaviour and trends to a global or semi-global economic and spatially-effective system, as outlined, among other hypotheses, by the Wallerstein model. In addition, there is the critical question of the chronology of these industries, and the light that their market arrangements can shed on the notion of core-periphery characteristics and on the validity of the chronology of Wallerstein's model of a world-system. The link between two very different scales of behaviour and experience is not easy to make, and it seems that one of the basic questions to be asked is how the production system of, say a cloth-producing family in West Yorkshire in the sixteenth and seventeenth centuries might have been bound to the fluctuations of regional, national and international markets. The scale factor is a very important one, as also is the question of cyclical economic fluctuations and their effects on industrial dynamics at global, national and regional scales.

As it is impossible in the space available to draw on examples of proto-industrialization from the whole of England and Europe, examples will be drawn from just a few English regions, and will then be used as a basis for broader comparisons and contexts.


## 3.   The English Experience

The existence of rural regions of industrial production in England, producing for distant markets, is well documented, though not all such industries and regions have been exposed to equally-intensive levels of investigation. The historiography of research in this area is well known, starting with the important general insights into the nature of industrial change in modernizing Britain produced in the first half of the twentieth century by such scholars as Clapham (1926) and Ashton (1948), and intensifying in relation to the more recent specifically proto-industrialization debate in the work of such scholars as Thirsk (1961), Berg (1985), Gregory (1982), Hudson (1986) and Clarkson (1985). New questions, particularly regarding the conventional associations with poor or marginal environments, the symbiotic existence of more 'commercial' industries, and the importance of complex social and institutional factors, are currently being asked and investigated. One interesting development in all this has been the rediscovery of the importance of the regional dimension to questions of early and full industrialization. The recent book edited by Hudson (1989) on regional perspectives on the industrial revolution in Britain is an important addition to the literature, stressing the context of increasing dissatisfaction with aggregate national levels of analysis and the growing attention to the early occurrence of industrialization within regions rather than nations, and the additional need to consider social and cultural as well as technological and economic factors.

In essence the balance of the theoretical and specific empirical work undertaken suggests a period of active rural industrial activity between the seventeenth and the early nineteenth centuries. The predominant rural industries were textiles, but there were in addition significant areas of metal-working and other craft industries. The specific characteristics of one or two of these regions will be mentioned later, with particular reference to the questions of labour and capital input, to the nature of social and organizational contexts, and to the subsequent experience of transition to full industrialization (or deindustrialization). The relationship between the sources of capital and the organization of the labour force is thought to be of particular importance, especially in cases where the capital came from outside the region.

One important point to stress at the outset is that industry in early modern Britain was almost without exception not a separate and distinctive economic activity but part of a complex set of economic and social relations or 'ecologies' bound up together at both small and larger scales, that is at the household and regional level, with other activities such as agriculture and urban change. If one includes in the term 'industry' various trades and crafts, in addition to manufacturing, then the complexity deepens.

The demographic background to early industrialization is complex, but at a national level involved two phases: rapid population growth, low mortality and falling wages in the period 1548-1640; followed by a period of decline and then slow growth to the 1720s, with rising wages in the period 1641-1743, and after a brief fall in the 1720s an increase again for the rest of the eighteenth century. The total population of England grew from just over 3 million in 1550 to 5.5 million in 1650, to 6 million by 1750, and to 7.91 million by 1801 (Wrigley and Schofield, 1981; Smith 1990), the mid-eighteenth century witnessing in effect the beginning of the demographic transition, related to reduction in mortality through improved diet, improved hygiene, and increasingly effective control of epidemic disease (Lawton, 1990, 287).

While the aggregate description of overall population trends affords interesting contextual material, it does little to help us understand the highly complex interactions between demographic behaviour and regional and local economies which are of critical importance in understanding the nature of early industrialization.

In terms of patterns of trade, the experience in England was roughly as follows. Internal trading patterns and institutions reveal a steady increase in the scale of commercial activity in the sixteenth and seventeenth centuries, with larger and more complex areas of production being bound into capitalist trading systems, particularly in the textile industries, and notably through circulating rather than fixed capital, with a consequence of steady change in the fortunes of specific industries in their spatial locations. One hypothetical argument that could be advanced, therefore, is that the influence of national and international markets produced an unsteady system with important spatial consequences (Dodgshon, 1990). At the outset of the early modern period, however, it is clear that what might be seen as the integration of the space-economy in relation to capital, goods, and even information, was very imperfect and uneven, partly due to the poorly developed inland transport systems. By the mid- to late eighteenth century a more sophisticated system had developed, with price-fixing markets, national and international, determining the tighter integration of the space-economy and the expansion or demise of proto-industrial regions.

An important and obviously relevant feature of proto-industry is its connection with non-local markets, sometimes far removed from the place of production. By the mid-eighteenth century, the national economy was led by export industries, with a very high percentage by value of exports being of cloth (Langton, 1978, p.174). Of particular importance were the markets in the British colonies, but there was intensive interaction with European countries also. Rowlands (1989) has pointed to the extension of the overseas markets for metallurgical goods from the West Midlands from 1660-1760, including nails, plantation hoes, cane cutters, oxchains and slave collars sent to the West Indies from the 1650s, while similar goods were linked with the triangular trade of slave traders, and ironware and brassware were exported to Sweden and Russia. Such links not only facilitated the development of regionally-based early industries in England, but also in some cases effected their demise. Thus in the case of the Wealden region of Kent and Sussex in south-east England, the success of such early industries as cloth-making, iron-making and glass-making was tied to a form of dependent capitalism through entrepreneurs who were

> "guided by national or international market forces, and the social structures, processes and landscapes which they did so much to create were in tune with national and international, rather than regional, movements. Glass, cloth, ordnance and timber products were dispatched to metropolitan or foreign destinations and were controlled largely by monopolistic mercantile or landed wealth" (Short, 1989, 162).

Short's account of the demise and disappearance of all the Wealden industries except for timber and timber products from about 1650 to 1820 places emphasis on the international market competition from Swedish iron and internal competition from the Midlands, South Wales, and Scotland, poor communications and shortage of water in the decline of the iron industry, and similar factors, including the uncertainty of external markets, in the cases of the cloth and glass industries. There were also important effects of changes in capital flow, Thus :

> "With little identity of interest between merchant and industrial capital, those regions, such as the Weald, formerly receiving metropolitan mercantile investment capital were henceforth less favoured as government stocks and foreign investment became more attractive".

## 3.1. The Textile Industries

Most accounts of proto-industrialization in England stress the importance of the textile industries, and not without justification, for the output of the textile industries increased substantially between the early sixteenth and mid-eighteenth centuries. There was a striking increase in the early modern period in the numbers of people employed in England in crafts and trades, that is from about 25% in the early sixteenth century to 40% by the late seventeenth century. What is especially important is the way in which competition, changes in supply of raw materials, and changes in technology, effected dynamic changes in the nature of success, and thus in the rise in importance of some producing regions at the expense of others. As Kerridge, in his major study of the textile industries of early modern England has clearly indicated,

> "Originally textile manufactures had been dispersed throughout the length and breadth of the kingdom. Even in the sixteenth century the geographical division of labour was incomplete. Alongside enterprises whose products sold all over the world, there were workers who served solely local consumers. Homespun wool was entrusted to a local weaver, then to a fuller and a

shearman, and finally to a tailor.....These activities all had their place in the
scheme of things"(Kerridge, 1985, p. 14).

In late medieval times the major producers were Wiltshire and
Gloucestershire, which produced broad cloth for export, West Yorkshire,
producing lower grade cloth for the home market, the worsted-producing region in
East Anglia, centered on Norwich, the traditional cloth regions of Essex and
Suffolk, and the cloth-producing regions of Devon and Somerset. The subsequent
experiences of these established regions varied enormously, as did those of the
newer regions of the north of England.

## 3.2   Early Industrial Lancashire

It will perhaps help to focus attention if particular consideration is given to
one proto-industrial region, namely Lancashire. The process of industrialization in
Lancashire has been in some respects reasonably well documented, though much
more detailed work is required. In a recent essay, Walton (1989) has indicated that
a wide variety of proto-industries existed in Lancashire in the early modern
period, and that there were major shifts over time in the pattern and degrees of
success of such manufacturing industries, but he also alludes to the basic
difficulties of establishing when the phase of proto-industrialization begins and
ends, and to the problem of definition of a region for this particular purpose. The
latter problem

> "provokes argument because it becomes so protean and elusive whenever a
> precise frame of reference is needed for empirical enquiry. The area which
> became the Lancashire cotton district is problematic because different parts of
> it evolved in different ways, so that what was recognizable as a distinctive
> economic region by 1840 had been a jigsaw of contrasting experiences less
> than a century earlier" ( Walton, 1989, p.43).

The basis of the production system in Lancashire was textiles, though there
were also in the south-west of the county mining and metalworking industries. The
textile industries offer a complex and changing narrative, for

> "Different areas specialized in different cloths using different raw materials;
> but there was considerable overlap between the textile industries, and the
> boundaries of their areas of predominance changed over time" ( Walton, 1989,
> p.45).

The history of the Lancashire textile industry , examined in detail in what is
still the outstanding account - by Wadsworth and Mann (1931) - has its roots in
the fourteenth and fifteenth centuries with the expansion of rural domestic
production, especially of woollen and linen cloth from locally grown flax and
wool. Coarse, light and cheap woollen cloth was increasingly exported to Spain,
Portugal and France during the course of the sixteenth century, with the necessity
arising of supplementing local wool with coarse wool from Ireland and the
Midlands. Linen was mainly produced for the home market, but local supplies also
had to be supplemented, in this case from Ireland. Need for the entrepreneurial

and credit-facilitating skills of middlemen, to assist in sales to distant markets and in the obtaining of adequate supplies of wool was expressed by the smallholding peasant producers in the later sixteenth century in petitions against state prohibitions of middlemen wool dealers. The 'undressed' (i.e. unfinished) woollen cloths of these small producers were brought to weekly markets where it was bought by dealers who arranged for its export via seasonal fairs, the London cloth market, or merchants at ports such as Chester, Liverpool and Hull. Wadsworth and Mann see in these links the beginning of capitalistic organization, though the extent to which the dealers and merchants actually organized production is not at all clear.

In the late sixteenth and early seventeenth century there were important new developments, including the introduction of cotton, change towards a capitalist system of industrial organization, and the increased proportional contribution of spinning and weaving to the economy of the small landholder. The main products at the beginning of the seventeenth century were a greater variety than hitherto of coarse cheap woollen cloths and the half-worsted bay cloths, light and cheap new draperies introduced to England by Flemish immigrants in the mid-sixteenth century and both of these types produced for the export market, linen cloth for the home market, and the new cotton, initially combined with flax for the making of fustians, an industry very much earlier and more strongly established in south Germany. The fustian industry grew rapidly, and an active export market developed to western and south-western Europe. The growth of bays and fustian production have been linked to changes in taste and fashion - away from the coarser wool to higher quality 'foreign' types of cloth for clothes, and also to vicissitudes in shipping and trade opportunities, including the fall of Antwerp, the entrepôt of the south German market. Hence

> "the South German and Italian industries which had supplied Europe with fustians for centuries were undermined, and, with the catastrophic assistance of war, were displaced by the countries of the Atlantic seaboard and the North Sea" ( Wadsworth and Mann, 1931, p.23),

that is Britain, France and Holland. This view gives perhaps too great a sense of permanence to the temporary blockage of cloth exports to northern and western Europe by war, for in the late sixteenth and early seventeenth centuries the restoration of trade between England and France and Spain, and the temporary restoration of trade with the Dutch, seems to have given rise to a short-lived small increase in the export of the older style of unfinished cloth to the finishing industries of the Netherlands and Germany, and a more rapid rise in the export of the new draperies, especially to France and Spain.

This system of domestic production in Lancashire structured by merchant capitalism had very specific spatial consequences, with increased regional specialization of production becoming evident. A major woollen industry region existed in the north and east in the Pennine valleys, which was a shrunken version of a formerly more extensive wool region. The linen area, essentially that of the Lancashire Plain, had expanded eastwards in the form of fustian production onto the moorlands between Blackburn and Bolton, and in the area immediately north of Manchester, with most of the rest of the county given to pure linen production.

These production systems and regions were characterized by the classic symbiosis of rural industry and agriculture, especially in the eastern upland and valley pasture regions, where the predominance of smallholdings with some form of rural industry, usually wool spinning and weaving, was assisted and extended by the sub-division of holdings through inheritance and the practice of colonization of moorland, sometimes through enclosure. On the western edge of the Pennines, therefore, the number of farmer-craftsmen/manufacturers grew in the seventeenth century. Hey cites the interesting example of the honour of Clitheroe where

> "the fines were certain, and the copyholders had taken advantage of the right
> to enclose their commons at 6d. per acre in proportion to their customary
> rents. In addition to these new intakes, many old tenements had been divided
> by partible inheritance, with the result that by 1662 there were half as many
> holdings again as there had been in 1608. Of the 654 householders recorded in
> part of the honour in the poll-tax returns of 1660, 406 held land worth less
> than £5 per annum, 145 had holdings valued between £5 and £10, 68 had
> tenements worth between £10 and £20, and only 30 had land valued at over
> £20" (Hey, 1984, p.68).

A classic dual economy prevailed. The picture is, inevitably, more complex than this, for in some of the marginal pastoral areas the system of inheritance was not partible inheritance but primogeniture, though Walton points out, in the case of the Rossendale woollen region, this did not prevent the sub-division of estates via sale or mortgage (Walton, 1989, p.53). In the Rossendale area, in the east of the county, the spread of weavers' smallholdings was rapid and marked on the upland benches above the valley floors, with members of families contributing to the carding and spinning of wool into yarn which the men spun into cloth, marketed in the town of Rochdale. Thus Rochdale itself

> "was the headquarters of merchants, chapmen and woolstaplers and the centre
> for fulling, dyeing and finishing, while the manufacturing processes of
> carding, spinning and weaving were carried on in the surrounding country
> district, where the industry had free scope for expansion owing to the absence
> of any urban corporation seeking to monopolize the trade" (Tupling, 1927,
> p.169).

The organization of the textiles industries continued to change in the course of the seventeenth century. Between the increasingly important urban merchant families of clothiers, mercers and drapers, who were large commercial capitalists dealing with both raw materials and finished goods, and the rural producers, employing the handloom, was a network of middlemen dealers, who sold raw materials on credit and who bought the finished goods. The fustian weavers were on the whole more closely dependent on these middlemen than were the woollen clothiers, whose supplies of wool were more ubiquitous than cotton and the structure of whose industry was, in consequence, much looser. The relations with the urban network and the generally unregulated nature of Lancashire industry contrasted markedly with the experience of other textile regions such as the West of England and with the restrictions on French industry and of that of Swiss, German and Dutch towns.

Major expansion occurred from the late seventeenth century, for

"Lancashire industry was well equipped to take part in, and adapt itself to that fierce competition in textile goods that was now entering on a new phase with the expansion of the Atlantic trade, and the permeation of Western markets by the textiles of the East" (Wadsworth and Mann, 1931, p.71).

The means by which it was so equipped included an increasingly intensive system of middlemen and a putting out system involving more and more dependent spinners and weavers, notably in the fustian industry, a more elaborate credit mechanism, and the introduction of two technical innovations, the Dutch loom and adapted silk-throwing machines. The Lancashire industries also adapted well to the changing market requirements, which involved in essence a rapid decline in the demand from northern Europe for the old draperies and a rapid rise in demand for the new draperies in the markets of the Mediterranean.

The major changes in the Lancashire textile industries in the late seventeenth and eighteenth centuries involved the expansion of the area producing fustians, and the development from it of the beginnings of the cotton industry. The geography of textiles in the early eighteenth century involved three distinctive production regions: the wool area of the eastern uplands and valleys, with packhorse transport and a dual economy being characteristic features; the linen area of the west of the county, whose products were marketed at Warrington, Preston and Ormskirk, and the third region was the fustian region of the centre and southeast, with an extension into the Manchester lowland, on account of the large amounts of imported cotton available from Manchester via the Mersey-Irwell Navigation. Although handlooms were still employed into the nineteenth century, the eighteenth century witnessed an increase in water-driven machinery. A greater specialization of occupation was a consequence of the increase in production in the eighteenth century, and new machinery included the flying shuttle. More and more families were devoting all their time to domestic manufacture. The spectacular changes which were to transform the area to a major manufacturing region in the period of full industrial revolution only date from the late eighteenth century, and thus cannot concern us here. What is important, however, is the effect of external and national demand on the growth of the Lancashire textile industries. The early British colonies such as North America and India became markets for textile exports, including coarse cloth, but the inappropriateness of such cloth for hot climates led to the exploitation, through regulated trade, of the Indian market by imitation of Indian production of cotton and calico, and to the development of colour printing on cloth.

Some comparison with production conditions of European producers is possible and necessary. A particularly important feature of textile development was the attitude of government to regulation. Wadsworth and Mann are helpful on this, especially in the comparison of England and France:

"In England, the system of regulated industry had almost died out, and the Government was ceasing to take an active part in industrial

development...After 1720, the Board of Trade and Plantations, which had
been active twenty or thirty years earlier in projects to establish new
industries, took no further steps in this direction....France presented a
complete contrast...French industry worked under a bureaucratic hierarchy
responsible to the central government...There was an intense and partially
successful struggle against the trade corporations and the enforcement of
regulations on the manufacture of goods, and a movement towards the removal
of commercial barriers, but for new inventions and their development industry
tended to rely more and more on Government assistance. The system of
inspection and regulation assisted in producing this result" (Wadsworth and
Mann, 1931, pp. 196).

There was similar regulation for protection of new industries in Prussia, and
doubtless elsewhere.

If one looks, therefore, for the key factors in the development of an
increasingly varied proto-industrial complex in Lancashire in the early modern
period, they would include geographical influences on the availability of raw
materials, that is wool, linen, and cotton; the long tradition of domestic industry
in, but not entirely confined to, areas of marginal agriculture; the pattern of
population increase over the period, the changes and vicissitudes in external
markets, the changing nature of organization and regulation of the industries,
technological innovations, and the changing nature of the political and trade
relations with near European neighbours and more distant colonies. An important
addition to this list is the fact that England was politically and territorially united,
and free from war on home territory, with fewer restrictions to the movement of
goods than many European territories, and a coastline and river system that
facilitated the movement of finished products. We will assume that the area under
consideration does qualify for consideration as a proto-industrial region, though
Walton has indicated that the demographic patterns and seasonalities of production
experienced in parts of this area do not correspond to the classic proto-industrial
model.

## 4.    Theory and Practice: the Proto-industrial and the Wallerstein Models

How can these regional features and experiences shed light, if any, on the
question of the relations between proto-industrialization and the core-periphery
model of Wallerstein's world-system theory? A cynical view would be that it is
both too easy and too difficult to prove a connection between two explanatory
models, that is world-system theory and proto-industrialization, easy because of
the level of abstraction from reality involved, and difficult because of the
extraordinary complexity of the actual experience of industrialization under widely
differing geographical, political, technological and economic circumstances. If one
accepts that differing combinations of these and other factors could lead to
industrialization, then it is difficult to make the connection between the two
models. Sabel and Zeitlin, in their study of the historical alternatives to the
classical theory of the break-through to mass industrial production in Europe in the
nineteenth century, offer one of a series of criticisms of the standard proto-

industrial model on the grounds that it is insufficiently accommodating to variety and flexibility of origins, characteristic, and development (Sabel and Zeitlin 1985; Butlin, 1986; Clarkson, 1985). Thus

> "explanations of the way an industrializing region develops are as difficult to reduce to a list of purely 'economic' causes as explanations of which regions industrialize. The Lyonese silk industry was, for instance, the result of a programme of import substitution executed by a seventeenth-century alliance between urban guilds and a mercantile state. The calico industry in Mulhouse was founded in mid-eighteenth century by the local Protestant patriciate with the financial help of Swiss bankers... Birmingham and Sheffield grew without state aid as the result of a favourable combination of local raw materials and a prosperous pastoral agriculture that afforded numerous smallholders the leisure and capital to put industrial skills to productive use" (Sabel and Zeitlin, 1985, p.168).

The standard literature on proto-industrialization is of little help in the establishment of useful connections between the two sets of theories. In the major theoretical statement of the subject, Kriedte, Medick and Schlumbohm's *Industrialization before industrialization* (1977,1981), only one reference is made to Wallerstein, (admittedly only one of his volumes had been published when the book first appeared in German) : a short critical comment on his apparently unqualified and expansive use of the term 'capitalism' in the first volume of his *Modern World-System* series. In a more recent British publication (Hudson, ed., 1989, p.35), Wallerstein is given similarly short treatment : brief reference solely to his assertion that long-term periods of economic expansion create the conditions for the consolidation of core-periphery structures.

Wallerstein's own perspective on proto-industrialization is more helpful. He asserts, as we know, that there came into existence in the long sixteenth century, that is the period 1450-1640, an all-pervading and complex system, the European world-economy, which structured the space, economy and society of the world influenced and increasingly appropriated by European powers. This sixteenth-century European world-system was

> "constructed out of the linkage of two formerly more separate systems, the Christian Mediterranean system centering on the North Italian cities and the Flanders-Hanseatic trade network of north and north-west Europe, and the attachment to this new complex on the one hand of East Elbia, Poland, and some other areas of eastern Europe and on the other hand of the Atlantic islands and parts of the New World" (Wallerstein, 1974, p.68).

By 1640, the following areas were to be at the core of the world-economy : England, the Netherlands, and "to some extent" Northern France (Wallerstein, 1980, pp.108-9).

How does the concept and chronology of the world-system help explain the development of rural industry in parts of England from the sixteenth to the mid-eighteenth century ? Wallerstein's thesis is that the change in the geographical distribution of industry from about 1550 onwards involved, *inter alia* , the expansion of rural textile industry in England on the basis of a search for external markets, the domestic market being deemed too small, at times of economic crisis, to sustain production. In spite of its exposed position by extent of reliance on foreign trade, this was offset by the creation of new overseas trading companies, an increase in efficiency of production, greater political unity in England than the Netherlands, and less oppressive taxation than in such old areas of production as Flanders and northern Italy, the administrative revolution of the Tudor period, internal peace, and the strong role of London (Wallerstein, 1974, p.230 seq.). He is particularly enamoured of K.W. Taylor's simile, which describes the spatial consequences of peace and the new geography of world trade on the changes in the distribution of population (which, with prices, was increasing rapidly) and industry in the sixteenth century :

> "Like a potted plant, long left undisturbed on a window-sill and then transferred to an open garden, the economy of England threw out leaves and branches" (Wallerstein, 1974, pp. 260-619).

Crises of trade in the early seventeenth century were solved by the extension of production of the 'new draperies' and the development of a re-export trade. According to Dodgshon (1977, p.13) this argument that the pattern and experience of the growth of rural industry in Britain, the Low Countries and France in the sixteenth and seventeenth centuries, thought to be the by-product of an expanding world market and in part the withdrawal of industrial competition from eastern Europe, does not conform easily with the conventional explanation of the rise of such industries in terms of an impending crisis of subsistence through overpopulation in agriculturally marginal areas. This objection is partly overcome when one looks to proto-industrialization in areas that were not marginal , and to those areas, as Dodgshon admits, where there was a putting-out system, involving merchant capitalists.

My own view is that I am less than convinced for the market forces/world economy thesis in relation to the sixteenth and early seventeenth centuries, and more convinced of it for the later seventeenth and the eighteenth centuries when the scale of external trade in, for example, English textiles, to Europe and to British colonies, increased substantially, notably the Atlantic trade at the beginning of the seventeenth century. I am, therefore, partially supporting Robert Dodgshon's 1977 thesis that one really has to look to the eighteenth century and beyond for the convincing evidence of a world-system, related to the forces of a self-regulating market system, and also his more recent proposal (Dodgshon, 1990), that one has, in effect, to look to a period of increased growth of fixed capital in industry and related infrastructures, that is the eighteenth and nineteenth centuries, for a fuller integration of space into a major capitalist economy with clear global proportions. The supposed transition of England from semi-peripheral to core status during the second half of the seventeenth century awaits, I think, more rigorous conceptual and empirical examination, especially the nature of a supposed semi-peripheral status. I am happier with analyses of earlier English rural industrialization which rely essentially on the changing nature of European trade, rather than the more mystical notion of the early state of a world-system.

Similar arguments could be marshalled for other areas of textile production in England. The textile industry of the East Midlands, for example, exhibits a chronology of 'proto-industrial' development which is characteristic of the period from the late seventeenth through to the nineteenth century. As Levine (1977) has shown in his study of Shepshed in Leicestershire, the growth of the hosiery framework knitting industry dates essentially from the late seventeenth century, with the movement of hosiers from London to avail of lower labour costs and to avoid the control of production by the Company of Framework Knitters. Thus,

> "The expansion of the industry throughout western Leicestershire was extremely rapid. In 1660 there were only 50 frames in the whole county, whereas by 1795 an estimated 43% of the county's population depended on some branch of the trade" (Levine, 1977, p.18).

By quite early in the eighteenth century the framework knitters of Shepshed had, it seems, been reduced to the status of wage labourers, and therefore subject to the vicissitudes of market fluctuations and changes in fashion. The trade with North America was particularly vulnerable. An extremely important factor influencing the geographical distribution of the wool hosiery industry in Leicestershire was the matter of social control, with the industry having developed most strikingly in 'open' parishes with unregulated settlement, especially on commons, and relatively weak lordship, in contrast to the 'closed' parishes, which had a few large landholders who placed restrictions on the number of cottagers or rural labourers resident in their parishes.

## 5.    The European Experience

Very obviously there were many and complex regions of rural industry in early modern Europe, again mainly based on textiles, but also including metallurgy and handicraft activities. One thinks of the many studies of the proto-industrial regions of Flanders, the Rhineland, the Pays de Caux, Languedoc, Hesse, Lower Saxony, Baden-Württemberg, Silesia, and very many more. The studies of these industries on the whole tend to be region specific, and on the whole, as in the British case, their comparative perspectives could be strengthened. Particular difficulties arise when one attempts to combine their greatly varied characteristics into a broad thesis on the relations between the notion of a world-economy and early industrialization. Part of the difficulty attaches to the use, hitherto, of too fixed and rigid a conceptual model of proto-industrialization, which tends to marginalize conceptually the industries of urban areas, non-marginal agricultural (i.e. arable) areas, and metallurgy and mining. It is perhaps also urgently necessary to question some of the assumptions on which such studies are based, notably the assumed cheapness of labour, the seasonality of production, and the matter of related and consequent demographic characteristics, including early age at marriage. One might usefully proceed, therefore, by expanding the notion of proto-industry (or abandoning it for the simpler notion of early industry), in order to incorporate a more complex view of pre-factory industrial structure into the world-economy thesis. This point has particular validity in view of Wallerstein's over-simplistic view of the contrasts in the roles of entrepreneurs in, for example, the English proto-industrial system. His sections in his second volume, on the reasons for the decline of rural industry in what he calls the dorsal spine of Europe, are however, interesting, and merit further critique.

More larger-scale regional studies, such as Viazzo's recent study of the environment, population and social structure of the Alps since the sixteenth century (Viazzo, 1989) would be helpful in understanding world-systems and early industrialization. Wallerstein does not figure in his text, but there is interesting information on proto-industrialization. His use of the term proto-industrialization is broader than normal, and seems to include what he calls 'autarkic' communities, presumably meaning self-supplying rather than producing for distant markets, and he also includes mining as a proto-industrial activity. The focus of the study is the demographic history of the Alps, and the broader theories engaged are from that quarter. Of particular interest are his observations - of comparative relevance to the earlier observations in this essay on open and closed communities in Leicestershire - on the effects of closed corporate communities on the development of early industry. Thus:

> "It is interesting to note that in the late sixteenth century Riva d'Agordo [in the mountains of Veneto] displayed all the marks of the closed corporate community, but this scarcely enabled the members of this community to stem the massive immigration of miners which was to alter the economy and demography of the valley so radically... On the other hand, attempts by urban-based merchants to develop commercial pastoralism in the uplands or to establish a putting-out system in rural areas, although supported by powerful commercial interests, were not directly endorsed by central governments... this was the reason why cottage industry gained a foothold in the highlands of Zurich rather than in the villages of the flatter districts" [which had rigid, closed corporate communities] (Viazzo, pp.280).

## 6.    Conclusion

More detailed investigation, at both smaller and larger scales, of the phenomenon known as proto-industrialization are needed before we can confidently make useful connections between this phenomenon and broader general theories such as that of Wallerstein. Further detailed examination and exemplification of the internal logic and structure of the Wallerstein thesis is also needed. Comparative studies of proto-industrial regions of Europe and other regions of the semi-periphery and periphery will serve both to refine and develop the original thesis and to broaden our understanding of an important element of the complex processes of economic, social, and geographical change.

# References

Ashton, T.S. (1948) : The Industrial Revolution 1760-1830. Oxford: Oxford University Press

Berg, M. (1985) : The Age of Manufactures. London

Butlin, R.A., (1986) : Early industrialization in Europe: concepts and problems, The Geographical Journal, 152, pp.1-8

Clarkson, L. A., (1985) : Proto-industrialization: the first phase of industrialization? London : Macmillan

Clapham, J.H. (1926) : An Economic History of Modern Britain. The Early Railway Age 1820-1850. Cambridge: Cambridge University Press

Dodgshon, R.A., (1977) : The Modern World System. A Spatial Perspective, Peasant Studies, VI, 1, pp.8-19

Dodgshon, R.A., (1990) : The changing evaluation of space'. In: Dodgshon, R.A. and Butlin, R.A. (eds.) : An Historical Geography of England and Wales, 2nd edition. London : Academic Press

Gregory, D. (1982) : Regional Transformation and the Industrial Revolution : A Geography of the Yorkshire Woollen Industry. London : Macmillan

Hey, D. (1984) : Yorkshire and Lancashire. In: Thirsk, J. (ed.), The Agrarian History of England and Wales, Vol. V.1. 1640-1750. Regional farming systems. Cambridge: Cambridge University Press, pp. 59-88

Hudson, P. (1986): The Genesis of Industrial Capital : A Study of the West Riding Wool Textile Industry c. 1750-1850. Cambridge : Cambridge University Press

Hudson, P. (ed.), (1989) : Regions and Industries. A Perspective on the Industrial Revolution in Britain. Cambridge : Cambridge University Press

Kriedte, P., Medick, H. and Schlumbohm, J. (eds.), (1977, transl.1981) : Industrialization before Industrialization. Cambridge : Cambridge University Press

Kerridge, E., (1985) : Textile Manufactures in Early Modern England. Manchester : Manchester University Press.

Langton, J. (1978) : Industries and towns 1500-1730. In: Dodgshon, R.A. and Butlin, R.A. (eds.), An Historical Geography of England and Wales, first edition, pp.173-198

Lawton, R. (1990) : Population and Society 1730-1914. In: Dodgshon, R.A. and Butlin, R.A. (eds.), An Historical Geography of England and Wales, 2nd edn., London : Academic Press, pp.285-321

Levine, D., (1977) : Family formation in an age of nascent capitalism. Cambridge : Cambridge University Press

Rowlands, M (1989) : Continuity and Change in an industrialising society : the case of the West Midlands industries. In: Hudson, P. (ed.), Regions and Industries. A perspective on the Industrial Revolution. Cambridge, : Cambridge University Press, pp. 103-131

Sabel, C. and Zeitlin, J. (1985) : Historical alternatives to mass production: politics, markets and technology in nineteenth-century industrialization, Past and Present, 108, pp.133-176

Smith, R.M. (1990) : Geographical aspects of population change in England 1500-1730, in Dodgshon, R.A. and Butlin, R.A. (eds.), An Historical Geography of England and Wales, 2nd edn., London : Academic Press, pp.150-179

Short, B. (1989) : The de-industrialization process : a case study of the Weald, 1600-1850: In: Hudson, P. (ed.), Regions and Industries. A perspective on the Industrial Revolution in Britain. Cambridge : Cambridge University Press, pp.156-174

Thirsk, J. (1961) : Industries in the Countryside. In: Fisher, F.J. (ed.), Essays in the Economic
    and Social History of Tudor and Stuart England. Cambridge : Cambridge University Press,
    pp.70-88
Tupling, G.H. (1927, reprinted 1965): The Economic History of Rossendale. Manchester : The
    Chetham Society
Viazzo, P.P., (1989) : Upland Communities. Environment, population and social structure in the
    Alps since the sixteenth century. Cambridge: Cambridge University Press
Wadsworth, A.P. and Mann, J.de L. (1931, re-printed 1965) : The Cotton Trade and Industrial
    Lancashire 1600-1780. Manchester : Manchester University Press
Wallerstein, I., (1974): The Modern World-System I. Capitalist Agriculture and the Origins of the
    European World-Economy in the Sixteenth Century. New York and London : Academic
    Press
Wallerstein, I. (1980) : The Modern World-System II. Mercantilism and the Consolidation of the
    European World-Economy, 1600-1750. New York and London : Academic Press
Walton, J. K., (1989) : Proto-industrialization and the first industrial revolution: the case of
    Lancashire. In: Hudson, P. (ed.), Regions and Industries. Cambridge : Cambridge
    University Press, pp.41-68
Wrigley, E.A., and Schofield, R., (1981) : The population history of England 1541-1871. London:
    Edward Arnold

# 9

# Industries of the Western Regions of France from the Fifteenth to the Nineteenth century

Jacques Pinard

Translation of the French manuscript by R. A. Butlin

The provinces of the West of France - Brittany, Normandy, Maine, Anjou, Poitou - seem to have experienced a measure of economic decline in the recent past which could not have been envisaged at the beginning of the Modern Period. Admittedly, their economy at the end of the Middle Ages was similar to that of other regions of the Kingdom, but the development which they experienced during the following centuries could have led to an expectation that they would become important poles of development in the same manner as the regions around the more northerly coastal ports of the North Sea, the Channel or the Baltic. What were the factors and events which caused this change in their evolution, and what is the evidence for their former status which allowed them to be placed in the semi-peripheral zones of I. Wallerstein's model, or allowed them to be viewed as regional 'cores' of the intermediate zone of the world system according to the criteria laid down by F. Braudel, and which they had actually been for a short period in their history?

## 1.    The Undifferentiated Economy up to the Sixteenth Century

At the end of the Middle Ages and in the early decades of the sixteenth century, these regions lived off the produce from farming and fishing which had also, in the better-placed, given rise to several pre-industrial activities.

The lands on the most fertile soils produced cereals and vines, the wine from which was bought by the traders who frequented the coastal ports, especially to the south of La Rochelle, for transportation to and sale in the Nordic countries. These merchants also sought the salt provided by the saltmarshes on the flat , sun-exposed shores south of the mouth of the Loire, and which was in great demand in the North for the preservation of fish caught in the adjacent seas and which was a very important foodstuff for the coastal populations of that period. The products of pastoral farming were still unimportant because those from poultry, pigs, and sheep were consumed locally and were not involved in any international trade. There were elsewhere regions which produced wine, salt and some cereals which were bought by foreign merchants, but the trade was limited. It was a number of ports of the Atlantic and Channel coasts that were mainly involved, such as La Rochelle, Nantes, St. Malo and Rouen, which also engaged in inshore fishing.

The raw materials on which an economic transformation might be based were still limited in number; other than the production in mills of flour from such cereals as wheat and rye, it was mainly concerned with textile fibres extracted from such cultivated plants as flax and hemp, or the wool from sheep fleeces which was spun domestically, that is in the farmhouses, and woven on simple handlooms owned by several skilled village weavers working for payment and who produced coarse cloth and finer draperies, according to the quality of the raw material used, to make up clothes and curtains. In the regions located at reasonable distances from the ports, cloth for sails and ropes for rigging of boats were also made. This was a speciality of Brittany, Poitou and Aunis.

The first mineral deposits, exploited at the surface of the sedimentary plateaux covered with the superficial deposits which originated in older neighbouring massifs, were the bases for the furnaces which produced pig iron and which was used in the production of utensils, tools, and sometimes of weapons: each region having several forests providing the fuel contained a number of modest forges which satisfied the needs of the local population. There was no long-distance trade.

Elsewhere it was the same workers, the peasants, who engaged in all these activities: farming; the extraction of ore and of wood cut in the forest; their carriage; and the process of smelting during the winter months. They worked for their lay or ecclesiastical lord, from whom they received lands which they exploited as a holding, rendering to him part of the produce as payment of dues.

The artisans did not engage in this work, for, often concentrated in the small towns, the only form of nucleated settlement in these regions of the West, they made small objects which were sold to the rest of the population. These people were blacksmiths, carpenters, woodworkers, and weavers, Only the towns had trade gilds of butchers, bakers, confectioners and the like. The society of that time was relatively undifferentiated and non-hierarchical, apart from the nobles who could not demean themselves by manual labour, and the clergy who prayed, studied, and sometimes engaged in work in the fields, bringing new land into cultivation around their monasteries or draining marshes for cultivation.

All regions experienced much the same economic and social conditions and none at that time had really taken much initiative: just a few places could sell or exchange the products of their land or their workshops. Transport overland was very difficult in these hedged regions with sunken roads on sediments like clay or marl in times of rain, and were only exceptionally supported by a firm sub-soil or with a stony surface, as in the chalk regions.

## 2.   The Increase of Regional Development Nodes in the Seventeenth and Eighteenth Centuries

The opening of new commercial horizons with the discovery of the New World and the increase in new sea routes to Africa and Asia provided Western Europe, and especially the Atlantic seaboard, with enormous possibilities for development, because into their ports came raw materials, and notably precious metals, which favoured economic expansion and considerable monetary inflation. It was particularly through the ports of Brittany and Poitou that there penetrated and circulated the Spanish gold coins which facilitated the acceleration of trade.

This expansion was evidenced through an increase in food consumption which influenced the bringing into cultivation of new land in the West reclaimed from the coastal sands and marshes. In the sandy regions ('landes') of the Breton peninsula or of Poitou reclamation was followed by paring, burning and manuring before the introduction of cereals and buckwheat which provided food for what was often a large rural population. Dutch engineers, already well-known for their expertise in hydrology, were brought in for the draining of the Breton, Poitevin and Saintongeais marshes of the West, which was followed by the introduction of cereal cultivation and the rearing of dairy cattle. The construction of new windmills, on water or tidal locations, in the estuaries and bays of Brittany, rapidly increased and is an indicator of the increase in the supply of and demand for wheat.

Vine cultivation in the Charentes expanded because of the discovery of the means of preserving and transporting wines of mediocre quality over long distances: it sufficed to distill them in stills to obtain the spirits which, when aged in oak casks, acquired a distinctive aroma and became the celebrated Cognac. Foreign merchants became increasingly interested and even came to buy property in the region in order to plant vines and engage in this enterprise on their estates, of which some famous names - such as Hennessy - still exist.

Fishing also underwent major expansion when the sailors undertook voyages to distant places from the Breton (St. Malo) and Charentes (La Rochelle) ports in order to search for cod off the coast of North America. Catches were dried and salted before being sold in local or foreign markets. The trading ships from the provinces went particularly to Spain and Italy when they sailed to the Mediterranean to search for agricultural and industrial products which they re-exported, on their return, to the countries of Northern Europe, products such as the wines and fruits of Andalusia, wool from Castille, iron from the Basque country or oil from various southern regions. A truly triangular trade developed from the ports of the West whose sailors acquired the reputation of being great sea-traders and its cities of being major centres of international commerce.

These always more intensive agricultural and coastal activities stimulated the initiation of manufacturing enterprises by furnishing their equipment or processing the products which they transported to them before the merchants finally redistributed them to the centres of commerce which had also become centres of control of all kinds of activity.

To be sure the naval bases, which originated in the need to protect the most important harbours, were still quite small, having only a few teams of carpenters working with wood from the inland forests, in search of which they had increasingly to go farther afield when the nearest stands became rarer - and thus more expensive - with the competition from increasing numbers of users and with the newly established large arsenals (at Rochefort, Brest, Lorient, Cherbourg and

Le Havre) becoming major consumers for which the better quality large trees were
reserved.

It was particularly the furnaces and forges processing the iron ores
discovered in numerous places in these regions of the West which multiplied at the
initiative - at least in the early cases - of the lay or ecclesiastical lords, sometimes
by the great names of the French nobility such as de Broglie, Cond, and Rohan,
who wanted to realize the value of their forests and their estates and to provide
mills for the streams which ran through them and whose paddle wheels drove the
bellows of the furnaces and the hammers of the forges. They provided the arms
and munitions of the king's armies and the rigging, cannon and cannon-balls of the
Royal Navy and other fleets. These metallurgical industries began in the most
remote and distant parts of central Brittany, such as Paimpont, or on the borders
of its neighbouring regions such as Normandy, Maine and Anjou, and even in
Poitou or Angoumois. Here pieces of artillery for the Navy made in the forges
were transported by boat on the Charente to the arsenal at Rochefort created by
Colbert in 1666 and for which a royal foundry was to be established at Ruelle near
to Angouleme in the middle of the following century. These establishments could
be located in relatively thinly peopled zones because all the inhabitants were
'required' to carry out the different tasks which the managers required, such as
woodcutting in the forest, the transport of wood, and smelting, sometimes
undertaken for payment or more often as services which still could be imposed by
an estate-owner on the tenants of his estate.

The development of trade had introduced into the textile industries raw
materials from distant places, such as wool from regions having large flocks of
sheep, for example Champagne and Berry, and also imported from other
countries, like Spain, or cotton imported at first from the oriental countries and
subsequently from the English colonies in North America. Several regions of the
West of France which already had a tradition of processing local raw materials had
become specialized in the new type of textiles: there were in Brittany, for
example, regions developing the production of coarse cloths from hemp, at Vitre
and in the Rennes region, and alongside them there were places specializing in fine
linen of high value such as Leon and the Quintin-Loudéac region. One also saw
the appearance in Normandy, in the rural villages furthest removed from the Seine
valley such as those of the Pays de Caux or Pays de Bray, of the production of
coarse fabrics using local wool, while the towns situated near the river or its
tributaries and which already had weavers' gilds in the Middle Ages specialized in
fine cloth made with imported wool or with cotton, especially as the style of
painted or printed cloths expanded in the course of the eighteenth century and
because the urban workshops alone could use these new techniques. This provided
another way in which cities like Rouen, Elbeuf and Louviers could exert control
over the surrounding countryside and the workshops located there. Other changes
were to give more power to those towns which became small centres of regional or
local industry over several decades and attracted to them the inhabitants of the
surrounding areas, at least for several years.

The importation of exotic materials into the ports which engaged in trade with the West Indies, in effect the Antilles, began to develop processing industries on their quaysides. This was the case with sugar imported by the ports of Nantes, La Rochelle and Rouen. The capital of Normandy had in the early part of the eighteenth century twenty sugar refineries, and La Rochelle in the early years of the following century had seventeen refineries. The wars of the eighteenth century dealt a fatal blow from which they never recovered, particularly with the advent of competition from sugar-beet in the nineteenth century.

The use of tobacco which started to increase in the seventeeth century led to the state taking control, leading quickly (by 1674) to a monopoly, and constructing the first tobacco factories. Ports like Dieppe and Le Havre in Normandy as well as Morlaix in Brittany were thus equipped, the raw material at first being entirely imported. Other trade such as that in coffee, tea, cocoa, spices and dyes, even if it did not result in major industrial activities at least accrued substantial rewards to the merchants, and gave great prestige to the port cities, with their mansions, public buildings and churches, many of which still survive, and which are the fruits of investment by the rich citizens, some aristocrats, and the state, especially in the Age of Enlightenment.

The discovery of printing two centuries earlier had generated major demands for paper which led many owners of mills, located in regions with water-courses with year-round flow, to transform the mills to paper-mills: this was the case in Angoumois where the re-emergent karst gives a relatively regular and abundant flow to several tributaries of the left bank of the Charente south of Angoulme. The first paper-mills with fitted fulling equipment in the sixteenth and seventeenth centuries - there were 80 in 1656 - were modernized in the eighteenth century with the arrival of Dutch entrepreneurs who introduced their famous 'pile' or stamping trough for careful pounding of old paper and rags to produce the best vellum. Already famous publishing houses like Elsevier from the Low Countries were supplied from Charentes where there were Dutch merchants, and where a distinctive society developed, comprising the owners of the mills, the merchants or 'farmers' who controlled them, the master-craftsmen and the labourers, each occupying a particular place in this well-established hierarchy. A royal factory was founded in 1734 at La Couronne south of the capital of Charentes, confirming the importance of the place, which only declined as a result of the crises at the end of the eighteenth century.

Thus on the eve of the French revolution these regions of the West comprised a series of development nodes, each one experiencing periods of intense activity (during times of war for some, times of peace for others), around which were rural market towns whose workshops and inhabitants lent their support by working for the merchant-manufacturers of the towns - the putting-out system of domestic industry was not solely to be found in the textiles industry. The trading contacts between these cities accelerated, sometimes by sea and coastal networks, sometimes by land routes because of the network of routes established and improved during the eighteenth century and which facilitated links between the regions, leaving isolated only a few small regions inhabited by populations whose poverty reflected their limited resource bases.

## 3.    Favoured Regions in the Nineteenth Century

The radical political, economic and technical changes of the end of the eighteenth and beginning of the nineteenth centuries were focussed on new industrial processes and products and effected considerable modification of the methods of production and ways of life in all regions, particularly in the West of France.

Many major landowners of aristocratic origin who were in the preceding century engaged in the administration of their estates, not only in the exploitation of forest and mineral resources but also (and principally) in the introduction of new farming methods, and making use of new land and improving crop rotations, were deprived of their possessions by the Revolution. Their successors did not always have the experience and knowledge necessary for the successful revival of these activities. On the other hand the often energetic exploitation of the mineral deposits and the forested areas for supplying fuel to the blast furnaces (increasingly needed during times of war for the manufacture of arms and munitions for the troops and the fleets involved) exhausted some of the sites: a number of metallurgical plants had to stop production on a number of occasions.

With the return of peace in 1815 trade resumed, the increased wealth of the more active classes enabled new investment, and the recent scientific discoveries in France and elsewhere were brought into use in newly-constructed mills. Thus steam-engines were gradually brought into use, alongside water-wheels, as power sources for the new machines used in the textile industries and which were also being used in England (spinning jennies and weaving looms). Coke smelting was slowly to replace charcoal, but in order to take full advantage it was necessary to find nearby coal deposits - in this sense the poverty of resource of the West soon became obvious other than at two or three small sites - and the old blast furnaces were no longer adaptable to this new fuel. The iron ores discovered at the end of the nineteenth century in Anjou and Lower Normandy produced only one metallurgical industry - near Caen.

The revival of inland and external trade depended, before the coming of the railway, on the development of a sufficiently economic means of bulk transportation which had already proved its efficacy in the previous century: the navigable waterway. Some rivers were improved and several canals were constructed in the West, such as that from Nantes to Brest in Brittany, linking these two economic nodes and also small industrial cities between them, at least for several decades. The construction of a network of railway tracks in the second half of the nineteenth century, in spite of the fact that it tolled the knell of the earlier means of communication, was to have one role among others, not only in the West but in the whole of France, of causing the flow towards the capital of foodstuffs produced throughout the whole area: primary products, consumables etc. a number of which would be unprocessed or only slightly modified when they left these regions which thus began to lose the foundations of their life and work. This was a modern version of the system of exclusive rights of the metropolis which had applied during the Ancien Regime to the detriment of the conquered and developed territories which were later called colonies.

However, several sections of these western regions experienced expansion by attracting several modern industries. This was the case for example in the valley of the Lower Seine around Rouen which had already prospered from the textile industries in the preceding century. Large quantities of cotton were brought by sea from the United States; henceforth it was processed in spinning and weaving mills located on small tributaries near the main river, while the older industries were forced back towards more isolated regions such as those further to the south where flax and hemp cultivation had begun to serve the cloth mills and ropeworks established at Angers, Cholet or Beaufort-en-Valle. The phenomenon of the concentration of the cotton industry in the Lower Seine was further enhanced after the treaty of 1871 with the arrival in that region of industrialists from Alsace who relocated in the West. On the other hand the older centres of cloth production in Brittany and in the bordering regions were not affected by the new types of textile industry, and underwent industrial decline.

The old remote metallurgical sites in the middle of the great forests of central Brittany and in the bordering regions of Maine and Anjou, which gradually ceased production in the course of the nineteenth century, did not experience the birth of new enterprises as in the case of the mills of the Eure basin in Normandy which specialized in wireworking and the manufacture of other everyday common objects in great demand such as needles. Nevertheless some important enterprises were to develop, in some cases somewhat by chance, for political or economic reasons, such as the arms factory at Châtellerault located in Poitou in 1819 as part of the relocation policy for the arms factories which were thought to be vulnerable near the northern and eastern frontiers of France, or the forges of Hennebout near Lorient in Brittany established to supply strong sheet-metal for the naval shipyards of the arsenal at that port, and also tinplate for the canning factories established along the western coasts of the Breton peninsula, or the steelworks at Trignac on the Lower Loire created to provide metal for the shipbuilding industries at Nantes and Saint-Nazaire. But the major metallurgical companies did not locate in the West - with the exception of one in Normandy established at Mondeville near Caen - because of the lack of sufficiently large deposits of iron ore and coal.

Before the demise of the activities which had provided employment in the past thousands of people, the Atlantic regions were to move towards the processing of products obtained locally. It was thus the products of fish landed in the Breton ports which first gave rise to canneries thanks to the invention of the sterilization process by Nicholas Appert (1809) with the use of the autoclave: initially the canning of fish, then vegetables which the industrialists, especially from Nantes, having invested in these factories, encouraged the peasants to cultivate to provide supplementary produce outside the fishing season. The custom also of preparing dried and salted meat for the crews of naval vessels in their ports encouraged certain Breton cantons to develop the rearing of pigs and to prepare pork products such as pat and ham initially on an artisan and later an industrial basis: it was local capital in particular which was to be invested in this processing, finding opportunities both locally and very quickly also in the large cities including Paris when the rural exodus began to have severe effects on the population of central Brittany.

The other products of livestock rearing, especially milk products, experienced important changes in consequence of two or three major events. It is true that the peasants having pasture land in Normandy had invested very early, certainly from the eighteenth century, in the acquisition of new pasture lands and livestock to supplement their birthright (which was transmitted by inheritance via the eldest member of the family), but it was the construction of the first railways which facilitated the export of local products - first butter and then the already famous cheeses - to Paris and the large cities. Brittany did not follow this until the twentieth century.

In the Charentes, the phylloxera crisis which, beginning in 1880, ravaged the vineyards providing the wines for distillation, stimulated the extension of pasture and the development of dairy farming. The milk, skimmed on the farm for the production of butter, took on a new significance with the creation of the first dairy co-operatives and the arrival of the centrifugal creamer a little later. Butter was from now made in the dairies with the milk collected from the farms, expedited by the refrigerated vans, the earliest being those used to send produce from the Charantes to Paris; the by-products were used to feed pigs and to produce casein. These rural areas were wholly modernized by the appearance of these modern techniques, allowing part of the rural populace to remain (which would otherwise have been forced to leave because of the loss of the vines) pending the revival of the vineyards under the aegis of the major merchant houses at Cognac whose influence was not really felt until the beginning of the century. A similar development was experienced in the later specialization in the high quality wines by several vineyards in the Loire valley and its neighbouring areas, and also by those areas conducive to the cultivation of vegetables on account of the coming of the railway.

Thus four or five nodes out of a group in the West of France were able to take the opportunity from the nineteenth century onwards of either maintaining older established activities or acquiring new ones which enabled the population to remain, and found new openings either locally or in external - national or foreign - markets. But between these more dynamic sectors regions went steadily into decline, on account of lack of raw materials and capital, their populations ( often numerous and educated, as in Normandy for example) initially leaving to go to the largest cities in their region, and then sometimes in a second stage to Paris, especially the second and third generations. Those regions outside the cities which kept their service and some basic industries would find themselves marginalized and would become peripheries, or at best, in F.Braudel's phrase, intermediate zones, 'brilliant seconds', working above all for their regional metropolis, and for the capital city, and finding themselves increasingly vulnerable to being left by the wayside when the whole of France found itself in a changed position because the centres of gravity of economic activity were moving globally towards new countries.

In a feebly-developed and highly fragmented economy (the result of ineffective means of transport), the regions having relatively modest needs reached self-sufficiency (an autarchic system) and became small economic cores whose role weakened with the expansion of commercial horizons, and with the progress following the technological innovations, which bestowed selective advantage to regions in relation to their potential. The least well endowed became marginal 'peripheries', and only a few intermediate, semi-peripheral, zones continued to participate in the general economy of the world-system: this was the experience of the regions of the West of France at the dawn of the twentieth century.

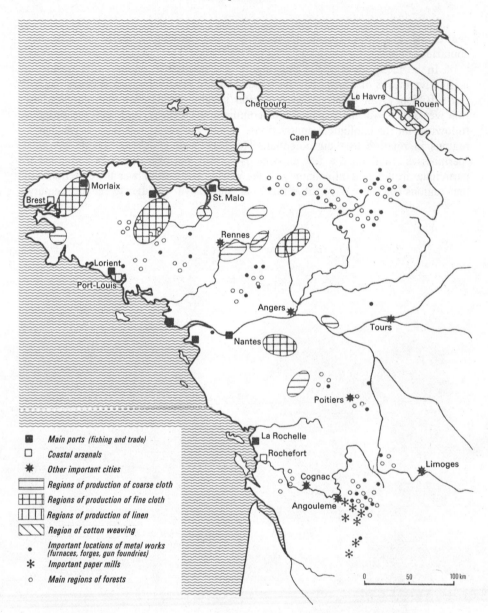

Legend:
- Main ports (fishing and trade)
- Coastal arsenals
- Other important cities
- Regions of production of coarse cloth
- Regions of production of fine cloth
- Regions of production of linen
- Region of cotton weaving
- Important locations of metal works (furnaces, forges, gun foundries)
- Important paper mills
- Main regions of forests

Main industrial activities in western France in the 16 th to 19 th centuries

# References

Beauchain R. (1982): L'industrie sucrière à Rouen jusqu'au milieu du XIXè siècle. Annales de Normandie 13, Caen, pp.315-319

Bourgin, H. (1920): L'industrie sidérurgique en France au début de la Révolution, Paris, 2 tomes

Chaunu, P. (1972): Réflexions sur l'échec industriel de la Normandie. In: L'industrialisation en Europe au XIXè siècle. C.N.R.S., Actes du Colloque de Lyon, Paris, pp. 285-304

Delafosse, M. (1949): La Rochelle et les Iles au XVIIè siècle. Revue d'Histoire des Colonies, Paris, pp. 215-224

Le Lannou, M. (1952): Géographique de la Bretagne, Rennes

Musset, A (1917): Le Bas-Maine. Etude géographique, Paris

de Planhol, X. (1988): Géographique historique de la France, Paris

Pinard, J. (1972): Les industrie du Poitou et des Charentes. Etude de l'industrialisation d'un milieu rural et de ses ville, Poitiers

Pinard, J. (1987): Les nouvelles activités manufacturières dans les campagnes du centre -ouest de la France du XVIe au XVIIIe siècle. In: H.-J. Nitz (ed.): The Medieval and Early Modern Rural Landscape of Europe under the Impact of the Commercial Economy, Göttingen, pp. 295-306

Sion, J. (1909): Les paysans de la Normandie orientale. Pays de Caux, Bray, Vexin Normand, Vallée de la Seine. Etude géographique, Paris, Rééd. G. Montfort 1978

Touchard, H. (1967): Le commerce maritime breton à la fin du Moyen Age. Ed. Les Belles Lettres, Paris

Woronoff, D. (1983): L'industrie sidérurgique en France pendant la Révolution et l'Empire, Paris

# 10

# Mining Regions and Metal Trade in Early Modern Europe

Dietrich Denecke

## 1. The Spatial Organization of Metal-Mining and Smelting in Central Europe

In the context of an "economic world-system" mining and metal trade is undoubtedly an outstanding example proving - and disproving - a general validity of the theory. This is especially true for the 16th century when a capitalist economic system was established going from exploration of minerals to mining and exploitation, to the smelting process, to the separation of pure metals, to metal trade, to profit made on the market, and finally, making the wheel come full circle, in new investment. From the geographical point of view the location of the different activities, their patterns of distribution, and moreover the functional coherence within the whole network of the extraction, processing, marketing, and investment chain is of specific interest. The economic management within the frame of capitalistic aims is the important process behind all mining and metal trading activities so that there are at least two different spatial economic systems involved: Firstly the pattern of economic development and production as evident in industrial regions and landscape and secondly the pattern of trading connections, of capital flow, and of investment strategies combined with economic and political policies. The first system and pattern located and investigated in the landscape can be reconstructed on maps. The pattern of economic strategies behind the scene however is difficult to obtain. This can be investigated by taking into account economic and political history or we may try to discover these strategies by analysing the impact they had on regional economic development.

Mining ventures always have to undergo certain stages of development, a centrifugal pattern of diffusion or, in the broader sense, a pattern of process and expansion. These stages are at the same time phases in time and clearly emerge from a core. All mining is bound to the deposits of minerals as such and all exploitation is bound to quality, stratification, accessibility and geographical location of the minerals. Vague information about deposits and an engagement in exploitation is the starting point to launch the economic and in general capitalistic venture of an initiative and investment of capital and labour for the exploitation of mineral deposits. Mining the minerals is just the first step. The highly technical processing of the minerals has to follow, often at a different place and under separate organization and management. The chain of operation continues with the organization of a transport system for heavy goods and the demanded establishment of a long-distance market. Most important for these economic activities in the Early Modern Period was the capitalistic economy within the framework of a credit and loan system.

The main centre of the European copper production at the end of the 16th century was the East Alpine region of the Habsburg territory, besides that Hungary and Slowakia in the second place, and the Mansfeld region in the third place. Between 1580 and 1610 the East Alpine region accounted for about 50% of the European production and for more than 100 years (1500-1620) this proportion remained stable between 40% and 45%. In 1520, at a time which can be characterized as a 'copper rush', 40 to 50.000 miners were employed in Tyrol.[1]

In 1420 copper mining began in the East Alpine region on the basis of local companies (*Gewerke*). The main mining areas (*Reviere*) were Falkenstein, Schwaz and Ringenwechsel, later Röhrerbichl (1540), Radmer (1547), Ahrntal (1491) and Samobor in Croatia. Production usually boomed in one place for a short time and then quite often declined rapidly because of growing expenses and technical difficulties. During the 16th century new mines were permanently opened to replace the declining ones so that between 1490 and 1620 a quite stable level of production was achieved.[2]

Between 1500 and 1530 there was an important change in ownership, management, investment, and market as copper mining and metal trade (also lead and silver) changed towards capitalistic forms of organization and holding. Offering or granting loans to the German Emperor, to the territorial sovereigns, and to the individual mining companies, more and more wealthy merchants from southern Germany, especially from the cities of Nürnberg and Augsburg, stepped in as shareholders and finally they had to take over the whole mining management to secure their interest and revenue.[3] This meant that capital investment and management from outside the region became more and more important as it steered the regional and economic development. Capital from the big European markets such as Nürnberg, Augsburg, Leipzig, Frankfurt, Lübeck, Hamburg, Venice, Genoa, Antwerp, and London was invested or withdrawn so that technical development, increase in production, influx of workers, supplying industries, agriculture, and the transport network became more and more subordinate to the overall flow of capital.

A key position in the process from mining the ore to selling the pure metal was the smelting process. This was taken over and organized especially by trading companies (*Saigerhandelsgesellschaften*). The main process was to extract silver from raw copper (*Schwarzkupfer*) by adding lead (*Saigerprozess*).

## 2.    The Periphery of European Copper Mining and the Capitalist Metal Trade Network (16/17th Centuries)

In Britain there were abundant copper deposits especially in Wales, Cornwall, and Cumberland. In the middle of the 18th century more than 6.000 tons of copper were produced annually in Britain.[4] At the beginning of the 19th century Britain contributed half of the world's copper production. However, up to the middle of the 16th century, when Central European mining regions already began to decline, there was no copper mining in Britain at all. Mining in Britain did not begin before 1567 and ended already after only 70 years of small-scale operation in 1640. The annual production of all Britain did rarely exceed 40 tons of copper. Copper mining in Britain was not taken up again before 1680 to develop rapidly during the 18th century.

This phenomenon of underdevelopment of copper mining in Britain during the Early Modern Period needs explanation which is not easy to obtain.

In Norway there were copper deposits in the northern region of Tröndelag and in the southern region of Osterdal and Gudbrandsdal.[5] Even in the 16th century the Norwegian king was interested in the promotion of copper mining in Norway. Especially in the 1530's and 40's he tried to attract miners from Germany to establish several mines. Some hundreds of miners came but after a short period of operation at about 1550 most of them returned home and mining more or less came to an end. A new start was made in 1630 when there was another influx of German miners which at this time was supported but not attracted or steered by the state. It rather seems to have been based on individual decision, i.e. German miners and experts moved from the declining mining areas in Central Europe towards the periphery looking for new ventures. Within a short period between 1630 and 1665 seven new mining districts, each comprising a number of mines, were opened up. Local labourers were integrated soon, whereas the main management remained in German hands. They transferred know-how and mining techniques from the German mining regions to these far off mining colonies. Mining operations were scattered and production was low, but nevertheless important for the Norwegian economy. The annual production of all Norwegian mines was on average between 240 and 320 tons of copper which was at that time only 1/6 of the production in adjoining Sweden. Although there was a decline after 1670 the mining operation in general continued. The market in Britain and Norway was of local nature and in Norway almost no additional import of copper took place.

Besides the Central European silver, lead, and copper imperium there was the Swedish copper mining which was geographically in the periphery and not included in the Central European trading system.[6] It developed and reached its peak during the 17th century.

Exploitation did not begin before 1575 increasing shortly after 1600. Most of the produced copper was exported via Stockholm, Gävle, and Tälje across the Baltic Sea to Lübeck, Stettin, Rostock, Danzig, Malmö, Helsingör, and Copenhagen, and across the Belt to the Netherlands and Britain.[7] In these harbours also copper from Saxony, Tyrol, and Hungary was traded. Denmark, the Netherlands, and the German lowlands completely depended on imports of copper.[8] This resulted in a competition between the Central European market system and Sweden for these markets.

Evidence for an economic 'World-System' can be seen in the fact that in many cases mining operations in peripheral areas were innovated or stimulated by entrepreneurs who came from central mining regions (e.g. Saxonia, Bohemia) or from cities engaged in metal trade, especially from southern Germany (Augsburg, Nürnberg).

In England the Keswick mine in Cumberland was established by Daniel Höchstätter in 1564/67.[9] After getting a prospectors warrant for mining in Devon, Cornwall, Lancashire, and Cumberland in 1563, he worked for a German company in Augsburg (Haug, Langnauer & Co.). To run his mining and smelting operation in England he organized a German-English joint stock corporation: 'The Society of the Mines Royal' and the 'Mineral and Battery Society' in London, the first corporations of this kind in England. At the same time his family and the German company were engaged in mining and metal trade in Hungary and in the Alpine region. Höchstetter was a business man and also an expert in practical mining. He brought German miners to Britain to erect the first copper mine in England at Keswick which was called 'Newlands', a typical name for a colonial venture. In 1576 a second mine was sunk at Caldbeck but as no profit had been made the German company withdrew and Höchstetter himself took over in 1580. From these two nuclei further mines were established with other Germans engaged: his son Emanuel, Mark Steinberger, Ulrich Frosse (at Merioneth in Cornwall), and Hans Ziegler from Nürnberg. This enterprise, initiated from a far off economic core area, did not succeed at last because of lacking support and investment from the centre and because it was not followed by local initiative. In 1642, at the time of the drastic decline of copper mining and metal trade in Central Europe, all copper mining in Britain came to an end for another fourty years.

Mining ventures, initiated from the centres of metal trade (Augsburg and Nürnberg), were not limited to Europe. There were already enterprises going to South America during the first half of the 16th century, especially to Venezuela, a territory, which, for a short period, was in the hands of the Welser family from Nürnberg. In 1529 Nikolaus Federmann went to Venezuela with 27 miners to search for precious metals and to start mining and smelting there. One year later Hans Seissenhofer followed with 26 miners. In 1528 the Welser company, which sent out Federmann and Seissenhofer for exploration, gained a smelting privilege for Venezuela and Santa Marta. However, after 1534 no more investment was made.[10]

In 1542 Hans Tetzel from Nürnberg went to Cuba to start a copper mine and copper smelting which was the very first smelting of copper in the New World.[11] The copper ore was discovered near Santiago de Cuba about 1530 but after some experiments the mines were given up. As copper was badly needed for tools and kettles in the developing Cuban sugar industry the young patrician from Nürnberg took the initiative to establish a copper industry in this far off region. In 1546 he secured his venture by a contract with the Spanish Crown. He even founded a mining company, 'St. Jacob de Cuba', which was joined by Lazarus Nürnberger from Sevilla, his two brothers, and two brothers-in-law. Tetzel's uncles were the most prominent copper merchants and *Saigerhütten*-masters in Nürnberg. Tetzel was engaged in the copper trade in Antwerp and Sevilla before he left to start his own venture which was supported by capital from the European metal market. He carried on until he died in 1571. Tetzel produced copper for the Cuban market, but his main aim was to export copper to the West Indies and Europe (especially to Spain), which was only successful for a short period.

Although there was already a vision to go overseas to expand the network of metal production and the import to Europe towards the newly discovered continents, these outposts or bridgeheads of the European economic world-system had to fail because of lack of supply and support, lack of skilled workers, hardship, and an unfavourable environment.

## 3.    The Spatial Organization of Metal Trade and Capital Flow

The metal trade was well organized in a network of trading agencies. One of the wealthiest merchant families, the Fugger family from Augsburg, had organized a European-wide network of agencies and trading places with offices and storehouses.[12] The central places in the mining areas where the raw copper (black copper) and also pure copper (red copper) and silver were brought together were Neusohl (Banska Bystrica) in Slovakia, then Upper Hungary, and Hall and Innsbruck in the East Alpine mining region. The maritime trading places in the Mediterranean were Venice (Italy) and Zengg (Yugoslavia). From these two harbours connected with the mining areas in Hungary, Bohemia, and the Alps a number of trading places in the East Mediterranean were supplied with copper.

The maritime trading centres in the Baltic region were Lübeck and Danzig. In Western Europe they were Antwerp and Hamburg. The Spanish and overseas' markets were served from Lisboa and Sevilla. The most important harbour and maritime trading place in the 16th century undoubtedly was Antwerp being connected with the centres in southern Germany by the main Gent-Frankfurt-Augsburg trade route.

The centres of business and trading management, the capitals of capital with a number of famous merchant families, were Augsburg and Nürnberg.[13] About 80% to 95% of the metal trade of the three leading European copper mining regions was controlled by merchants from Augsburg and Nürnberg. Augsburg, with 53% to 66%, was the most prominent trading centre. Certainly only a small proportion of the metal was brought to the market of these cities, but capital, mercantile decisions, and directions were concentrated here.

It is interesting to note a quite clear division of the influence in the mining regions between Augsburg and Nürnberg: Hungary was completely in the hands of Augsburg merchants and the Mansfeld production was completely controlled by the merchants of Nürnberg.[14] Only in the Alpine region both cities were engaged. Nürnberg had a share between 11% and 23%.

The money made with all the mining and metal trade accumulated in the treasury of the state or the sovereigns who had the prime mining rights, and increasingly, in the pockets of the merchants in the cities. State and capitalistic companies were in competition. Soon most of the sovereigns lost the game and became debtors or eventually ran into hopeless dependency.

Some income had been spent on the expenses for production and transport and some had to be invested to open up new mines, and to introduce new techniques and machinery. For the mining regions this was a pull factor for workers which led to an influx of population. In the course of regional economic development money was invested in infrastructure, construction, crafts, agriculture, transport, and the upgrading of the road system.

Not enough scientific attention has been paid to this process of rapid regional economic development in quite often very remote mountainous areas in the 16th century. There are almost no geographical studies which focus on the economic complexity of a mining region yet. The majority of the studies deal with the production and organization of mining and smelting and the economy of metal trade within the framework of the general capital market.

The surplus or usually the loan, which the sovereigns took up on their traditional mining rights and shares, was mainly used to pay for military equipment, wars and defence, and governmental expenses. The merchants in the cities invested in other trading ventures instead, in equipment and infrastructure for trade but also in the whole range of the cities' economy and culture, in property, housing standard, arts, and science. The high standard of urban culture in the 16th century in Central European cities, when cities in other parts of Europe were just established (e.g. in Sweden, Norway, and in some respect also in Britain), undoubtedly was a result of the flourishing European trade which was controlled by the cities' merchants and trading companies.

## 4. An Individual Economic System of Metal Production and Metal Trade: The Harz Mountains

The late 18th century trading organization for copper and lead produced in the Harz Mountains may serve as a regional example. The metal market for Harz-metals in the 16th century was already controlled by the sovereigns, the Dukes of Brunswick and later the Kings of Hannover.[15] During the 17th century the trade was leased to some merchants, however in 1714 the State of Hannover regained control and organized a special state administration for the metal sales of the territory, the so called *Berghandlungsadministration*. The towns of Goslar and Osterode were two central places in the Harz mining area where the products were brought together (lead, copper, zink and sulphur). Further trading posts (offices) were established in some big trading cities towards the east and to the north along the main roads leading to the sea. To the south there were only connections to the big centres of Frankfurt, Nürnberg and Regensburg. Export to other European countries was mainly organized by sea. Therefore all offices in foreign countries were located in harbour cities, especially Antwerp and Amsterdam. Here the state-organized metal market joined the by far more powerful capitalist metal market of the free merchants, which had decreased since the 18th century. The leading areas of lead production at that time were Scotland, England, and also Spain.

It is interesting to note that the state administration of metal trade in Hannover was also in charge of the 13 state owned grain storehouses, established to secure the food supply for the state and especially for the mining area. All profit and surplus went to the treasury of the state. Reinvestment of capital in the mining area, which was in decline during the 19th century, was low. Nevertheless we see that in the early 19th century there still existed a European trade system even for individual mining areas such as the Harz Mountains in Northern Germany.

## 5.    Summary: The International System of Copper Mining and Copper Trade

Drawing together the main results of the preceding chapters we may summarize as follows: The international system of copper mining and copper trade consisted of:
1. A wide range and long-distance management of production and trade,
2. an establishment and an organization of infrastructure for production, processing and trade,
3. the organization of companies, cartels, and monopolies to compete on a multinational European market.
4. Within a long-distance trade network lay cores of a hierarchy of trading cities.
5. An important impact of capital concentrated in the market cities can be observed on the cities themselves and on the production areas as long as there was any prospect of succeeding.
6. Concerning the 'world-system' I tried to draw the attention not to a 'core-periphery model' of regions but to the spatial organization of economic systems.

## 6.    Some Comparative Remarks About Iron Production and Iron Works in Early Modern Europe

Copper production and copper trade is undoubtedly one of the most evident examples for an economic world-system of Early Modern times which might support Wallerstein's theoretical concept. This is different in the small regional worlds of European iron production and iron industry in the Early Modern Period.[16] The iron industry of the Harz region, the Oberpfalz region in Bavaria[17] or the Dill region in Hessia are good examples for this. Let us take a closer look at the Dill region during the 18th and 19th centuries as a selected example:[18]

The iron deposits, mines and furnaces were located in the upper Dill region in the County of Nassau-Dillenburg. Here pig-iron was produced and offered for sale. Customers were a great number of hammerworks and smithies within the boundary of the territory (not yet mapped) but also in neighbouring territories, along the rivers and tributaries of the Sieg, Eder, Lenne, Bigge, Ennepe, Wupper, and the Ruhr.

The number of customers decreased with growing distance. This iron industry and trade was a typical system of supply for subsequent treatment. No far-reaching trading organization was established, there were no trading companies, there was no capital investment to make a profit on trade, and no transport system was organized. Instead it was a kind of 'self service system', i.e. the individual hammer works had to order the quantum of iron needed and then they had to pick up the iron themselves with their own horse carts. Many of the owners of the hammerworks bought on credit and fell in debt so that the capital flow was very slow and obstructed. These regional systems of iron production declined with the development of the modern iron industry in the 19th century.

Besides these regional systems of iron production and iron manufacturing there was also a long-distance trade especially from Sweden and Britain to the Netherlands.[19] The picture here is not very well investigated as it interferes with the regional markets and because of its instability. Undoubtedly Dutch entrepreneurs were heavily engaged in Swedish iron production.

Thus it remains a future task to study the international flow of iron from those source regions to the Netherlands as the former core of the Early Modern European world-system.

## Notes and References

1.  Wolfstrigl-Wolfskron, M. V. (1903): Die Tiroler Erzbergbaue 1301-1665. Innsbruck.
    Wenger, M. (1931): Ein Beitrag zur Statistik und Geschichte des Bergbaubetriebes in den österreichischen Alpenländern im 16. Jahrhundert: In: Montanistische Rundschau 23.
    Wiesner, H.(1951): Geschichte des Kärntner Bergbaus. Teil 2: Buntmetalle. Klagenfurt.
    Atzl, A. (1957):Die Verbreitung des Tiroler Bergbaus. In: Der Anschnitt 9 , pp. 42.
    Hildebrandt, H. (1972): Augsburger und Nürnberger Kupferhandel 1500-1619. Zzeitschrift für Wirtschafts- und Sozialwissenschaften 92, pp. 1-31.
2.  See Pickl, O. (1977): Kupfererzeugung und Kupferhandel in den Ostalpen: In: Kellenbenz, H. (ed.): Schwerpunkte der Kupferproduktion und des Kupferhandels in Europa, 1500-1650. Köln, pp. 119.
3.  See Pickl, O. (1977): Kupfererzeugung und Kupferhandel in den Ostalpen, in: H. Kellenbenz (ed.): Schwerpunkte der Kupferproduktion u. des Kupferhandels in Europa, 1500-1650. Köln pp. 117-147.
    Scheuermann, L. (1929): Die Fugger als Montanindustrielle in Tirol und Kärnten. Ein Beitrag zur Wirtschaftsgeschichte des 16. und 17. Jahrhunderts. Studien der Fugger Geschichte 8. München.
    Kunnert, H. (1962): Der Nürnberger Ratsherr Paul (II) Behaim als steirischer Gewerke. In: Der Anschnitt 14, pp. 20-27.
    Unger, E. E. (1967): Die Fugger in Hall in Tirol. Studien zur Fuggergeschichte 19, Tübingen.
4.  Kellenbenz, H. (ed.) (1981): Precious metals in the age of expansion, Stuttgart.
    Hammersleg, G. (1981): Technique or economy? The rise and decline of the early English copper industry, ca. 1550-1660. In: Kellenbenz, H. (ed.): Precious metals in the age of expansion. Stuttgart, pp. 1-40.
5.  Treite, S. (1981): Die norwegische Kupfererzeugung vor 1700. In: Kellenbenz, H. (ed.): Precious metals in the age of expansion. Stuttgart, p. 260.
6.  Kumlien, K. (1981): Staat, Kupfererzeugung und Kupferausfuhr in Schweden 1500-1650.In: Kellenbenz, H. (ed.): Precious metals in the age of expansion. Stuttgart, pp. 241-259.
7.  Thierfelder, H. (1958): Rostock-Osloer Handelsbeziehungen im 16. Jahrhundert, Weimar.

North, M. (1980): Bilanzen im Lübecker Schwedenhandel 14.-16. Jahrhundert. In: Gotlandia Irredenta. Festschrift für Gunnar Svahnström, ed. by Robert Bohn, Sigmaringen.

8.  Van der Wee, H. (1981): Worldproduction and Trade in Gold, Silver and Copper in the Low Countries, 1450-1700. In: Kellenbenz, H. (ed.); Precious metals in the age of expansion, Stuttgart , pp. 83.

9.  Hammersleg, G. (1981): Technique or economy? The rise and dechine of the early English copper industry, ca. 1550-1660. In: Kellenbenz, H. (ed.); Precious metals in the age of expansion, Stuttgart, pp. 1-40.

10. van Klaveren, J. (1960): Europäische Wirtschaftsgeschichte Spaniens im 16. und 17. Jahrhundert. Forschungen zur Sozial- und Wirtschaftsgeschichte 2, Stuttgart.
    Walter, R. (1986): Nürnberg, Augsburg und Lateinamerika im 16. Jahrhundert. Die Begegnung zweier Welten. In: Pirckheimer Jahrbuch 2, pp. 45-82.

11. Werner, T. G. (1961): Europäisches Kapital in ibero-amerikanischen Montanunternehmen im 16. Jahrhundert. Vierteljahrschrift für Wirtschaftsgeschichte 48, 2.
    Werner, T. G. (1961): Das Kupferhüttenwerk des Hans Tetzel aus Nürnberg auf Kuba und seine Finanzierung durch europäisches Finanzkapital (1545-1571). Vierteljahresschrift für Wirtschaftsgeschichte 48, 3.

12. Strieder, J. (1929): Die Entstehung eines deutschen frühkapitalistischen Montanunternehmertums im Zeitalter Jacob Fuggers des Reichen 1459-1525. Beiträge zur Geschichte der Technik und Industrie 19.
    Somerland, B. (1938): Die Faktorei der Fugger in Leibzig. Schriften des Vereins für die Geschichte Leibzigs 22.
    Pölnitz, G. von (1949/51)): Jakob Fugger. Tübingen.
    Pölnitz, G. v.on (1960): Die Fugger. Frankfurt.

13. See Hildebrandt, H. (1972): Augsburger und Nürnberger Kupferhandel 1500-1619. Zeitschrift für Wirtschafts- und Sozialwissenschaften 92, pp. 1-31.

14. Mück, W. (1910): Der Mansfelder Kupferschieferbergbau in seiner rechtsgeschichtlichen Entwicklung, Eisleben.
    Möllenberg, W. (1991): Die Eroberung des Weltmarktes durch das mansfeldische Kupfer. Studien zur Geschichte des Thüringer Saigerhüttenhandels im 16. Jahrhundert, Gotha.
    Möllenberg, W. (1915): Urkundenbuch zur Geschichte des Mansfeldischen Saigerhandels im 16. Jahrhundert. Geschichtsquellen der Provinz Sachsen 47, Halle.
    Westermann, E. (1971): Das Eislebener Garkupfer und seine Bedeutung für den Europäischen Kupfermarkt 1460-1560, Wien.

15. Westermann, E. (1971): Der Goslarer Bergbau vom 14. bis zum 16. Jahrundert. Forschungsergebnisse - Einwände - Thesen. Jahrbuch für die Geschichte Mittel- und Ostdeutschlands 20.
    Henschke, E. (1974): Landesherrschaft und Bergbauwirtschaft. Zur Wirtschafts- und Verwaltungsgeschichte des Oberharzer Bergbaugebietes im 16. und 17. Jahrhundert. Schriften zur Wirtschafts- und Sozialgeschichte 23, Berlin.
    Irsigler, F. (1985): Über Harzmetalle, ihre Verarbeitung und Verbreitung im Mittelalter. Ein Überblick. In: Meckseper, C. (ed.): Stadt im Wandel. Bd. 3, Stuttgart , pp. 315-321.
    Kraschewski, H.-J. (1985): Heinrich Cramer von Clausbruck und seine Handelsverbindungen mit Herzog Julius von Braunschweig-Wolfenbüttel. Zur Geschichte des Fernhandels mit Blei und Vitriol in der 2. Hälfte des 16. Jahrhunderts. Braunschweiger Jahrbuch 66, pp. 115-128.
    Kraschewski, H.-J. (1990): Quellen zum Goslarer Bleihandel in der frühen Neuzeit (1525-1626). Veröffentlichungen der Historischen Kommission für Niedersachsen und Bremen 34, Hildesheim.

16. Sprandel, R. (1968): Das Eisengewerbe im Mittelalter. Stuttgart.

Kellenbenz, H. (1974): Schwerpunkte der Eisengewinnung und Eisenverarbeitung in Europa 1500-1650. Köln/Wien.

17.　Ress, F. M. (1950): Geschichte und wirtschaftliche Bedeutung der oberpfälzischen Eisenindustrie von den Anfängen bis zur Zeit des 30-jährigen Krieges. Verh. der Hist. Ver. von Oberpfalz und Regensburg 91, Regensburg.

18.　Einecke, G. (comp.) (1932): Der Bergbau- und Hüttenbetrieb im Lahn-Dillgebiet und in Oberhessen, Marburg.

Born, M. (1985): Die industrielle Entwicklung des Dillgebietes in ihren Beziehungen zur Territorialgeschichte. In: Hessisches Jahrbuch für Landesgeschichte 8, pp. 150-170.

Dösseler, E. (1968): Eisenhandel im südlichen Westfalen und in seiner Nachbarschaft in der vorindustriellen Zeit. Westfälische Forschungen 21, Münster.

19.　Wernstedt, F. (1935): Järnkrämare i Stockholm 1651-1890, Stockholm.

# Part IV   The European Periphery

# 11

# Spatial Changes in Poland under the Impact of the Economic Dynamics of the 16th and 17th Centuries.

Anna Dunin-Wasowiczowa

This paper will concern itself with the spatial changes in the Polish territory which occurred under the impact of the economic dynamics of the 16th and 17th centuries, and through the evolution of the large manorial estates with corvee labour. These changes should be studied within the frame of a complex approach, thus the paper is based on interdisciplinary research. Certain aspects of the spatial pattern of Poland were studied on the basis of source material on large scale, but the reconstruction of rural landuse has been based on detailed source material recorded in the 16th - 17th centuries, and also on plans of settlements from the 18th and 19th centuries.

## 1.   Area, Population, Economic Relations

By the end of the 16th century the area of the Polish-Lithuanian State amounted to ca. 815.000 sq km, and Central Poland (the Crown lands) ca. 278.000 sq km.

Table 1: Central Poland

| Region | area in sq km |
|---|---|
| Great Poland | 57.000 |
| Royal Prussia | 24.000 |
| Mazovia | 35.000 |
| Kuyavia | 8.000 |
| Little Poland | 55.000 |
| Ruthenia | 86.000 |
| Podlachia (since 1569) | 12.525 |
| | 277.525 |

The population of 16th century Central Poland with the exclusion of Royal Prussia and Ruthenia is estimated at ca. 3-3.5 mill. inhabitants. The average population density was about 15 inhabitants per sq km. On this area there existed about 22.500 settlements, of which only 3% were towns.[1]

The 16th century was a period of ever increasing political power of the gentry and, at the same time, of restrictions of liberty and economic deterioration of the other classes, i.e. peasants and burghers. On landed estates of the Crown (the Kingdom of Poland), of the nobility and of the Church an economy developed based on serfdom as the predominant form of agricultural production. A precondition for the existence and development of the manor was the international system of market relations, which provided incentives for large-scale production of grain.

In the 16th century the position of Poland in the international trade system changed, when Turkey (the Ottoman Empire) conquered the ports of the Black Sea: Poland lost her position as intermediate trader of luxuries. Despite this her traditional position in the Baltic trade region changed very little, but increased quite considerably in volume with the growth of the international grain market which centered on Amsterdam from where Dutch merchants tapped the Polish grain resources through Gdansk. Thus Poland's access to the Baltic Sea (since 1466) permitted her to take part in the transit trade to the West. The structure of trade changed according to the demands of the European market. In the 16th and 17th centuries Poland was among countries exporting mainly raw materials (grain, timber, naval stores, cattle), first of all to the Netherlands, and to a lesser extent to Sweden, Portugal, Spain, France, England, and other European countries[2]. Since the political union of Poland and Lithuania, Poland's role as a region of transit trade, as well as the role of Polish merchants as intermediate traders between Lithuania and southern Germany, had been strengthened. Poland became part of what Wallerstein has defined as the dependant periphery of the early modern European world-system or, alternatively, part of the remotest zones of the Thünen-system as Nitz has shown in his article in these proceedings.

## 2.    Regional Markets and Production

The early 16th century was a period of formation of regional markets which specialized in one or several branches of handicraft or (proto-industrial) cottage industry in which even characteristics of early capitalism emerged in the form of the putting-out system. In Great Poland there developed structures of commercial manufacturing which were mainly based of cloth weaving. In Little Poland mining, iron and steel production as well as linen weaving dominated. In the mountains to the south animal husbandry prevailed.[3]

By the end of the 16th century a growing limitation of independence of producers and a decline of handicraft and rural industry, connected with the development of manors based on  serfdom, was to be observed. The regional pattern of the mining and iron industry depended on the limitations of investments. The old places of salt mining (Wieliczka, Bochnia) expanded (for the location of the various branches see Fig.1).[4] Some branches of mining shifted their locations. For the first time the indigenous resources of marble, limestone and sandstone were exploited and used for Renaissance and Baroque buildings and sculptures.[5] Quarries were to be found around Krakow (the Polish capital until 1609), in the

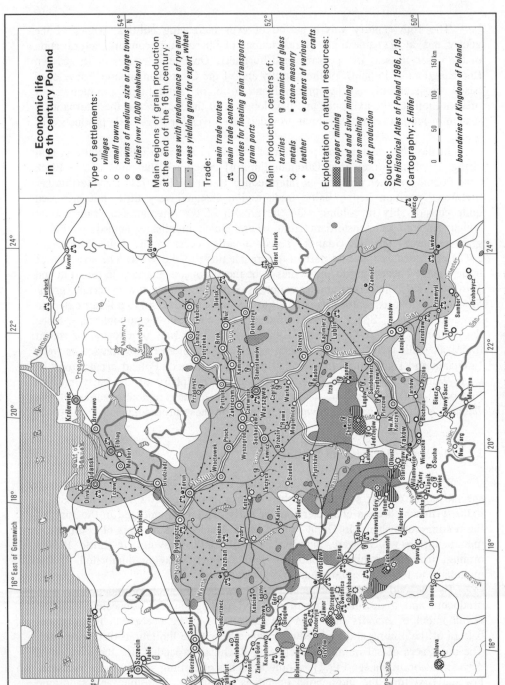

*Fig.1: Economic life in 16 th century Poland*

Góry Switokrzyskie (Hills of the Holy Cross) and in Lower Silesia. Due to newly opened iron mines the traditional exploitation of local iron deposits in morasses, which were to be found everywhere in Poland, was abandoned. Now iron mining was concentrated in Little Poland, especially in the Old Iron Basin and around Bedzin.[6] By the end of the 17th century a partial regression of mining can be observed which was caused by exhaustion of resources, by water problems in the pits, or by lack of capital.

## 3.    Impact of Economic Changes in Poland on New Spatial Patterns

### 3.1. Towns

By the end of the 16th century in the Crown territories there existed more than 1.000 towns. During the 16th century and the first half of the 17th century about 190 new towns were founded on these territories. The number of new towns in Lithuania and Ruthenia is not included because their exact number is not known. The majority of the new towns were founded by the high ranking magnates and by the rich nobility. Two thirds of the 16th and 17th century new towns had been privately founded[7] (see table 2 and 3).

Table 2: Growth of the proportion of new "located" (founded) towns of the nobility[8] in the 15th-17th centuries in % of the total number of new towns

| Region | 15th century % | 16th century % | 17th century % |
|---|---|---|---|
| Great Poland | 65 | 68 | 70 |
| Little Poland | 56 | 64 | 68 |
| Mazovia | 41 | 57 | 58 |
| Royal Prussia | 16 | 16 | 18 |
| Podlachia | 33 | 43 | 55 |

Inspite of this new wave of foundation of central places, medium sized and small towns were in an unfavourable economic situation. Small towns built *en masse* at the beginning of the 16th century in medium nobility estates often turned, at the end of the 16th and in the 17th centuries, into villages, and their inhabitants were forced into labour on manorial estates. When originally established they were intended to serve as local collection points for  export but actually the grain merchants bypassed them, and the peasantry of the villages was too poor - and still became poorer - as to use the facilities of the towns.

Table 3: Degree of urbanization of Central Poland in the 16th and 17th century

| Region | number of towns at the beginning of the 16th century | number of new towns in the 16th c. (increase in %) | number of new towns in the first half of the 17.th c. (increase in %) | number of towns per 1000 sq.miles |
|---|---|---|---|---|
| Gr. Poland | 263 | 31 (12) | 8 (3) | 4,5 |
| Lt. Poland | 164 | 46 (28) | 18 (12) | 3,5 |
| Mazovia | 88 | 40 (45) | 1 (3) | 3,2 |
| Royal Prussia | 36 | - - | 1 (3) | 1,8 |
| Warmia | 12 | - - | - | 3,0 |
| Podlachia | 13 | 12 (92) | 2 (15) | 2(?) |
| Red Russia | 110 | 105 (95) | 38 (18) | 3,9 |
| Central Poland | 686 | 234 (34) | 68 (10) | |

In contrast, the large towns flourished during the 16th century e.g. Danzig (as the main grain port) and so did all the towns located along the rivers due to their grain trade (for instance: Torun, Wloclawek, Plock, Wyszogrod, Zakroczym, Warszawa, Kazimierz on the Vistula; Pultusk, Lomza, Wizna, Tykocin on the Narew; Kamienczyk, Brok, Nur, Drohiczyn on the Bug; Krzeszów and Sieniawa on the San).[9] Other towns developed as seats of the regional parliaments of the nobility (Warsaw, Lublin, Piotrków, Grodno) and of jurisdiction (Lublin, Piotrków), of international fairs (Poznan, Lublin) and of the courts of the king (Kraków, Warsaw) or of big magnates (e.g. Zamosc). The towns of the larger manors served as their administrative centres: the courts of the magnates were also located at these places. In eastern Poland they were fortified against the invasions of the Tartars (for instance: Zamosc, Brody, Zólkiew, Stanislawów, Brzezany, Sluck, Stary Bychów, Nieswiez).[10] In addition, the towns of the magnates served as local centres of commerce and of crafts (for instance: Lesko and Przeworsk in Red Russia, Wegrów in Podlachia, Rawicz and Leszno in Great Poland).[11] Some towns which were located along international trade routes to the West functioned as transit markets, e.g. Nowy Sacz, Biecz in Little Poland, which also supplied the surrounding local markets.[12] Another example is Zamosc which was founded by Chancellor Jan Zamoyski in 1580. Like Przemysl farther south, this town was located on the cattle transit route to the markets of western Central Europe at a point where several cattle trails from Moldavia and the Ukraina converged. The influence of the distributional pattern of towns on trade with rural commodities has been studied in the context of the local market economy. The Polish traders of the 17th century did not specialize in a single commodity only. Jerzy Tymowski of Nowy Sacs was engaged in big international trade, mainly cloth, and some Hungarian wine, hardware, and spices which he sold to the burghers of his home town, to members of the Polish nobility, and merchants of

North Hungary.[13] At the same time he bought grain and other agricultural commodities on the rural markets in the vicinity.

During this period in several towns new quarters were laid out. In Great Poland "New Towns" for artisans and traders were founded as extensions of older urban cores, for example in Rawicz in 1638. In Podlachia (Lithuania, from 1569 Poland) new towns, which has been established under the Agrarian Reform of King Zygmunt August were true agricultural towns, e.g.Suraz.[14] They were intended to serve as the market centres especially for grain which was expected to be produced by the newly established villages with their three-field systems which had been formed by a complete reorganization of the traditional settlement pattern of dispersed forest hamlets. Frequently the new towns were laid out with a grid of straight streets lined by long rows of strip plots which contained the homestead on the frontage and the agricultural land in the rear. Towns of this eastern region could also include new urban elements such as the palace of the magnate who gave great importance to a representative appearance of his "capital", as well as to the defence of the town. If it was located close to the eastern border, where one had to be aware of the attacks of the Tartars, earthen ramparts and other fortifications like bastions were built after Italian or Dutch models, e.g. Zamość in Red Russia and Głowów in Little Poland. Sometimes the fortifications circumscribed parts of the arable land of the towns, too. Towns which had their economic basis on river navigation, especially those on the Vistula and their tributaries, had well-developed ports with large granaries, which served as intermediary collection points for the grain sent from the estates to Gdansk.[15]

## 3.2. Villages

The new early modern wave of agricultural colonization (after the earlier wave of medieval colonization) was largely directed by the great magnates[16] and has partly - as in Lithuania - to be seen in the context of the extension of grain production for the international market. New settlement networks evolved.

By the end of the 16th century, in those parts of the territories inhabited by ethnic Polish stock on an area of 175.000 sq km (Great Poland, Little Poland, Kuyavia, Mazovia, Podlachia), there existed arround 22.500 settlements.[17] Only 3% of them were towns, and they together represented about 1/4 of the total population. The average settlement density was about 1 per 8 sq km. Mazovia was the most densely settled region with 1 settlement per 5 sq km[18], while the lowest density was to be found (apart from mountainous and densely forested tracts) in Royal Prussia (1 per 12 sq km).[19]

It is very difficult to estimate the increase of colonization in Poland as a whole in the 16th and 17th centuries. In comparison with the general trends of development we can estimate approximately the trends for regions where colonization was stronger than elsewhere.[20] In the Carpathian Mountains new settlements were founded based on the Walachian Settlers' Law (or statute) with immigrants who as pastoral people centering on sheep herding originated from Walachia (now part of Romania) and expanded from there all along the Carpathian

Mountains, part of whom finally reached the Beskides of Southern Poland and even settled in the plains.[21] The coarse wool produced by them was of no importance for export.

Colonization which in earlier centuries had been strongest in central Poland, in its new wave expanded into peripheral regions which so far were only thinly populated with scattered settlements along the valleys. This is especially true for Lithuania where colonization was initiated by the royal Agrarian Reform of the 16th century, and marginal areas of Mazovia, Pomerania and Silesia where colonization also restored the losses of the wars of the 15th century. It was not only through peasant and estate colonization that the area under forest was reduced during the 16th century: in the basin of the Vistula wooded tracts were heavily exploited for export of timber, potassium and naval stores, and attracted iron and glass works, too. After extraction of wood the exploited areas were spontaneously settled by small peasants.

All these activities had strong ecological implications. The originally most important tree species of the forest - the yew-tree - as well as species of wild animal like aurochs and wild horse, disappeared. In the private forests overexploitation caused the lowering of the water table. In the hills erosion increased which led to flooding in the valleys and plains; so soils deteriorated. To some degree impoverishment of soils is thought to have been caused by continuous application of the three-field course on soils of low quality. Large estates dug fish ponds which made use of uncultivable land.[22]

The evolution of estates based on corvee labour led to changes in the traditional field pattern of former peasant villages. The fields of the demesnes expanded: their owners occupied wasteland and, forcibly, bought out the peasants of their traditional property rights and drew their lands into the demesne. Most of the plots which the farmers were permitted to retain were located in the less fertile parts of the fields. Nevertheless, part of the demesne fields and peasant fields were still intermingled. This type of demesne village predominated in the Polish territory and was very characteristic of villages completely in the hands of the nobility.[23] The areal extension of demesnes (dominal lands) in individual villages as well as in the country as a whole has yet to be studied.[24] In the manorial inventories of the villages the area of the demesne is given as an estimate of the amount of grain sown. The fields of the peasants are given in "mansi" (Latin, medieval land unit).

The distribution of other types of settlement was more limited. Demesnes as separate, isolated, settlements are to be found in Royal Prussia and in Silesia[25], to a lesser degree also in central Poland.[26] The settlements of the "petty nobility" east of the Vistula mainly consisted of hamlets. The villages of Lithuania which had been created through the colonization of the Agrarian Reform, with the aims to expand the grain export to the world-market, especially via Riga,[27] were laid out in a very schematic way with a regular street pattern (straight street villages).[28] Up to 1567 almost 1 million hectares of the territories of the princely state of united Poland-Lithuania had been surveyed. The royal manors served as the model which was soon copied by the magnates and the nobility of Lithuania.[29]

Villages which developed spontaneously on the basis of manufacturing (e.g. cottage industries in the forests including those where naval stores for export were produced) display irregular patterns. In the mountains the layout of the new settlements follows the traditional pattern of row villages. An important aspect of future research will be to study those villages which have been founded under the Walachian Settlers Statute in Red Russia, in southern Little Poland, in the Duchy of Cieszyn and in Silesia. Colonization of the Dutch type used the characteristic pattern of isolated farms with compact strip holdings arranged in rows and located in the moist river lowlands.[30] This new type of colonization - with Dutch immigrants as colonists - was first invented in the delta of the Vistula (Polish "Zulawy", German "Werder") and expanded south in the marshy tracts along the Vistula almost as far as Warsaw. Whilst the German term "Hollaenderei" (Polish "olederskie" or "oledry") is applied to this type of settlement, the settlers were not necessarily Dutch but had settled "iure Hollandico". This model was so attractive that it even expanded beyond the lowlands. At the same time colonization also spread to the regions between the Warta and Notec rivers. The economic role of these new colonies in the river lowlands is not yet clear. They concentrated on intensive animal husbandry. In the delta of the Vistula they supplied the large city of Gdansk with dairy products and meat. Whether the settlers along the Vistula and its tributaries did use navigation for supplying markets of distant towns like Warszawa, is an open question.

## 4. The Peripheral Character of the Polish Economy in the European World-System

### 4.1. Volume and Regions of Grain Export

Around 1600 on the manors of the Crown the economy was already based on corvee labour.[31] Certain trends in the evolution of this new type of manorial estates in east central Europe (east of the Elbe) are still discussed with some controversy.[32] The basic condition of its evolution is a system of market links which strongly encouraged the production of grain. A lot of research has already been devoted to the evolution of the dual international economy in Europe with the east central European countries, especially those of the Baltic zone, supplying grain and raw materials to the western countries.[33]

A reconstruction of the location of landed estates which at that time sent grain by barge is possible due to registers of "water tariffs", duties collected for commodity floated down the Vistula in Włocławek. For the remaining areas use was made of monographic research, connected with individual estates, studies of the range of the market, on the basis of grain prices and the composition of crops.

Of great importance were the routes by which grain was exported to the West.[34] The most important regions of grain cultivation for export were areas

connected with the Baltic ports, i.e. Gdansk (Danzig) and to a lesser degree
Königsberg, Elbing, Riga, Stettin and Kolberg (Fig.2).

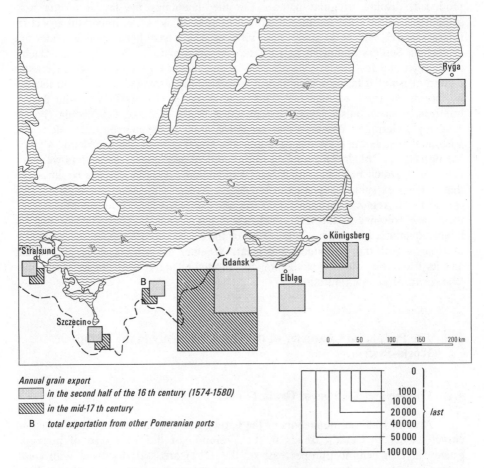

*Fig.2: Annual grain export in the second half of the sixteenth and in the mid-seventeenth century*

    The total area under cultivation in Poland is estimated to have been ca. 3,5
mill. hectares - about 12% of the total area of the Crown (Central Poland). The
annual harvest of the main varieties of grain (rye, wheat and barley) was around
1,3 mill. tons.[35] Concerning the amount of grain from Poland cleared at the
Danish Sund toll station, about 10% of the harvested rye was exported on ships,
but very little over land. Towns located on the border of Great Poland could
increase their importance in overland grain export as royal toll stations
(Międzyrzecz, Zbąszyn, Grodziec, Kiebłów, Kopanica, Wschowa, Poniec,
Jutroszyn, Sulimierzyce, Ostrzeszów, Krzepice, Częstochowa, Siewierz.[36] More
important however for grain export was the network of rivers in the Vistula basin
in the central section of her course (see Fig.1).

Table 4: Grain export (rye and wheat) from various regions through the toll station in Włocławek. Grain in lasts (ancient unit, ca. 2000-2500 kg)

| Region | 1537 lasts | % | 1546 lasts | % | 1556 lasts | % | 1575 lasts | % |
|--------|-----------|-----|-----------|------|-----------|------|-----------|------|
| Kuyavia | – | – | 44 | 1.1 | 1,697 | 12 | 828 | 6.6 |
| Mazovia | 2,372 | 90.6 | 2,935 | 76.9 | 6,327 | 44.9 | 6,677 | 53.2 |
| Podlachia | 70 | 2.6 | 20 | 0.5 | 962 | 6.8 | 926 | 7.4 |
| Gr. Poland | 2 | – | – | – | 506 | 3.6 | 562 | 4.5 |
| Lt. Poland | 120 | 4.6 | 672 | 17.7 | 2,508 | 17.8 | 2,850 | 22.8 |
| Red Russia | – | – | 52 | 1.3 | 1,752 | 12.5 | 424 | 3.5 |
| Lithuania | 58 | 2.2 | 97 | 2.5 | 327 | 2.3 | 80 | 0.6 |
|  | 2,621 | 100 | 3,820 | 100 | 14,080 | 100 | 12,347 | 100 |

Mazovia, which extends from the banks of the Vistula to the east, was the most important grain exporting region of Poland (see Table 4 and Fig.3). In the early 16th century more than 90% of exported grain that passed the toll station of Włocławek (on the Vistula) came from Mazovia, mainly from the region east of Warsaw and from the Voivodship of Płock, both close to the Vistula.[37] During the second half of the 16th century the demand for grain grew and the area under grain production for export expanded further east into the region along the Narew and Bug rivers (which are tributaries of the Vistula). The area under cultivation in Mazovia in the second half of the 16th century is estimated at as high as 20% of the total area of the region.

Podlachia is the adjoining westernmost province of Lithuania with access to the upper reaches of the Narew and Bug. From 1569 it was part of the Polish Crown. Research on this region is incomplete but it is known that grain was exported from the area around Tykocin. In Great Poland the Voivodships of Poznan and part of the Voivodship of Sieradz produced grain for export. So far nothing is known of grain shipped on the Warta river, which flows west into the Odra (Oder).

In the 17th century production of grain increased in the province of Kuyavia (northern Poland, along the Vistula). In the south, in Little Poland, the main aristocratic producers of grain were those of the Voivodships of Sandomierz and of Lublin (east of the upper Vistula). The natural conditions of the mountains (Carpathian Hills) did not permit the southernmost part of Poland to participate in the grain business.[38]

Insufficient research has been carried out on the easternmost parts of early modern Poland for which little is known about the progress of colonization.[39] Some scholars maintain that the Ukraine region did not contribute to grain export due to lack of navigable rivers flowing to the west. Other scholars, however, claim that part of the basin of the Dniepr river did send commodities to the west on the Dniepr tributaries (which would mean: transporting the goods upstream).

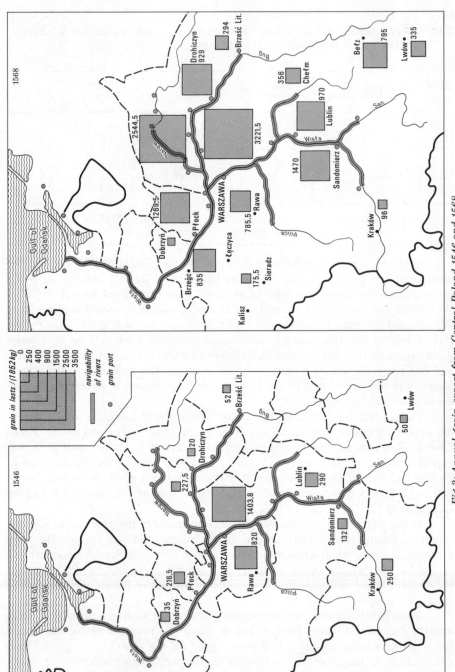

*Fig.3: Annual grain export from Central Poland 1546 and 1568*

The Grand Duchy of Lithuania exported to Riga in Livland and to Königsberg in (East) Prussia.[40]

Table 5: Annual grain export from Baltic ports of Prussia and Lithuania in lasts (ancient unit, ca. 2000-2500 kg).

| ports | 1577-1580 (annual average export) | 1585 | 1591-1600 (annual average export) | 1649 | 1665 | 1681-1690 (annual average export) |
|-------|-------|------|-------|------|------|-------|
| Danzig | 20,444 | 27,400 | 37,094 | 95,700 | 11,700 | 36,300 |
| Elbing | 10,372 | | | | | |
| Königsberg | 19,159 | | | | 1,200 | 10,700 |
| Riga | | | 6,679 (1593) | | | |

In Lithuania the expansion of the manorial economy was supported by the agrarian reform of 1557. The traditional hamlets and their small fields were completely reorganized and surveyed in a schematic way, in order to introduce the three-field economy as the modern system of the time.[41] Some parts of Lithuania had specialized in commodities other than grain, for example the province of Samogitia (in the northwest of Lithuania) grew flax and exported fibres and seed.[42] With its location off the large rivers of Dvina and Memel this region suffered from high transport costs for grain and therefore had to resort to the alternative of producing and exporting a high value, low weight commodity (according to the von Thünen law; see H.-J. Nitz, in this volume).

Silesia, in the southwest, was a classical agricultural region which mainly supplied the local markets of the densely populated proto-industrial regions - linen weaving in the foothills of the Sudetia Mountains, and mining.[43] With respect to Royal Prussia (in the north) and western Pomerania (in the northwest) the opinions of scholars diverge on the scale of the export economy[44].

## 4.2. Transit Trade of Beef Cattle

The export of beef cattle from the East was of great importance to Germany. During the first half of the 16th century cattle passed the toll stations of Great Poland (west central Poland), mainly through Poznan, Kalisz, Ostrzeszów and Wielun in the direction of Silesia. In the second half of the century cattle transit through southern Poland grew in importance. Cattle trails from the southeast, i.e. Moldavia, Walachia (now parts of Romania and Russia) and Podolia (now part of Russia), passed through Little Poland (southern Poland) where two great routes were available: one through Sandomierz to the northwest, and the other from Rzeszów through the southern part of Little Poland.

Table 6: Beef cattle export (passed by the toll stations)

| Toll station | 1585 (Head of beef cattle) | Toll station | 1586 (Head of beef cattle) |
|---|---|---|---|
| Poznan | 3,405 | Poniec | 829 |
| Sulmierzyce | 244 | Kieblów | 38 |
| Kopanica | 18 | Zbaszyn | 1,643 |
| Boleslawiec | 2,469 | Krzepice | 18,693 |
| Czestochowa | 2,177 | Oswicim | 7,932 |
| Przedbórz | 3,090 | Kozieglowy | 5,234 |
| Chrzanów | 1,553 | Myslenice | 306 |
| | | Mstów | 359 |
| | | Olkusz | 697 |
| | 12,956 | | 35,731 |

The annual number of cattle is estimated at ca. 60,000. Beef cattle were also drawn from the western and central parts of the Crown Lands and from Lithuania.[45]

Cattle which were exported, mainly beef, came first of all from the southeastern regions (Podolia, Ukraine, Ruthenia, Moldavia, Walachia) and were characterized by a large fleshy animal with grey fur - a steppe breed. Beef cattle production and export were carried out by big estates of magnates in Ruthenia and Podolia (including great families such as the Herburts, Lanckoronskis, Koniecpolskis, Zólkiewskis, Chodkiewiczs, Wisniowieckis and Zamoyskis among others.[46]

Beef cattle from Central Poland, mainly from Great Poland, the vicinity of Leczyca and Sieradz, were the subject of a small export. It was a red milk breed, not particularly big.[47]

## 5.   M. Malowist's Concept of Economic Zones of East Europe

All the aforementioned factors forming spatial changes in the Polish territory were, on the one hand, directly related to European economic changes, and, on the other hand, connected with the problem of the origin of the early modern economy in Central East Europe, treated as an independent economic unit; one may speak of a geographical-historical unit.

The divergence of the history of West and East has been affected by, firstly, a general crisis in the 14th and 15th centuries, and, secondly, demographical changes and the "price revolution" in Western Europe.

According to this idea, formulated by M. Malowist and his students[48] (it differed in the field under research from F. Braudel's theory[49] who used to divide Europe along parallels of latitude), this part of Europe comprised five zones: 1. a Baltic zone, highly urbanized; 2. a central region, having an agricultural base with an industrial subzone in Silesia and Great Poland; 3. an ore-bearing or mountainous zone; 4. a Balkan zone; 5. a Black Sea - Caspian zone.

This theory, developed by a research worker of European history, and based on data concerning, to a great extent, Central East Europe, has become to a certain degree, as regards the international trade position, a stimulus for I. Wallerstein's conception of world economic zones.[50] It should, however, be pointed out that M. Malowist's theory renounced ideas of treating a Central European manorial estate as a form of a capitalist firm, as well as of equating the estate owning Polish noblemen at the turn of the 16th century with the English "gentry"[51].

The problem needs further microanalytical spatial research, and it seems that the application of J.H. von Thünen's theory on functional spatial elements of the European economic world system[52] might offer new perspectives.

## Notes

1.    A. Wyczanski, Gospodarka polska XVI w. (Polish economy in the 16th c.), Zarys Historii Polski pod red. J. Tazbira (in: History of Poland ed. J. Tazbir), Warsaw 1979, p.189; B. Baranowski, S. Herbst, Historia Polski pod red. T. Manteuffel (in: History of Poland ed. T. Manteuffel), Warsaw 1957, vol. I, 2, chapter 8, p.416; I. Gieysztorowa, Ludnosc, w: Encyklopedia Historii Gospodarczej pod red. A. Maczaka, Population. In: Encyclopaedia of Economical History of Poland ed. A. Maczak), Warsaw 1981, vol. 1, p.432; T. Zarebska, A. Zaboklicka, Przemiany srodowiska geograficznego. In: Historia Kultury Materialnej Polski w zarysie, pod red. W. Hensla, t. 3 pp.27-61 (Changes of the geographical environment, in: History of Material Culture of Poland, ed. W. Hensel), vol. 3, Wroclaw 1978, pp.27-61.
2.    A. Rybarski. Handel i polityka handlowa Polski w XVI stuleciu (Polish trade and politics of commerce in the 16th century), vol. 1-2, Poznan 1928-1929; M. Bogucka, H. Samsonowicz, Dzieje miast i mieszczanstwa w Polsce (History of Polish towns and townspeople), Wroclaw 1986.
3.    J. Malecki, Zwiazki handlowe miast polskich z Gdanskiem w XVI w. i w pierwszej polowie XVII w. (Trade of Polish towns with Danzig in the 16th and 1st half of the 17th century), Wroclaw 1968; I. Baranowski, Przemysl Polski w XVI w. (Polish industry in the 16th century), 1919; A. Maczak, Sukiennictwo wielkopolski XIV-XVII w. (Industry of wool in Great Poland in the 14th-17th centuries), Warszawa 1955; S. Mielczarski, Rynek zbozowy na ziemiach polskich w 2 polowie XVI i 1 pol. XVII wieku. Proba rejonizacji (Marketing of grain in the Polish territory in the 16th and 17th centuries. An attempt at regionalization), Gdansk 1962; J. Malecki, Studia nad rynkiem regionalnym Krakowa (Studies on the regional market of Krakow in Little Poland), Warsaw 1963; B. Zientara, Dzieje malopolskiego hutnictwa zelaznego XIV-XVII w. (History of Polish metallurgy in the 14th to 17th centuries), Warsaw 1963.
4.    A. Keckowa, Zupy krakowskie w XVI-XVIII wieku (Salt-works of Krakow in the 16th to 18th centuries), Wroclaw 1969.
5.    Zarys dziejow gòrnictwa na ziemiach polskich pod. red. J. Pazdura (History of mining in Polish territory ed. J. Pazdur), Katowice 1960.
6.    D. Molenda, Kopalnie rud olowiu na terenie zlóz slaskokrakowskich w XVI-XVIII w. (Lead mines in Silesie and Little Poland in the 16th to 18th centuries), Wroclaw 1972.
7.    M. Bogucka, H. Samsonowicz, op. cit. (2).

8.    A. Wyrobiz, Rola miast prywatnych w Polsce w XVI i XVII wieku,  (The role of private
      towns in Polish territory in the 16th and 17th centuries), "Przeglad Historyczny" (Historical
      Review) 65 (1974),pp.19-46.

9.    See map "Economic Life in 16th Century Poland" by A. Zaboklicka (the present author),
      The Historical Atlas of Poland 1986, Department of the State Cartographical Publishers,
      Wroclaw, p.19.

10.   A. Wyrobisz, Miasta prywatne w Polsce XVI-XVIII w. jako inwestycje kulturalne (Private
      towns in Poland in the 16th-18th centuries), Kwartalnik Historii Kultury Materialnej,
      (Quarterly Review of History of Material Culture) 26(1978) pp.47-56, p.28; J. Burszta,
      Handel magnacki i kupiecki pomiedzy Sieniawa nad Sanem a Gdanskiem od konca XVII do
      polowy XVIII w. (Trade of the magnates and merchants between Sieniawa town on the San
      and Danzig from the end of the 17th to the 1st half of the 18th century) Roczniki Dziejów
      Spolecznych i Gospodarczych (Annals of Economic and Social History) 16, 1955.

11.   A. Wyrobisz, Miasta prywatne... op. cit. (10), p.37.

12.   T. Zarebska, Les transformations spatiales des villes polonaises à l'epoque de la renaissance
      et du baroque, Acta Poloniae Historica 34 (1976) pp.189-218; S.Herbst, Zamosc, Warszawa
      1953; A.Milobedzki, Ze studiów nad urbanistyka Zamoscia (Studies on the town of
      Zamosc), Biuletyn Historii Sztuki (Bulletin of History of Art) 15 (1953) n.3-4.

13.   A. Dunin-Wasowicz, Kapital mieszczanski Nowego Sacza na przelomie XVI i XVII
      w.Wplyw na ekonomike miasta i zaplecza (Capital of the townpeople of Nowy Sacz in the
      16th and 17th centuries. Its influence on the economy of the towns and agricultural
      environs, French summ.), Warszawa 1967; H.-J. Nitz, in this volume.

14.   G.Demidowicz, Planned Landscapes in North-East Poland: The Suraz Estatc 1550-1760,
      Journal of Historical Geography, 11 (1956), pp.21-47.

15. W.Kalinowski, City Development in Poland up to the Mid-19th century, vol.1-2, Warszawa
      1966; T.Zarebska, Les transformations spatiales... op.cit. (12); A.Milobedzki,
      Budownictwo militarne miast polskich w okresie nowozytnym (Military architecture of
      Polish towns in the modern era), Kwartalnik Historii Kultury Materialnej (Quarterly Review
      of History of Material Culture), 1978, p.1.

16.   L.Zytkowcz, Studia nad gospodarstwem wiejskim w dobrach koscielnych w XVI w.
      (Studies on farming in Church owned properties in the 16th century), Warszawa 1952;
      J.Topolski, Gospodarstwo wiejskie w dobrach arcybiskupstwa gnieznienskiego od XVI do
      XVIII w. (Farming on the properties of the Cathedral in Gniezno in the 16th-18th
      centuries), Poznan 1955; J.Leskiewicz, Dobra osieckie w okresie gospodarki folwarczno-
      panszczyznianej XVI-XIX w. (Economy of manorial estates with corvee in Osiecko 16th-
      19th centuries), Wroclaw 1957; Al. Tarnawski, Dzialalnosc gospodarcza Jana Zamoyskiego
      (1572-1605), (Economic activity of chancellor Jan Zamoyski), Lwów 1935;
      A.Wawrzynczyk, Rozwòj wielkiej wlasnosci na Podlasiu w XV i XVI w. (Development of
      big manors in Podlachia in the 15th and 16th centuries), Wroclaw 1951; A.Wyczanski,
      Studia nad folwarkiem szlacheckim w Polsce w latach 1500-1580 (Studies on manors of the
      nobility in Poland in 1500-1580), Warszawa 1960.

17.   T.Zarebska, A.Zaboklicka, op. cit. (1).

18.   Atlas Historyczny Polski, mapy szczególowe XVI w., Mazowsze w drugiej polowie XVI w.
      pod red. W. Paluckiego (Historical Atlas of Poland, Mazovia in the second half the 16th
      century), vol.1: maps 1:250 000, vol.2: commentary, index, Warsaw 1973.

19.   Atlas Historyczny Polski, Prusy Królewskie w 2. pol. XVI wieku (Historical Atlas of
      Poland, Royal Prussia in the 16th century) ed. M. Biskup, L.Koc, vol: commentary, vol. 2:
      maps 1:500 000, Warsaw 1961.

20.   T.Zarebska, A. Zaboklicka, o.c. (1).

21.   Wl. Semkowicz, Materialy zródlowe do dziejów osadnictwa górnej Orawy (Sources of
      colonisation in Orawa environment), Zakopane vol. 1-2, 1932-1938; S.Kuras, Osadnictwo i

zagadnienia wiejskie w Gorlickiem do polowy XVI w. (Colonisation of the Gorlice environment - 16th century), Nad rzeka Ropa,3, Kraków 1968, pp.61-90.

22.  W.Szczygielski, Dzieje gospodarki stawowo-rybnej od XVI do konca XVIII w. (History of the economy of fishery in the 16th-18th c.), "Rocznik Lódzkiy".

23.  H.Szulc, The impact of the evolution of large estates on the rural settlement of Silesia and Pomerania, in this volume; J.Burszta, Wies-uklad przestrzenny (Layout of the village). In: Encyklopedia Historii Gospodarczej Polski pod red. A. Maczaka (Encyclopaedia of Economic History of Poland, ed. A.Maczak), vol.2, Warszawa 1981, p.492-499; H.Szulc, Les changements de formes de l'habitat rurale en Pologne). In: Recherches de Geographie Rurale. Hommage au Professeur Frans Dussart, Liège 1979.

24.  S.Orsini-Rosenberg, Rozwój i geneza folwarku panszczyznianego w dobrach katedry gnieznienskiej w XVI w. (Beginning and development of farming with corvee in the manors of the Cathedral in Gniezno in the 16th century), Poznan 1925.

25.  H.Szulc, Typy wsi Slaska Opolskieg na poczatu XIX w. i ich geneza, (Types of rural settlements of Opole-Silesia at the beginning of 19th century and their origin), Warszawa 1968; H.Szulc, Osiedla podwroclawskie na poczatku XIX w. (Eng.summ.: Studies of the scttlement in the Wroclaw suburban area from the end of the XVIIIth till the first half of the XIXth century), Wroclaw 1963; H.Szulc, Morfogentyczne typy osiedli wiejskich na Pomorzu Zachodnim (Eng.summ.: Morphogenetic types of rural settlements in Western Pomerania), Prace Geograficzne Nr 149, Wrocklaw 1988; Atlas Historyczny Polski, Prusy Królewskie op. cit (19); R.Heck, Uwagi o gospodarce folwarcznej na Slasku w okresie Odrodzenia (Economy of manorial estates with corvee labour in Silesia in the epoch of Renaissance), Sobótka 11 (1956) n.2.

26.  Atlas Historyczny Polski, Mazowsze... op. cit. (16).

27.  A.Zaboklicka, The Range of Riga Trade in the 16th century, The Historical Atlas of Poland, Warszawa-Wroclaw 1986, p.19, II.

28.  G.Demidowicz, Planned Landscapes... op.cit. (14).

29.  L.Kolankowskik, Pomiara wlóczna (Land surveying in "mansi" - Latin medieval land unit), Ateneum Wilenskie 4 (1927) n.13; M.Kosman, Pomiara wlóczna na Polesiu pinskim (Land surveying in Polesie on the Pina), Roczniki Dziejów Spolecznych i Gospodarczych (Annals of Economic and Social History), vol.31; W.Conze, Agrarverfassung und Bevölkerung in Litauen und Weißrussland, I.Teil, Hufenverfassung. Deutschland und der Osten, Bd.15, Leipzig 1940.

30.  W. Rusinski, Osady Olderów w dawnym wojewódzlwie poznanskim (Dutch villages in the Voivodship of Poznan), Poznan 1939 - Kraków 1947; W.Rusinski, Deutsche Siedlungen auf polnischem Boden im 16.-19.Jahrhundert, Acta Poloniae Historica XLVII (1983) pp.209-256.

31.  A.Wyczanski, Studia nad folwarkiem szlacheckim w Polsce w latach 1500-1580 (Studies on the manors of the nobility in Poland in 1500-1580), Warszawa 1960; J.Popolski, Gospodarstwo wiejskie ... op.cit. (14); L.Zytkowicz, Studia nad gospodarstwem...op. cit. (14).

32.  B.Zientara, Z zagadnien spornych tzw. "Wtórnego poddanstwa" w Europie srodkowej (Problems of "second serfdom of peasants" in Middle Europe), Przeglad Historyczny (Historical Review) 47 (1956); Wl. Rusinski, Drogi rozwojowe folwarku panszczyznianego (Development of manorial estates with corvee labour), Przeglad Historyczny (Historical Review) 46 (1955); W.Kula, Teoria ekonomiczna ustroju feudalnego (An economic theory of the feudal system), Warszawa 1962. English translation from the Italian edition: An Economic Theory of the Feudal System. Towards a Model of the Polish Economy 1500-1800. Foundations of History Library, London 1976 (Presentation by F.Braudel).

33.  M.Malowist, Wschód a Zachód Europy, Warszawa 1973 (East and West of Europe); J.Topolski, Narodziny kapitalizmu w Europie XIV-XVII w. (Birth of capitalism in Europe

in the 14th to 17th centuries), Warszawa 1965; A. Maczak, H.Samsonowicz: La zone baltique; l'un des elements du marche européen, in: Acta Poloniae Historica 11, (1965) pp.71-99.

34. A.Maczak, H.Samsonowicz, La zone baltique... op.cit. (33); Regestra thelonei aquatici Vladislaviensis saeculi XVI (Register of the river customs at Wloclawek in the 16th century), ed. S.Kutrzeba, Fr.Duda, Kraków 1915; I.Gieysztorowa, Études cartographiques de l'histoire de Pologne, Acta Poloniae Historica 2 (1959) pp.71-99; T.Chudoba, Z zagadnien handlu wislanego Warszawy (Problems of the Vistula trade in Warsaw), Przeglad Historyczny (Historical Review) 50 (1959).

35. S.Trawkowski, Przemiany spoleczno-gospodarcze w 1 pol. XV w. (Social economic development in the first half in the 15th century), in: Zarys historii Polski pod red. J.Tazbira (History of Poland ed. J.Tazbir), p.124-129; A.Wyczanski, Gospodarka Polska XVI w., Zarys Historii Polski, op. cit. (1).

36. A.Rybarski, Handel i polityka handlowa Polski... op. cit. (2).

37. For the location of the grain exporting regions of the 16th century see the map of A.Zaboklicka, Economic life in 16th century-Poland. The Historical Atlas of Poland, Warszawa-Wroclaw 1986, p.19; A.Rybarski, Handel i polityka handlowa Polski... op.cit. (2); A.Dunin-Wasowicz, Geografia historyczna Mazowsza XVI-XVII w. (Historical Geography of Mazovia in the 16th-17th centuries) ,Rocznik Mazowiecki (Annals of Mazovia) 7, pp.57-67.

38. H.Madurowicz, A.Podraza, Regiony gospodarcze Malopolski Zachodniej (Economic regions of west Little Poland in the 18th century), Wrocław 1958.

Additional comment by the editor:

The largest supply for Danzig's grain export came from Royal Prussia in the Vistula delta, the immediate hinterland of the port. This region is not included in A.Dunin-Wasowiczowa's study which analyses the grain export of interior Poland passing the toll station of Wloclawek more than 230 km from Danzig. Based on data collected by A.Wyczanski for the second half of the 16th century, M.North (article quoted at the end of this comment) calculated the average annual surplus of grain in Royal Prussia at about 70,000 t which is very large compared to Mazovia as the main supplier from interior Poland with an average annual surplus on 15,720 t. The total annual surplus of Poland (including Royal Prussia) may have, in the editor's calculation with reference to A.Dunin-Wasowiczowa's data for the late 16th century, amounted to about 100,000 t. Of this, according to the quantities calculated by Wyczanski, on average about 71,000 t were exported via Danzig (the difference must have been consumed by the Polish towns, including Danzig). Taking the export data of the late 16th century presented by A.Dunin-Wasowiczowa (Tab.3) which amount to about 20,000 to 25,000 t of grain exported from interior Poland to Danzig, we would arrive at a share of about one third of the town's total export, and two thirds from its immediate hinterland, i.e. Royal Prussia.

This uneven spatial distribution of grain surplus for the market can, according to North, for the most part be explained by the high soil fertility of the delta which permitted intense corn-growing which in turn was a reaction of the demesnes and the peasant farms to the stimulating effect of the nearby export market conveyed through Dutch and Danzig corn traders. Even during phases of recession of international demand the immediate hinterland of the export centre was in a more favourable position to sell its corn which it could offer at a cheaper price than the interior because of lower transport costs. This guaranteed a stable demand for corn in the delta.

In addition, as North has shown, the average size of peasant farms in the delta was three to four times larger than in interior Poland. This would have permitted a larger share of surplus for the market. Peasants had less corvee obligations to the feudal lords who to a considerable extent engaged wage labourers. In the editor's opinion this more expensive

labour system was possible for them because they saved transport costs compared to distant regions of interior Poland, which in turn could be invested in wage labour.
See Michael North, Getreideanbau und Getreidehandel im Königlichen Preußen und im Herzogtum Preußen. Überlegungen zu den Beziehungen zwischen Produktion, Binnenmarkt und Weltmarkt im 16. und 17.Jahrhundert (Corn-growing and corn-business in Royal Prussia and the Duchy of Prussia. Considerations regarding the relations between production, home market and international market in the 16th and 17th centuries. In: Zeitschrift für Ostforschung 34 (1985), pp.39-47.

39. T.Zarebska, A.Zaboklicka, Przemiany srodowiska... op. cit. (1).
40. G.Jentsch, Der Handel Rigas im 17. Jahrhundert, Riga 1930.
41. A.Wawrzynczyk, Rozwój wielkiej wlasnosci na Podlasiu w XV i XVI w. (Development of large manors in Podlachia in the 15th and 16th centuries), Wroclaw 1951; W.Conze, Agrarverfassung und Bevölkerung... op.cit. (29); L.Kolankowski, Pomiara wlóczna... op. cit. (29); M.Kosman, Pomiara wlóczna na Polesiu pinskim... op.cit. (29).
42. L.Zytkowicz, Rozwarstwienie chlopstwa a gospodarka na Zmudzi w polowie XVII i w XVIII w. (Social differentiation of the peasants in Samogitia in the 17th and 18th centuries) Studia Staropolskie ed. A.Wyczanski, vol.2, Warszawa 1979, pp.229-308.
43. H.Szulc on Silesian estates, in this volume; S.Inglot, Wies i rolnictwo (Village and agriculture). In: Historia Slaska, vol.I no.3 (History of Silesia, from 16th century to 1763).
44. Atlas Historyczny Polski, Prusy Królewskie, op. cit (19); A.Maczak, Miedzy Gdanskiem a Sundem, Studia nad handlem baltyckim od polowy XVI do polowy XVII w. (Between Gdansk and Sund. Studies of Baltic commerce in the 16th and 17th centuries), Warszawa 1972; Cz. Biernat, Statystyka obrotu towarowego Gdanska w latach 1651-1815 (Statistics of the turnover of goods in Gdansk 1651-1815), Warsaw 1962; M.Bogucka, Handel zagraniczny Gdanska w pierwszej polowie XVI w. (Gdansk's foreign trade in the first half of the 17th century), Wroclaw 1970; A.Manikowski, Zmiany czy stagnacja? Z problematyki handlu polskiego XVII w. (Change is stagnation? Problems of Polish trade in the 17th century), Przeglad Historyczny (Historical Review) 60 (19719; B.Wachowiak, Przemiany gospodarczo-spoleczne i ustrojowe na wsi zachodnio-pomorskiej (Economic and social transformation in the villages of the West Pomerania). In: Historia Pomorza (History of Pomerania), ed. G.Labuda, vol. II part 1, Poznan 1976, pp.698-726.
45. R.Rybarski, Handel i polityka handlowa Polski... op. cit. (2); J.Baszanowski, Z dziejów handlu polskiego w XVI-XVIII w. Handel wolami (History of Polish trade in the 16th to 19th centuries. Trade of beef cattle), Gdansk 1977.
46. J.Baszanowski, Z dziejów handlu polskiego... op. cit. (41); See also the discussion in the contribution of H.-J.Nitz, in this volume.
47. B.Baranowski, Chów bydla w drugiej polowie XVII i XVIII w. w Leczyckiem i na terenach sasiednich (Breeding of cattle in the second half of the 17th and in the 18th centuries in Voivodship Leczyca), studia z dziejów Gospodarstwa Wiejskiego (Historical studies on the rural economy), vol.1, ed. J.Leskiewiczowa, p.198-258; B.Baranowski, T.Sobczak, Regiony gospodarki hodowlanej na ziemiach polskich w XVI-XIX w. (Regions of cattle breeding in Polish territory in the 16th to 19th centuries), ibidem, vol.8, ed. J.Leskiewiczowa, Warszawa 1966, pp.177-186.
48. M.Malowist,Wschód a zachód Europy... op. cit. (33); M.Malowist, Croissance et regression en Europe, Paris 1972; M.Malowist, Continual Trends and Social Developments in Central Europe. The Baltic Countries and the Polish-Lithuanian Commonwealth. In: State and Society in Europe from the XVth to the XVIIIth Century (ed. J.Pelenski), Warsaw 1981; M.Malowist, The Problem of the Inequality of Economic Development in Europe in the Later Middle Ages, The Economic History Review 19 (1966); M.Malowist, The Problems of the Growth of the National Economy of Central Eastern Europe in the Late Middle Ages, Journal of European Economic History 3 (1974); Central Europe in Transition

from the XVth to the XVIth Century (ed. A.Maczak, H. Samsonowicz, P.Burke), Cambridge 1985; A.Manikowski, Zmiany czy stagnacja, op. cit. (44). The articles of Malowist in English are quoted after Igor Kakolewski and Krzysztof Olendzki, Strefa poludnika 20: Mit czy rzeczywistosc? Wokól problematyki badan porównawczych Mariana Malowista nad historia Europy Srodkowowschodniej (The zone of the 20th meridian: Myth or reality? On the problem of the comparative studies of Marian Malowist on the history of East Central Europe), Przeglad Historyczny 81 (1990) pp.301-311.

49.   F.Braudel, Mediterranée et le monde mediterranéen à l'epoque du Philippe II, Paris 1949; F.Braudel, La dynamique du capitalisme, Paris 1985.

50.   I.Wallerstein, The Modern World System I-II New York 1974-1980, chapters II, V, VI.

51.   J.Topolski, Narodziny kapitalizmu... op. cit. (33); J.Topolski, La regression économique en Pologne du XVIe au XVIIIe siècle, Acta Poloniae Historica 7 (1962) pp.28-49.

52.   H.-J. Nitz, The European World-System. A von Thünen Interpretation of its Eastern Continental Sector. In this volume.

# 12

# The Impact of the Evolution of Large Estates on the Rural Settlements of Silesia and Pomerania

## Halina Szulc

The beginnings of the manorial economy with coerced peasant labour, which date back to the turn of the 16th century, at the same time exerted an influence on the formation of new settlement structures. Studies of this process by the present author have been geographically limited to the areas of Silesia and West Pomerania (Hinterpommern). But the question under discussion concerns the whole area of Central and Eastern Europe, the territories located to the east of the Elbe. Processes related to the manorial field-service economy took place earlier in Eastern than in Western Europe, and earlier in Poland than in Silesia, Pomerania, Bohemia or Meklemburgia.

The concept of duality in the economic development of Eastern and Western Europe is covered by numerous publications, both from Poland and abroad.[1] It is proposed that the main reason for introducing the manorial field-service economy in Eastern, earlier than in Western, Europe, was a lower level of economic development in the former. Thus, for instance, F. Bujak[2] expresses the opinion that transition towards the field-service manor system was caused by an agricultural crisis which occurred in these areas at the end of 15th century. Further potential causes elicited are:
- a great number of deserted villages,
-limitations in personal freedom of peasants, leading to the creation of the prerequisites for field-service manors,
- a low level of urbanization.

Why was the course of events related to the transition towards the manorial field-service economy different in the territories west of the Elbe? W. Rusinski[3], in trying to answer this question, refers to the theory of higher profit rates in industry than in agriculture, both in England and in the Netherlands. This entailed population outflows from rural to urban areas, so that fields were being turned into pastures.

Silesia and Pomerania represent two different regions of the manorial field-service economy. In Silesia manors had already existed for a long time as the economic centres of feudal estates ("grangiae"), but they were not numerous. An increase in the number of manors occurred only in the 16-17th centuries, when noblemen's field-service manors developed. This was principally a result of the demographic growth and the development of towns, while the export to highly industrialized countries of western Europe had a lesser influence. Therefore Silesian manors were located mainly on fertile soils, near towns, urbanized settlements and industrial villages, typical of the Sudeten foothills (Vorsudeten).

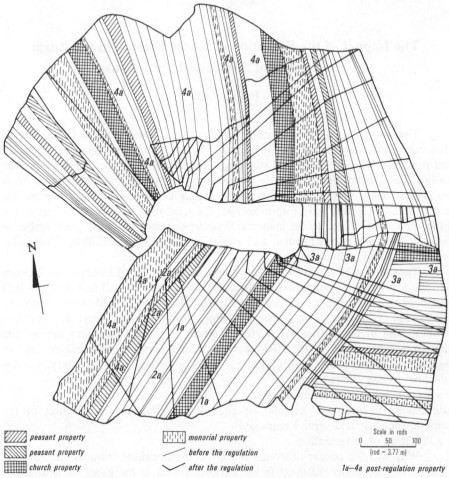

Scale in rods
0      50      100
(rod = 3.77 m)

peasant property

peasant property

church property

monorial property

before the regulation

after the regulation

1a–4a post-regulation property

Fig.1: Goworowice (Gauers, former Kreis Grottkau). Copy of a map of the village reorganised
in the consolidation process at the beginning of the 19th century in a star shape.
Manorial property before consolidation was mixed with peasant fields.

Coal mining developing in the Wałbrzych Basin in Upper Silesia was the
cause of the rapidly growing population in towns and industrial villages that were
a good market for grain and livestock. In the author's investigations of the growth
of the urban population, she took the example of the villages around Zabrze where
the development of industrial enterprises set up in the late 18th century was
accompanied by a fourfold increase of inhabitants between 1787 and 1840.[4]

T. Ładogórski[5], using principally late 18th century[6] statistical data, included
great weavers' villages lying at the foot of the Eagle Mountains (Eulengebirge) as
town-like industrial settlements.[7] It follows from his studies[8] that by the end of
the 18th century the territories to the south-west of Złotoryja (Goldberg),

Strzegom (Striegau) and Dzierzoniow (Reichenbach) were the most densely populated regions in Silesia. They were a lucrative market for cattle and grain produced in numerous manors, established mainly on Silesia's fertile soils and forming a long belt on the left bank of the Odra from Legnica (Liegnitz) to Głubchyce (Leobschütz), and on the right bank around Trzebnica (Trebnitz).

The present author has studied the problem of the spatial distribution of manors in the early 19th century taking the suburban area of Wrocław (Breslau) as an example.[9] There she distinguishes two types of manorial villages oriented at grain production and livestock breeding. Free-gardener (Freigärtner) villages were located in the direct neighbourhood of Wroclaw. They were situated on fertile soils to the south of the town, and were oriented at intensive vegetable and herb cultivation, a characteristic feature of the suburban areas. These villages were, until secularization, the property of Wrocław monasteries and afterwards of the King. Some of them belonged to the municipality.

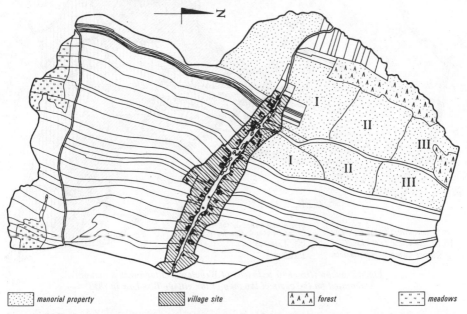

| manorial property | village site | forest | meadows |

Fig.2: Lubiatów (Lobedau, former Kreis Grottkau). Copy of a map of the village in 1825.
Row village with compact strip farms (Waldhufendorf), founded according to the 'German Law'
about 1300. Manorial property in blocks formed out of about five strip farms.

At the same time, the manorial villages located in the Odra river valley were engaged in livestock breeding. By the end of the 18th century in the area near Wrocław, 25-30 km from the old town, manorial villages constituted nearly half of the study area. It follows from J.Jańczak's[10] research that by the end of the 18th century sheep played an important role in Silesian animal husbandry. Closely connected with it was production of wool as a basis for advanced cloth making. In the area near Wrocław there were 120 sheep per 100 ha of arable land. The manorial and manorial-peasant villages oriented at grain production were located on black soils to the south of Wrocław.

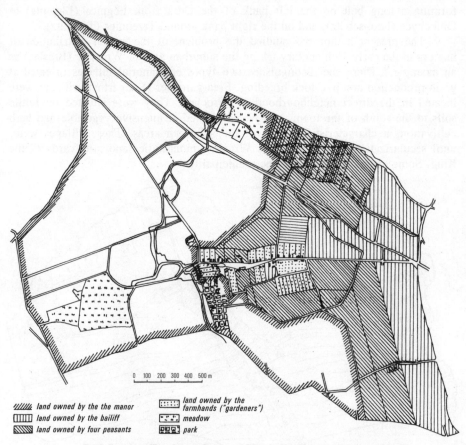

land owned by the the manor
land owned by the bailiff
land owned by four peasants
land owned by the farmhands ("gardeners")
meadow
park

Fig.3:Brochow (Brockau, a district of Wrocław): Village with a manor
Elaborated on the basis of the map of the village Brochów in 1818.

In Pomerania the spatial distribution of manors was quite different.[11] Manors were established there principally in the 15th and 16th centuries, this being related to the export of grain to Western Europe and Sweden. The question of the origin of manorial economy in Western Pomerania has not yet been fully answered. B. Wachowiak[12] quotes the most comprehensive approach of C.J. Fuchs[13], who claims that the reason behind the growth and spread of manors was the participation of the Pomeranian gentry in wars. Therefore the noblemen's land had to be cultivated by villeins which was facilitated by new privileges granted to the gentry by the dukes, for war service. An important factor contributing to the development of manors were numerous wastelands, which were largely created as a result of destruction in war, plagues decimating the population, and peasants escaping to the towns. Export of grain was the main determinant of the character and increase in manors which were established on peasants' land after eviction of the owners, on wastelands and clearings. The spatial distribution of manors in Pomerania is very characteristic. Manors were set up mostly in the areas which up to the 15th/16th centuries had been sparsely populated with small irregular villages which had come into being in a slow evolution. The manors were located on soils of various classes, mostly unfertile, like sand and clays, on culminations of terminal moraines, kames, eskers and outwashes covered with forests. In Pomerania manors did not form any compact settlement region. They were principally situated to the east of a line running from Kołobrzeg (Kolberg), Gryfice (Greifenberg), and Chociwel (Freienwalde) to Dobrzany (Jakobshagen). An exception is the area between Koszalin (Köslin) and Słupsk (Stolp), and the area west of Bytow (Bütow) and Szczecinek (Neustettin), where no manors were established. Here, on fertile soils, large regular peasant villages had been set up during the times of medieval colonization. Manors established on poor soils were oriented at livestock breeding and forestry economy, not at grain production.

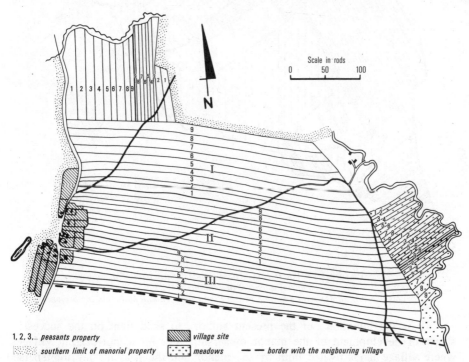

Fig.4:Zabno (Saaben, former Kreis Stolp). Copy of a map of the village in 1797, copy in 1820. Circular village with a three-field system. The village was inhabited by 9 peasants; the manorial property to the north has not been marked on the village map.

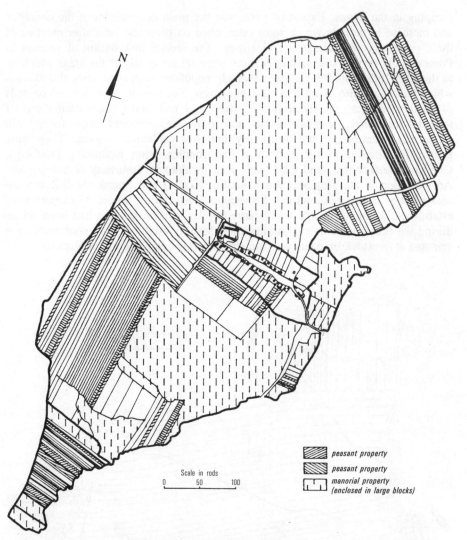

*Fig.5:Przydroze Wielkie (Gross Schnellendorf, former Kreis Falkenberg).*
*Copy of a map of the village in 1826.*
*A regular peasant village with a manor, peasant land reduced and shifted to the periphery.*

It is the intention of the present author to shed light on the successive stages of development of the manor ownership, i.e. on manor-peasant villages, where village lands owned both by the manor and the peasants lay interspersed with each other, manor-peasant villages with the manor and peasant land-ownership clearly separated from each other, and manor villages with small "gardeners" property (labourers of the manor owning just a cottage and a garden).

Examples will be taken from the authors' own studies related to the morphogenesis of villages in Silesia[14] and Pomerania.[15]

The development of the manorial property took place in various ways. At the beginning of the period in question the size of the manor was usually increased by taking into cultivation new areas from the so-called "landlord's reserve" of forested land. When this source had been exhausted, then the commons of the village, or farms abandoned by fleeing peasants (serfs) were taken. They had left their villages because of the growing burden of the field-service system. There were also cases of buying-out of bailiff properties or of larger peasant farms. As a consequence of these processes there were changes in the settlement structure of villages. Manor centres were established in the cores or peripheries of previously pure peasant settlements, thus transforming peasant villages into manor-peasant villages. Manor lands were spatially mixed with the peasants' fields. A good example of a manor-peasant village with spatially mixed manor and peasant land-ownership is the village of Gaworowice (Gauers)[16] (Fig.1).

A further stage in the evolution of the manor-peasant village was separation of manorial land from the peasants' land. Comparing the structure of the row village (Waldhufendorf) of Lubiatow (Lobedau)[17] in 1300 (only peasant strip farms) with that of 1825, it can be seen that the estate land was formed out of several strip farms and took the shape of compact blocks (Fig.2). Spatial organization of the manorial peasant villages, as well as ownership and socio-agrarian classes, as they existed at the beginning of 19th century, were studied for the village of Brochow (Brockau), now a quarter of Wrocław (Breslau)[18] (Fig.3). In this case only the bailiff and four peasants remained as farmers, while the majority was degraded to the status of "gardeners" (owning just a house and a garden).

The process of compacting of manorial lands by separating them from the peasants' lands, which was started in Pomerania in conjunction with reconstruction of villages after the Thirty Years War, led to frequent changes in the shape of village sites and field outlines. This process took a more rapid course in landed properties belonging to the Crown, according to the regulations introduced under Prussian rule, starting in the second quarter of the 18th century. In landed estates belonging to noblemen, it did not occur until after the Seven Years' War (1756-1763).

In the manuscript cadastral plans of villages, dating from the beginning of 19th century and drawn for the purpose of consolidation, often only peasant land-ownership is presented, appearing in three common fields cut into strips, as for example in the village of Zabno (Saaben)[19] (Fig.4), while manorial land was already enclosed in the form of large blocks. Przydroze Wielkie (Groß Schnellendorf)[20] (Fig.5) is an example for villages where the landlord was able to consolidate his manorial property by shifting the reduced peasant fields to the periphery. In circular villages one can sometimes state whether a given village originated as a circular one, or a manor was an element of a so-called incomplete circular village, which, until the manor was added had the shape of, for example, a curved row village. Such a village is to be found at Zelazo (Selesen)[20] (Fig.6). Only by studying villages in precise detail is it possible to draw conclusions about the ancient spatial structure of the village.

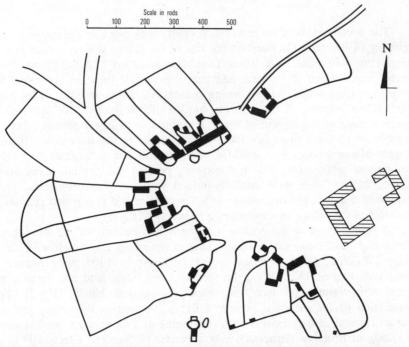

Fig.6: Zelazo (Selesen, former Kreis Stolp). Copy of a map of the village site in 1818.
Circular village. Manor houses with estate farm buildings in the east of the village-green.

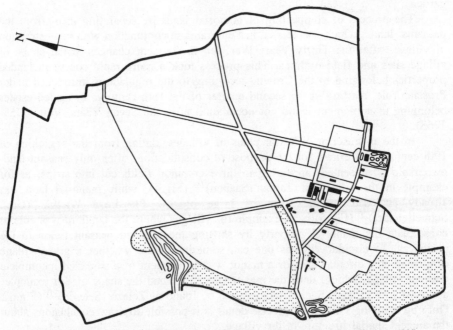

Fig.7: Poswietne (Lilienthal, former Kreis Breslau). Copy of a map of the village in 1835.
Manorial village: 'gardener' houses along a short street and manorial buildings.

The process of separation of manorial ownership away from the areas of peasants' land took an evolutionary course in some villages, requiring a long period of time, while in the others it occurred in just one step, being simultaneous land reforms connected with affranchisement of peasants at the beginning of 19th century.

Further development of the manorial land-ownership system resulted in the almost complete eradication of peasant ownership from villages. In manorial villages the whole area, excepting a small piece of land left to dependent small farmers, belonged to the manor. In such villages the central place was occupied by the manor which was composed of the owner's residence, together with the adjacent park and the farm buildings, e.g. in the village Poswiętne (Lilienthal)[22] (Fig.7). Groups of "gardeners" who worked in the manor clustered along a short road (Fig.8). That is why H. Schlenger[23] regards manorial villages as road villages. Manorial villages can be easily recognized on a topographical map due to their characteristic spatial structure, which is quite different from the layout of medieval peasant villages, and due to the course of their bounds: Such villages appear as if "cut out" of the areas of other villages, and for this reason their boundaries are often straight lines, while the village site is not located centrally. An example of this is Krzyki (Krietern), presently a quarter of Wrocław (Breslau), (Fig.8).

At the beginning of the 19th century all types of manorial villages were represented: manor-and-peasant villages with intermixed land-ownership pattern, manor and peasant villages with separated manorial and peasant land, and finally manorial villages with small gardener parcels. It is not until the land reform laws of the first half of 19th century that radical separation of the two kinds of land-ownership was carried out in the villages of the first type.

In Pomerania the noblemen's property constituted more then half of all landed properties. The medium sized noblemen's one- or two village property prevailed. In contrast, the group of village gentry usually owning just some "mansi" (medieval unit of farm land, ca. 16.5 hectares) was relatively small. F.Engel[25] has shown that, in terms of area, the second in rank was the ducal property, principally in the south-western part of Pomerania along the coast, but also around the towns of Bytow (Bütow), Szczecinek (Neustettin), to the north of Czaplinek (Tempelburg) and Dobrzan (Jakobshagen). As these areas coincide with those in which peasant villages were not so much affected by the formation of manors we may conclude that it was the Dukes' policy to keep their peasantry as tax payers. Church property, however, after the Reformation occupied only relatively small areas. The property of the Chapter of Kołobrzeg (Domkapitel Kolberg) was the largest. Around the towns, peasant villages in municipal property could also be found.

The problem of the feudal property in Silesia has been elaborated upon by J.Wosch[26]. According to his accounts, by the end of the 18th century in Silesia (as in Pomerania) the noblemen's property prevailed - it accounted for 66.8% of the total area.[27] Moreover, noblemen owned approximately 80% of the total number of manors in Silesia which means that there were as many as 1.16 manors per village. In noblemen's estates there were twice as many manors as in the royal domains, and two and a half times that in the Church and municipal estates. This condition persisted until the beginning of the 19th century, that is, until the secularized Church domains were sold. Noblemen's property dominated in the eastern part of Upper Silesia and in the northern and north-western areas of Lower Silesia.

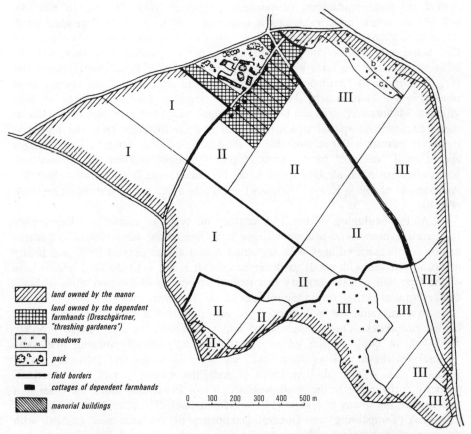

*Fig.8:Krzyki (Krietern, now a district of Wrocław). Copy of a map of the village in 1809. Manorial village (over 90% of the village area is occupied by the manor)*

Second in rank, after the noblemen's property, was, in late 18th century Silesia, the Church property, which accounted for some 15% of Silesia. It contained, however, only a small number of manors, and their distribution over Silesia was uneven. Manors of the Church were concentrated mainly in central Silesia, on fertile areas of early medieval settlements. Third in rank were the royal properties, accounting altogether for some 11% of the area of Silesia. Here also, the number of manors is low, for they constitute just 5% of all the Silesian manors. There were, on average, only 0.58 manors per village, a fact from which, as in Pomerania, it can be concluded that the Crown (and the Church as well) preferred tax paying peasant villages. After royal properties, in the fourth rank, there was municipal landed property, accounting for approximately 5% of Silesian territory.[28] Spatial distribution of municipal property was naturally uneven and mainly concentrated around towns located along the Odra and on the left bank, the richer part of Silesia. Municipal estates formed compact areas located in the direct vicinity of towns, to which they belonged. Bourgeois property was poorly represented in Silesia, constituting only 2% of its area. As can be conclude from the studies of J.Wosch,[29] large landed estates outnumbered the smaller ones in late 18th century Silesia. These large estates were concentrated in the hands of a relatively small group of owners, including big magnates, the Crown and the Church. The spatial distribution of large landed estates is uneven. It is most often in areas of poorer soils, with extensive forest cover, and pre-existing villages forming sparser settlement networks where the large estates were mainly concentrated. It was here that the nobility could easily create large demesnes through forest clearings and suppression of the peasantry of small villages. Thus it was here that latifundia appeared in the form of compact areas stretching over tens of kilometers.

## Notes

1.    This problem is discussed in more detail by J. Topolski, Uwagi o przyczynach powstania i rozwoju gospodarki folwarczno pańszczyźnianej na Pomorzu Zachodnim (Remarks on the cause of emergence and development of the manorial field-service economy in Western Pomerania), Studia i Materiały do dziejow Wielkopolski i Pomorza, vol. IV, nr 2, pp.109-122, 1960) who considers publications by W.Bruneck, G.F.Knapp, C.J.Fuchs, G.Below, M.Sering, F.Rachfahl, H.Bechtel, M.R.Krzymowski and G.Mortensen (German authors) and J.Rutkowski, B.Zientara, W.Rusiński and B.Wachowiak (Polish authors).

2.    F.Bujak, Z dziejow wsi polskiej (The History of Polish Villages), Studia historyczne, Krakow 1908.

3.    W.Rusiński, Drogi rozwojowe folwarku pánszcźyznianego (Courses of evolution of the field-service manor), Przegląd Historyczny, 47, 1956.

4.    H.Szulc, Zabrze, Czasopismo Geograficzne, 30, 1959, pp.179-194.

5.    T.Ładogorski, Gęstość zaludnienia, Komentarz i mapy zamieszczone w: Atlas Historyczny Polski, Sląsk w koncu XVIII wieku (Population Density, a commentary and maps. In: Historical Atlas of Poland. Silesia by the end of the 18th century), edited by J.Jańczak and T.Ładogórski, Polish Academy of Sciences, Institute of History, Wrocław 1976.

6.    F.A. Zimmermann, Beiträge zur Beschreibung von Schlesien, vol.1-13, Brieg 1788-1796; J.G. Knie, Alphabetisch-statistisch-topographische Übersicht aller Dörfer, Flecken, Städte und anderer Orte der Königl. Preuss. Provinz Schlesien, Breslau 1845.

7.    Bielawa (Langenbielau), Pieszyce (Peterswaldau), Piława (Gnadenfrei), Walim (Wüstewaltersdorf).

8.    T.Ładogórski, op. cit. note 5.

9.    H. Szulc, Osiedla podwrocławskie na poczatku XIX w. (Eng. sum.: Suburban settlements in the vicinity of Wrocław at the beginning of the 19th century), Monografie Sląskie Ossolineum, Wrocław 1963, on map D. Settlement Zones at the Beginning of the 19th Century.

10.   J. Jańczak, Rolnictwo (Agriculture), a commentary and maps.In: Silesia by the end of the 18th century, op. cit. note 5. By the end of the 18th century the number of draught oxen and cows was quite sufficient but there was a shortage of beef cattle. Therefore great numbers of oxen were driven to the popular fairs in Brzeg (Brieg), Namysłów (Namslau), Wrocław (Breslau), Legnica (Liegnitz) and Swidnica (Schweinitz), principally from the Polish Common wealth, Moldavia and Walachia (J. Jańczak, commentary p.96).

11.   H.Szulc, Morfogenetyczne typy osiedli wiejskich na Pomorzu Za chodnim (Eng.sum.: Morphogenetic types of rural settlements in Western Pomerania), Prace Geograficzne IG i PZ PAN, no. 149, 1988, on map "Settlement network of Western Pomerania against the background of hydrography and forests in the first half of the 19th century".

12.   B. Wachowiak, Gospodarcze położenie chłopów w domenach Księstwa Szczecińskiego w XVI i w pierwszej połowie XVII w. (The economic situation of the peasants in the domains of the dukedom of Szczecin in the 16th and the first half of the 17th century), Szczecin 1967.

13.   C.J. Fuchs, Der Untergang des Bauernstandes und das Aufkommen der Gutsherrschaften, Strassburg 1888.

14.   H. Szulc, Suburban settlements..., op cit.; H.Szulc, Typy wsi Śląska Opolskiego na początku XIX wieku i ich geneza (Eng. sum.: Types of rural settlements in Opole-Silesia at the beginning of 19th cent. and their origin), Prace Geograficzne IG PAN, nr. 66, 1968.

15.   H. Szulc, Morphogenetic types..., op. cit.

16.   Goworowice (Gauers, Gauwald, Kreis Grottkau), copy of a plan of the village in 1819. Scale ca. 1:5000. Archiwum Państwowe (AP) Wrocław, Kom.Gen. Gr.5.

17.   Lubiatów (Lobedau, Kreis Grottkau). Copy of a plan of the village in 1825. Scale ca. 1:4000, AP Wrocław KG Gr.

18.   Brochów (Brockau, now a district of Wrocław). Copy of a plan of the village in 1818. Scale ca. 1:5000, AP Wrocław, KG Wr. 133.

19.   Żabno (Saaben, Kreis Stolp). Copy of a plan of the village in 1820. Scale ca. 1:5000, AP Słupsk, UK Miastko.

20.   Przydroze Wielkie (Gross Schnellendorf, Kreis Falkenberg). Copy of a plan of the village in 1826, scale ca. 1:4000, AP Wr., KG Nie.

21.   Żelazo (Selesen, Kreis Stolp). Copy of a plan of the village in 1818. Scale ca. 1:4000, AP Słupsk, UK Słupsk 597.

22.   Poświętne (Lilienthal, Kreis Breslau). Copy of a plan of the village in 1835, scale ca. 1:2500, AP Wrocław, Wydz. Geodezyjny MRN Wr.

22.   Przydroże Wielkie (Gross Schnellendorf, Kreis Falkenberg). A copy of a plan of the village from 1826, scale ca. 1:4000, AP Wr., KG Nie.

23.   H. Schlenger, Formen ländlicher Siedlungen in Schlesien, Veröff. der Schlesischen Gesellschaft für Erdkunde und des Geogr. Instituts der Universität Breslau, Nr. 10, Breslau 1930.

24.   Krzyki (Krietern, presently a quarter of Wrocław). Copy of a plan of the village in 1809, scale ca. 1:4000, AP Wr. m. Wr. 608.

25.   F. Engel, Historischer Atlas von Pommern, Neue Folge, Lieferung 1, Veröff. der Historischen Kommission für Pommern, Karte 1: Besitzstandskarte von 1628, Karte 2: Besitzstandskarte von 1780, Köln/Graz 1959.

26.   H.Wosch, Commentary "Feudal Landed Property", pp.65-88, and on the basis of the maps: "Rozmieszczenie feudalnej własności ziemskiej i folwarków według grup stanowych w latach 1783-87" (The Distribution of Feudal Lands and Manorial Property in the Years 1783-1787), scale 1:500 000, and "Rozmieszczenie feudalnej własności ziemskiej według grup wielkościowych w latach 1783-1787" (The Distribution of Feudal Landed Property According to Size in the Years 1783-1787), scale 1:500 000. In: Historical Atlas of Poland. Silesia by the end of the 18th century, op. cit. note 5.

27.  For instance, according to late 18th century data, Count von Schaffgotsch owned 11 1/2
     villages without manors and seven villages with manors in the Jelenia Góra district
     (Hirschberg), and 15 villages without manors and villages with four manors in the Lwówek
     district (Löwenberg). The total extent of his estate was 36 1/2 villages and 11 manors
     totalling an area of 49,160 ha. Another example of noblemen's estates ranging between large
     and medium sized property: according to the 1782 data of the Lublinitz district of Upper
     Silesia the landlord von Stürmer owned 11 villages and  the Zamkowa commune
     (Schlossgemeinde), with 16 manors, the total area of which was 16,291 ha. - All statistical
     data on the landed property in Silesia are taken from Dr. J.Wosch's unpublished studies.
28.  According to the 1885 data published in the "Gemeindelexikon für Schlesien 1887" the
     property of the town of Jelenia Góra (Hirschberg) constituted five villages without manors
     and two villages with four manors, total area 4,787 ha.
29.  J. Wosch, op. cit. note 26.

# 13

# Ireland in the World-System 1600-1800

Kevin Whelan

## 1. Introduction

At the beginning of the seventeenth century, Ireland was a very lightly settled, overwhelmingly pastoral, heavily wooded country, whose economy was characterized by its quasi autarchic state, its lack of integration, its weak urbanism and technological archaisms. This was reflected in the unsophisticated and unprocessed nature of its exports - hides, fish, and timber in particular - and by its low population of about one million.[1] By the end of the century, all that had changed and the most rapid transformation in any European seventeenth century economy, society and culture had been effected. Two processes were central to this transformation; the initial subjugation, subsequent colonization and final integration of Ireland into the expanding English mercantilistic state, and the concurrent enhancement of Ireland's location, with the rapid articulation of the North Atlantic commercial world. After 1685, that Atlantic world was increasingly an English one,[2] and the fusion of the two processes accelerated the transformation of Ireland. From being an island behind an island on the rainy rim of Western Europe, Ireland now became the last European stepping-stone to America, and a very close neighbour of the increasingly powerful and consumptive England. Thus, Ireland's new location gave it easy access to two growing markets for livestock and livestock products. The rapacious foreign demand for pastoral products created a commercial agricultural sector; sustained economic diversification was generated around this initial export base.

## 2. Integration of Ireland into the First British Empire

Aggressive post-Elizabethan English expansion led in Ireland as in America to the spread of what Meinig has called "the ragged, bloody edge of empire".[3] Throughout the seventeenth century, Ireland (like Scotland) was increasingly integrated into the large political and economic entity evolving around London and its imperial interests.[4] The centralization of wealth and power in London, the emergence of the First British Empire, and the economic modernization of Ireland and Scotland are all concurrent and closely linked processes. By the late seventeenth century, London's economic and demographic dominance was unequalled in Europe, and necessitated the ingestion of resources from Ireland. Indeed, in the first half of the eighteenth century, half the English army was carried on the Irish establishment; in a situation where London's massive growth was absorbing English migration and in a sluggish demographic regime, the Irish and Scots were also needed to people British North America. London's growing power was also evident in the English legislation which increasingly curtailed Ireland's trade, and which dictated that Irish prosperity would depend heavily upon trade with Britain itself, and with her Atlantic colonies. This colonial dimension was copperfastened by the acts forcing Ireland's American trade to be channelled through English ports, and by the fact that Ireland's emerging manufacturing base was primarily dependent upon the trading connection with Britain and with British North America.[5] In other words, the Ireland-English relationship assumed a political economy of dependency. Simultaneously, the emerging world economic system of the seventeenth century impacted on what had been a remote, peripheral and backward island economy; in response, the landscape itself, settlement patterns, farming systems, land use, demography and regional economies were all transformed.[6]

## 3.    Economic and Social Transformation

### 3.1.  A New Landed Class

Central to these processes was the creation of a new landed class (primarily of non-Irish origin) in the seventeenth century, by confiscation and plantation. The positive dimension of this change was that Ireland acquired a precociously modern estate system, commercially oriented, with a rent-paying leasehold tenantry. This was ideally suited to create a system of agricultural production which would be notably sensitive to the dictates of external demand, as mediated through commodity price-responsiveness. The total destruction of the Gaelic lordships (with their introspective economies and their endemic political instability) and their replacement by a commercial estate system was therefore a key component of the enforced modernization of the Irish economy.[7]

As well as a new landlord class, the seventeenth century witnessed unprecedented immigration into Ireland from England and Scotland; over 100,000 had arrived prior to 1641 and by the early eighteenth century, perhaps 27% of the Irish population were of immigrant stock.[8] This was the largest single immigration into any European country in the seventeenth century, and was accompanied by the doubling of population between 1600-1700, again the highest rate of change in contemporary Europe.

Associated with the new landlord class and the demographic surge was massive environmental change. In particular, there was a sustained onslaught on the great Irish oak woodlands, which until 1600 had occupied 13% of the countryside.[9] Woodland depletion was a classic example of the way in which the Irish economy was being incorporated into the English one. As the iron-smelting industries of the English midlands began to run out of local supplies of oak (and therefore of charcoal), Irish supplies were substituted in a way which did not structurally enhance the Irish economy. It generated only impermanent exports (some timber, pipe staves, iron) and no forward linkages.

The other great by-product of the colonization process was the creation of a whole wave of town and village formation.[10] Ireland, hitherto a dominantly rural society, was now endowed with almost four hundred new seventeenth century towns and villages, initiated by the plantation grantees and new landlords. They wished to provide a nucleus for their estates, a seedbed from which settlers could be planted in the surrounding countryside, and a commercial, legal, cultural and military centre of gravity in a potentially hostile environment. These towns and villages (along with many other rural sites) became the sites of fairs, which, in a dominantly pastoral environment, quickly became the main centres of economic interchange. More crucially, the establishment of a dense, islandwide network of fairs was critical in articulating a cohesive national economy, which was geared primarily to the gradual west-east movement of cattle from the breeding and rearing lands in the hilly and boggy west to the fattening and finishing lands on the sleek pastures of North Leinster. So dense was the fair network that no drove roads emerged as in contemporary Scotland, or Denmark.

In turn the pervasive presence of the fairs was symptomatic of the penetration of the cash economy; by the end of the seventeenth century, Ireland was a heavily monetarised region by European standards. This is demonstrated by the fact that rents were overwhelmingly paid in cash, not in kind. The network of fairs and the associated commercial cattle production were ultimately geared to the expanding English market, itself flourishing as a result of the growth of London and the rise of the woollen industry. Irish cattle and sheep exports boomed in the 1600-1640 period and resumed in the post-war period at consistently high levels.[11] In the mid 1660s, 50,000 cattle and 100,000 sheep were entering England annually. Inevitably, this led to protests from English landed interests, fearful of the local depressant effect on rents (estimated at 20%) caused by this influx. Given the interest system of the English parliament, it was possible to get the Irish cattle trade banned completely in 1667 by the celebrated Cattle Act, frequently cited as an example of the mercantilistic orientation of the English state. Less frequently observed (and ignored by Wallerstein, for example) is the fact that the impact of the Cattle Act on the Irish economy was actually positive; it led to the growth of the victualling trades, and the shipment of beef and butter both to the continent and to the Atlantic economies.[12] This in turn exerted a positive influence on dairying and cattle rearing. Butter exports, for example, increased five fold in the twenty five years after 1665. The export trade fostered the rise of an indigenous mercantile class, and supplied local multiplier effects.

## 3.2 The Growth of Dublin

Dublin especially benefited from the increased orientation to the English economy. As a contemporary expressed it:

"it lyeing in Ireland as ye centre of a semi-circle, much of ye importacions,
especially those from England are carried out of Dublin and Leinster itself,
even with the other three provinces".[13]

Given the accelerating commercialization of the Irish economy, and given
Dublin's favourable location for links with England, it is not surprising to find
Dublin experiencing a demographic surge. It grew from a mere 10,000 in 1600, to
40,000 in 1660, to 60,000 by 1700 to a massive 180,000 by 1800 (Fig.2).[14] This
astonishing growth catapulted Dublin into being the second largest city of the
British empire, and the sixth largest city in Europe (after London, Paris, Vienna,
Naples and Amsterdam).[15] This population growth was triggered by Dublin's
ability to centralize the Irish economy around itself; this was reflected in the
emerging transportation system, its ever widening hinterland, and its
unprecedented accumulation of functions. Thus, Dublin monopolized tertiary
services (law, finance, education, publishing), it controlled foreign trade
(especially its remunerative wholesale and redistributive sectors), it was the centre
of a court and administrative sector, and it had a major manufacturing sector,
especially in luxury goods (sugar refining, drink, textiles). Dublin was both
national warehouse and national workshop.[16] In this sense, Dublin combining the
functions of a Glasgow and an Edinburgh; only Copenhagen could rival its sheer
range of functions. As the eighteenth century progressed, Dublin became, in
Werner Sombart's terms, a classic city of consumption, both a court city and an
aristocratic social centre, dominated by a gentry class who absorbed rural rent
increases in conspicuous consumption.[17] Therefore, Dublin was full of luxury
craftsmen, gold and silversmiths, bookbinders, silk weavers, watchmakers and
especially servants - 10% of the population in 1800.[18]

Yet, one could argue that Dublin's explosive growth was epi-phenomenal,
essentially a short lived phase while the city was playing the key integrative role
between the widening British economy and its Irish counterpart. Once this
integration was solidly achieved by the end of the century (symbolized in the Act
of Union of the two countries in 1800), Dublin inevitably shrank. It had
performed the function of integrating the national economy and linking it to
England's, thereby consolidating Ireland's role in the emerging world economic
system, as mediated through its English links.

In von Thünenian terms, Ireland was now being sucked into the London
orbit. In turn, this obliterated and overrode the earlier rings which had existed
independently around Dublin. The old tillage zone round the city was
metamorphosed in the late seventeenth century into a cattle finishing area. By
1758, the Irish parliament had introduced transport bounties on flour to encourage
the Dublin supply; by 1770, no fewer than fifteen counties (even as far away as
Cork and Galway) were supplying flour to Dublin. In turn, this actually depressed
population growth in the immediate Dublin hinterland, and nothing is more
striking in Ireland's demographic history than the way in which the density
distribution shifted from the seventeenth to the nineteenth centuries. (Compare
national maps of population density in 1660 and 1841 Fig.1)[19]

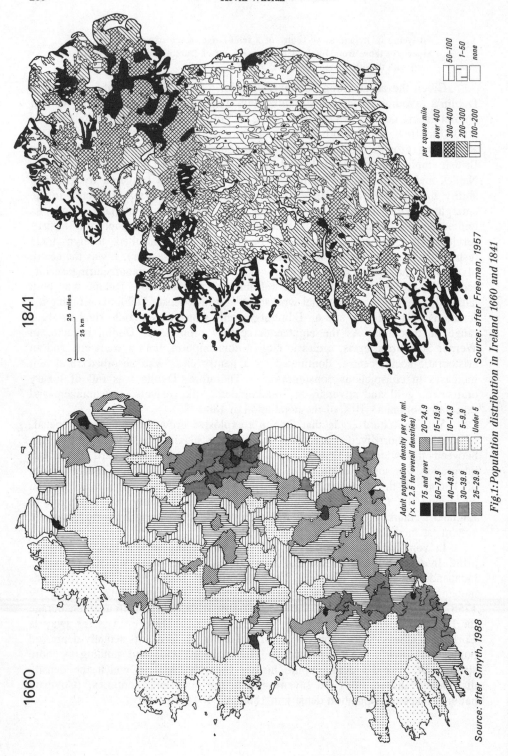

Fig.1: Population distribution in Ireland 1660 and 1841

Source: after Freeman, 1957

Source: after Smyth, 1988

1841

1660

per square mile

over 400
300–400
200–300
100–200

50–100
1–50
none

Adult population density per sq. ml.
(× c. 2.5 for overall densities)

75 and over
50–74.9
40–49.9
30–39.9
25–29.9

20–24.9
15–19.9
10–14.9
5–9.9
Under 5

0    25 miles

0    25 km

## 4.    Ireland and the Atlantic Economy

### 4.1    Food Processing Industries

Alongside growing integration with the English economy, the other key change in seventeenth-century Ireland was the quickening pace of the Atlantic economy and the increased articulation of Ireland with that economy. Given lower labour and production costs than in England, Ireland was ideally placed to take advantage of the emergence of the English North Atlantic after 1685. It was to curb this potential threat to English regional interests that the navigation legislation of 1660-1705 was enacted: these acts were essentially an attempt to reconcile the mercantilistic ambitions of England with Irish demands for participation in the North American trade. The 1663 Staple Act (and its subsequent modifications in 1671 and 1696) eventually completely barred direct imports from America to Ireland; by 1696, there was a full stop to two-way trade. This severely curtailed Irish trade, as the return leg of a bilateral trading system was legislatively amputated. The Irish provisions export trade had then to radically re-orient itself, to suit English self-interest.

Essentially transatlantic trade within the first British empire involved the movement of tropical and semi-tropical staples (sugar, tobacco, rice, indigo) from the plantation colonies of the West Indies and the North American mainland to Great Britain, where they were exchanged for an array of manufactured articles and specialized foodstuffs. Linked to this was a commerce to Iberian and Mediterranean Europe and the wine islands (Madeira, Azores, Canaries) in which colonial fish (especially cod) was traded for wine and salt. Within this broad system, trade between Ireland and the American colonies was essentially intercolonial commerce, with the activities of both partners carefully structured to serve the interests of England.  The growth of Irish trade was predicated on an increased dependence upon England, and Irish-American trade from 1660-1783 exhibited the extent of economic integration within the first British empire.[20]

There were two components in Irish American trade, one directed to the West Indies, the other to the North American mainland.[21] The Irish West Indies linkage is an especially clear example of the specialization associated with the emerging world economic system. Once the Caribbean islands began to move to a widespread sugar monoculture in the mid 1660s, provisions had to be imported to feed the planters.[22] The Irish provisions trade between 1670-1720 was the main source of food. The Governor of Barbados observed in 1675 re Ireland 'it being from thence wee feed so many mouths as must be used in the management of the sugaries'.[23] In the Caribbean trade, Ireland's comparative advantage lay in her ability to organize her agricultural resources into sophisticated food producing and food packaging industries, which effectively complemented the West Indian sugar industry, itself the most aggressively expanding sector of the colonial economy.

However, the Irish merchants involved did not act on independent initiative but rather as commission agents (or factors) of the London sugar industry. As a result, the Irish ports never developed entrepreneurial independence and were heavily dependent on British capital, British ships, British insurance and credit systems.[24] In this sense, the commerce of the provisions ports, like Cork and Waterford, was colonial in character. In turn, this meant that even after the repeal of the Navigation Acts, market forces perpetuated their thrust, and the pattern established under them had a life of its own that long outlived the acts themselves. Because the entire North Atlantic trade was focussed in this way, it had a hothouse effect on the financial sector of the British economy, which inevitably grew in scale, scope and complexity.[25] Sterling bills of exchange, drawn on both sides of the Atlantic against London trading firms, served as the primary financial instruments in the Irish-American trade. This pattern militated against the growth of independent financial services in Ireland. The net result was to create a very high degree of economic integration within the first British empire, an integration copperfastened by the Navigation Acts, which defined an appropriate set of trading activities, unified by a single financial services and marketing apparatus.

The problem for Ireland was that the consequent unilateral trading system did not sufficiently penetrate the economy to lead to self-sustained growth after the break up of the first British empire in the 1780s; when added to the coincident industrial revolution in England and the subsequent demise of nascent Irish industries, it led to a dangerous conjuncture in the Irish economy, whose tragic consequences were to find brutal expression in the famine of the 1840s, the last great subsistence crisis in Europe.[26]

These flows reinforced the major ports of the east and southeast coasts. By 1725, eight of the ten largest Irish towns were ports. Inevitably, those ports which commanded wide hinterlands, rendered accessible by navigable river systems, prospered. However, these developments did not benefit Ireland as a whole; instead, they created thriving regional economies on the east coast and in the Irish sea area, but depressed the far western ports; with their poor and weakly articulated hinterlands, they lost out to the blossoming ports of the south and east.

It was their ability to command rich agricultural hinterlands that catapulted Cork and Waterford onto the leading edge of the Atlantic economy, at the expense of Galway, Limerick, Sligo and Derry in the seventeenth century. Cork and Waterford ports achieved a deep market penetration across most strata of rural society. This is reflected in the sixfold growth in rents in the eighteenth century, by the commercialization of dairying and cattle rearing, and the sharpening and deepening of the social stratigraphy.[27] The output revolution (general exports from Cork increased by a factor of ten between 1700 and 1800) was based especially on a rapidly developing supply responsiveness, mediated by the leasehold system, which was highly sensitive to market trends. Given abundant land and labour, commercialized farming expanded rapidly to meet the growing demand.

## 4.2   The Social Costs of Modernization

However, there was a social cost to this economic expansion; the old independent small farmer ("gneever") class and partnership ("village") farmers were squeezed out and there was a massive rise in the number of impoverished agricultural labourers (or cottiers), accompanied by a narrowing of their diet towards a monotonous, and potentially dangerous, dependence on the potato. In other words, the big farmers prospered but at the expense of widening class divisions, and associated latent tensions which could frequently erupt into bitter agrarian disturbances. Thus, the main impact of the penetration of the cash economy in rural areas was to hasten the process of social differentiation; the class-based disturbances in eighteenth century Munster were most in evidence in the richest farming areas.[28] This social differentiation also had a settlement impact, especially seem in the dispersal of the potato-dependent cottier class to the fringes of the commercialized farms, where their mud cabins were frequently found in a necklace arrangement - social stratification mirrored in micro-spatial segregation.[29] Given the scale and extent of seventeenth century immigration and the inherently colonial structure of the Irish landowning system, these processes also tended to superimpose an ethnic and sectarian configuration on the social hierarchy, reified by penal laws. Thus, the majority native population were disadvantaged throughout much of the eighteenth century.[30]

This meant that Irish society was fissured vertically in a way which militated against social, cultural and ultimately political cohesion. In Ulster, for example, Catholics were almost invariably found at the base of the social pyramid, and remained a marginalized and disadvantaged culture group. Irish society, with these inbuilt lines of cleavage, was liable to fragment in periods of intense economic or political pressure, unlike the experience in England or Scotland. Indeed, a Scottish historian, Devine, discussing the high level of agrarian disturbance in Ireland as opposed to Scotland in the eighteenth century, has seen the contrast as a central explanation as to why Ireland was unable to fully modernize her agriculture in the ruthless fashion which was implemented in the culturally homogeneous Scottish lowlands.[31]

## 4.3   The Irish Provisions Ports

The Irish grazier areas sustained the victualling needs of Cork city, the greatest killing fields in Europe outside of Smithfield. It was Cork which supplied the salted beef and butter for the West Indies, and which victualled the British navy in the eighteenth century. As it became one of the great Atlantic ports, Cork's population trebled from 17,000 to 60,000 between 1700 and 1800 (Fig.2).[32] This was a striking example of the new potential generated by combining agricultural surpluses and the convenient navigation of Ireland with the expanding Caribbean and North American markets. Cork developed the most advanced provisions sector in the Atlantic economy, utilizing high quality curing, packaging and inspection. In particular, it dominated beef exporting, using specialization, low wages and advanced processing to become "the slaughter house of Ireland" and the centre of the most sophisticated meat packaging industry in the eighteenth century world.[33] As a result, Cork's shipping grew by a factor of six between 1700 - 1800 and its hinterland was transformed to service the trade.

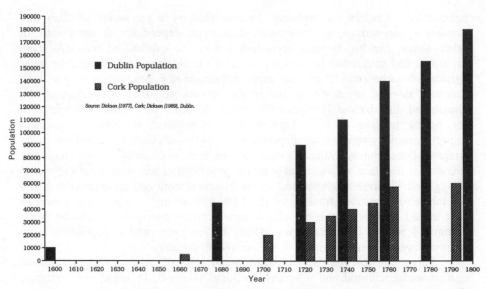

*Fig.2:Population growth 1600-1800,Cork and Dublin*

Waterford city, geared more to the Newfoundland trade, was also to respond positively to its newly acquired locational advantages.[34] After 1675, it became the main port both for provisions and for youngsters to prosecute the fisheries in the Newfoundland migratory cod fishery. This trade was regulated by West County English merchants who found a cheaper labour and provisions market in south east Ireland. By the 1770s, 5,000 people a year were moved from Waterford to the Grand Banks in this transatlantic seasonal fishery. As in Cork, however, the development of an indigenous Irish industry was militated against by the general lack of capital, and investment in shipowning (in contrast to Bristol, Liverpool, or Glasgow). Indeed, 70% of the total shipping in Irish eighteenth century trade was British owned, a dominance which intensified as the century proceeded.

## 4.4   The Irish Mercantile System

Alongside this dependency, another key feature of the Irish mercantile world in the eighteenth century was its ability to manipulate the seventeenth century dispersal of its old landed class to create and exploit a widespread network of family relationships in trade, especially in the Southern European ports, in the Caribbean and in British North America.[35] In France and Iberia, Irish merchants dominated the anglophone community - as in Cadiz (50 merchants in 1770), Bordeaux (70 merchants in the 1770s), Lisbon, Tenerife, Nantes, La Rochelle, Cognac, and St. Malo.[36] Ireland was therefore solidly embedded in the European Atlantic world (links also extended to Bruges, Rotterdam and inevitably, London), at least until the 1770s, after which its commerce became increasingly focused on

Britain. These expatriate Irish mercantile groups appeared wherever intensive shuttling of ships occurred, or where return cargoes were available. Essentially a tight mixture of faith, kinship and credit, these merchants were the ligaments of the economic system, bound by a common language, common financial mechanisms, a common set of service institutions, and common navigation laws (all radiating from London).[37] They were the key component in knitting together the Irish provisions industry, the West Indies economy, and the London sugar interest.

However, two crucial failings affected the mercantile Irish: (I) their inability to move beyond activity as commission agents, to become fully fledged independent merchants; (II) their reluctance (or inability) to establish large jointstock companies, or partnerships, which militated against their developing anything other than a perennially subordinate position in world trade.

## 5. The Linen Industry

### 5.1 Proto-Industrialisation

Beef and butter were two of the great Irish exports of the eighteenth century; linen was the third. Again, the growth of the linen industry in Ulster to one of the world's leading half dozen industries by the end of the eighteenth century is intimately related to the evolving world economic system. In the seventeenth century, Ireland had developed a nascent woollen industry; as with the Cattle Acts, regional interests in England (especially the Bristol and Exeter merchants) were able to force parliament in 1699 to prohibit the export of Irish woollens to Britain - the celebrated Woollen Act. Almost simultaneously duties on Irish linens entering the British market were abolished (in 1696). Free access to the British market was crucial in giving Irish linen a competitive advantage over its continental rivals.[38] After 1705, Irish linens were also allowed to be carried directly to the colonies (the only significant modification of the Navigation Acts until 1731). Just as beef and butter were the principal commodities going to the West Indies, so cheap, coarse linens became the principal Irish export to British North America in the eighteenth century. This export was similarly based around lenient and flexible credit terms, articulated through the London financial world. It was this credit flexibility which allowed for the massive export-led expansion of Irish linen production.

Linen production in Ireland in the late seventeenth century became possible for four reasons: (1) Ulster landlords were already aware by then that southern Irish agricultural production (sustained by better edaphic and climatic conditions) was superior. They were therefore keen to encourage their tenants to diversify into linen production, to keep rentals buoyant, especially in a period of population growth.[39] (2) Ireland (unlike Scotland) could grow its own flax and could therefore integrate its linen production in a more controlled fashion.(3) Landlords encouraged the rapid development of a marketing system, which allowed independent farmer/weavers to evolve. This system was well suited to Irish conditions, as opposed to a putting-out system. The consequent adaptability made Irish weavers more competitive than their urban-based Scottish competitors. (4) Irish linens were therefore cheaper to produce in a low labour-cost environment. Because they were sold via the Dublin/London nexus, they were marketed on flexible credit terms. Cheaper, custom-free Irish linens were therefore able to displace German and Dutch linen out of the English and then the American market.

Thus, linen production grew phenomenally: exports soared from one and half million yards in 1712, to eight million yards by 1740, to 17 million yards in 1760, to an astonishing 46 million yards by 1796. This expansion was sustained by technical innovation, via the application of chemicals and water power, especially in bleaching. Ulster pioneered the successful adaptation of the technology used to finish woollen cloth in water-powered tuck mills, so that it could be employed in bleach mills to wash and finish linen cloth; this enabled Ulster bleachers to process greater quantities of cloth more cheaply than anywhere else in Western Europe, but without spoiling the cloth.[40] Meanwhile, the overall development of the industry was carefully nurtured by the Linen Board established in Dublin in 1711. All these positive factors generated immense proto-industrialisation. As elsewhere, this was accommodated by massive subdivision of leaseholdings, intense population growth, and the incorporation of women and children into the workforce.[40] Thus, the Ulster countryside was festooned with a myriad of small weaver/farmer holdings. Like Pennsylvania, Ulster in the eighteenth century became "the best poorman's country".[41] In turn, the intensive proto-industrial area, especially in the famous linen triangle between Belfast, Dungannon and Newry created a prosperous aureole of agricultural supply (yarn, oats, potatoes, turf) in the surrounding region arcing from North Connaught to North Leinster.

Intensification of agricultural production via small, diligently laboured family farms, triggered massive demographic growth, in association with proto-industrialisation. By 1770, there were 42,000 weavers in Ulster: by 1820, when the proto-industrialisation phase was at its height, there was 70,000 weavers. The vast majority of these were never organized on a putting-out (Verlags-) system, but were independent farmer/weavers who sold their webs of cloth in the many provincial towns with brown linen markets, fostered by landlords eager to cash-in on the trade. The Ulster's industry good fortune was to develop precisely at the time when demand for cheap textiles was growing fast in Britain. By the 1720s, two-thirds of Irish linen exports were going to Britain and the industry was a classic export-driven one. After Britain, British North America was Ireland's second largest market for linens, absorbing especially coarse, cheap clothes.[42] By the 1770s, 20% of total exports (14.4 million yards) went to America. Basically, it was a trade carried on low prices and narrow margins, and therefore dependent on high volume marketing and low cost production. One side effect of its competitive edge was to depress any incipient textile industry in British North America in the eighteenth century, to such an extent that the Connecticut valley area even developed a specialist agriculture geared to supplying flaxseed to the Ulster market. By the end of the century, County Armagh had population densities in excess of 500 per square mile - perhaps the heaviest concentration of rural population anywhere in contemporary Europe. In 1821, Ulster's population of two million almost equalled the total population of Scotland.

## 6.  Conclusion

In summarizing Ireland's position in the world-system in the 1600-1800 period, one needs to make five key points. The first is the suddenness of its transition from utter backwardness to comparative modernity, without experiencing the long conditioning of other medieval societies. This great leap forward was predicated on its enhanced locational and strategic importance in the seventeenth century, especially the re-orientation of the North Atlantic world, and the growing significance of its nearest neighbour, England. However, this economic 'fast forward' was not achieved without great social and cultural trauma. Ireland was one of the only European countries to suffer the severe decline of its national language; there were inbuilt ethnic, sectarian and social fissures in the island; the country exhibited startling and frequently incompatible blends of archaism and precocious modernity.

Secondly, the period was marked by the sudden, but effective intrusion of the market economy. In the period of buoyant export-led growth, between 1740 and 1814, this intrusion was linked to integration with the British economy, and led to the intensification of livestock farming, the focusing of the export-trade more narrowly on the British market and upon the British Atlantic colonies, and the fostering of a dependence on England for manufactured goods.[43] The strength of the market economy can also be seen in the ubiquitous role of fairs, the equally pervasive role of the brown linen markets in the linen industry, and the extraordinarily busy traffic on the dense Irish road network. Arthur Young, for example, commented in the 1770s on the density of Irish road traffic, vis-a-vis its French counterpart.[44]

Thirdly, one must take cognizance of the ubiquity and strength of the landed estate system in Ireland, from the 1660s onwards. This heavily commercialized estate structure, regulated by a universal leasing system, was the key stimulus to the modernization of the system of agricultural production, and its consequent sensitivity to external demand. Therefore, with leasing acting as the pacemaker for the transformation of agricultural production, farming enterprises were price-responsive, receptive to external pressure, and therefore flexible in their choice of commodity production. The main external stimulus was provided by the Atlantic provisions trade, which became one of the principal engines of growth in the Irish economy in the eighteenth century. Thus, the estate system, functioning via a leaseholding, rent-paying tenantry, provided a market mechanism, while simultaneously orienting agricultural production to those commodities dictated by the emerging world-system.

Fourthly, all this change was constrained within a colonial relationship, mediated through London, which inhibited Ireland's independent development. This was manifested most acutely in the lack of financial infrastructure in the country - and the consequent dependence on London for bills of exchange, banking, insurance, credit, shipping and the development of large jointstock companies. Ireland's eighteenth century growth was always export-driven, and therefore vulnerable; its lack of freedom to manoeuvre as a small, peripheral island economy was exacerbated by the mercantile framework of the Navigation Acts, and the protectionism inherent in the 1667 Cattle Act and the 1699 Woollen Act. These ensured the rise of a port-based mercantile community, essentially passive in character, to act as the cardinal hinges between external demand and hinterland production.

Fifthly, given the complexity of its relationship with England, its large immigrant privileged minority, and its strategic significance, Ireland's political position was inevitably ambiguous and its political culture latently divisive. Ireland hovered uncertainly in constitutional terms, between being an independent kingdom (as defined in the Act of Kingship of 1541) or being a dependent colony, firmly clasped to the English imperial bosom.[45] Accentuating these uncertainties, there was the problem of accommodating (or overawing) a permanent disenfranchised Catholic majority. There was always a legitimacy question mark hanging like Damocle's sword over Ireland, a question which was frequently voiced at times of tension (as in the aftermath of the American and French revolutions). Therefore, the monopoly of political power exercised by the Anglican landlord class always rested on a queasy, and potentially brittle base, and was only guaranteed by the British connection; this allowed an historically rooted popular alienation from central government in Ireland, a dissent which was politicized and radicalized in the 1790s, subsequently erupted in the 1798 rebellion, and culminated in the Act of Union between the Irish and British parliaments in 1800.

England's mercantilistic model conceived of the empire's trading system as a giant wheel; ideally, trade moved up and down the spokes but was always focussed through the hub (England). As a particularly short spoke, there was little possibility of Ireland's trade moving around the wheel's rim, although Irish political rhetoric frequently indulged the illusion of Ireland's semi-independence. It is just this proximity to England, and its intriguing location vis-a-vis North America, that makes Ireland a striking, if complex, example of the changing role of national economies in the world economic system between 1600 and 1800.

## Notes and References

1.   Cullen, L.M.: An Economic History of Ireland since 1660 (London, 1972).
2.   Steele, J.: The English Atlantic 1675-1740 (Oxford, 1986). See also review by N. Canny "The British Atlantic World: towards a working definition". In: Historical Journal (forthcoming).
3.   Meinig, D.: The Shaping of America; a geographical perspective on 500 years of history. Vol 1. Atlantic America 1492-1800. (New Haven, 1986), p. 212.
4.   Cullen, L.M.: Scotland and Ireland 1660-1800; their role in the evolution of British Society. In: R. Houston and J. Whyte (ed.), Scottish Society 1500-1800 (Cambridge, 1989)
5.   Truxes, T.: Irish-American trade 1660-1783 (Cambridge, 1988).
6.   Andrews, J.H.: Land and People c 1780. In: T.W. Moody and W. Vaughan (ed.), A New History of Ireland. IV Eighteenth Century 1691-1800 (Oxford, 1986), pp.236-64.
7.   Duffy, P.: The Territorial Organization of Gaelic Landownership and its transformation in County Monaghan 1591-1640. In: Irish Geography XIV (1980), pp.1-26.
8.   Cullen, L.M.: The Emergence of Modern Ireland 1600-1900 (London, 1981).
9.   Mc Cracken, E.: The Irish Woods since Tudor Times (Newton Abbot, 1971).
10   Cullen, L.M.: Irish towns and villages (Dublin, 1979).
11.  Woodward, D.: The Anglo-Irish livestock trade of the seventeenth century. In: Irish Historical Studies (1973), pp.489-523.
12.  Wallerstein, J.: The Modern World-System II. Mercantilism and the Consolidation of the European World Economy 1600-1750 (New York, 1980), pp.265-6.
13.  Cited in Cullen, L.M.: Economic History, p. 19 (see note 1).
14.  Dickson, D.: The demographic implications of Dublin's growth 1650-1850. In: R. Lawton and R. Lee (ed.), Urban Population development in Western Europe from the late Eighteenth century to the early Twentieth century (Liverpool, 1989), pp.178-89. See also J. de Vries, European Urbanization 1500-1800 (London, 1984).
15.  Cullen, L.M.: The Growth of Dublin 1600-1900. In: F.H. Aalen and K. Whelan (ed.) Dublin City and County. From Prehistory to Present (Dublin, 1992), pp.252-77.
16.  Dickson, D.: The place of Dublin in the eighteenth century Irish economy. In: T. Devine and D. Dickson (ed.), Ireland and Scotland 1600-1850 (Edinburgh, 1983).
17.  Sombart, W.: Luxury and capitalism (Ann Arbor, 1967), esp. pp.21-38. See also Price, J.: Economic function and the growth of American port towns in the Eighteenth century. In: Perspectives in American History, VIII (1974).
18.  Whitelaw, J.: An essay on the population of Dublin (Dublin, 1805). For a specialist study of one of these trades, see M. Pollard, Dublin's trade in books 1550-1800 (Oxford, 1990).
19.  Smyth, W.J.: Society and settlement in seventeenth century Ireland: the evidence of the '1659 census'. In: Smyth, W.J. and Whelan, K. (ed.): Common Ground on the Historical Geography of Ireland (Cork 1988 ), pp.55-83. The map is on p. 57; T.W. Freeman, Pre-famine Ireland (Manchester, 1957).
20.  Menard, R. and Mc Cusker, J.: The economy of British America 1607-1740 (Chapel Hill, 1985).
21.  Truxes, op.cit. note 5, p.86.
22.  Dunn, R.S.: Sugar and slaves. The rise of the planter class in the English West Indies 1624-1713 (New York, 1972).
23.  Cited in Truxes, op.cit. note 5, p. 10.
24.  Mannion, J.: The maritime trade of Waterford in the eighteenth century. In: Smyth and Whelan (ed.), Common Ground, op. cit. note 19, pp.208-33.

25.    Menard, R. and Mc Cusker, J.: op.cit. note 20, passim.
26.    Mokyr, J.: Why Ireland starved: a quantitative and analytical history of the Irish economy
       1800-1845 (London, 1985).
27.    Dickson, D.: An economic history of the Cork region in the eighteenth century, unpublished
       PhD thesis, Triniy College Dublin, 1977. This is the best regional economic history to have
       appeared on Ireland.
28.    Power, T.: Land, politics and society in eighteenth century County Tipperary, unpublished
       PhD thesis, Trinity College Dublin, 1987.
29.    Whelan, K.: The famine and post-famine adjustment. In: Nolan, W. (ed.): The shaping of
       Ireland. The geographical perspective (Cork, 1986), pp.153-4.
30.    Whelan, K.: The regional impact of Irish Catholicism 1700-1850. In: Smyth, W.J. and
       Whelan, K. (ed.): Common Ground, pp.253-77.
31.    Devine, T.: Unrest and stability in rural Ireland and Scotland 1760-1840. In: Mitchison, R.
       and Roebuck, P. (ed.): Economy and society in Ireland and   Scotland 1500-1939
       (Edinburgh, 1989), pp.12-29.
32.    Dickson, D.: (1977), op. cit. note 27, appendix, table XXII.
33.    The quotation is from Bush, J.: Hiberna Curiosa (Dublin, 1765), p. 92.
34.    Mannion, J.: The Waterford Merchants and the Irish-Newfoundland Provisions Trade 1770.
       - 1820. In: Cullen, L. and Butel, P. (ed.), Négoce et industrie en France et en Irlande aux
       XVIII et XIX siècles (Paris, 1980), pp.27-43. See also Mannion, J.: A Transatlantic
       Merchant Fishery. In: Whelan, K. (ed.): Wexford: History and Society (Dublin, 1987),
       pp.373-421.
35.    Cullen, L.: The Irish Merchant Communities of Bordeaux, La Rochelle and Nantes in the
       eighteenth century. In: Cullen, L. and Butel P., op. cit. note 34, pp.51-63.
36.    Cullen, L.M.: The response of Ireland and Scotland to the navigation acts. In: Cullen, L.M.
       and Smout, T. (ed.): Comparative Aspects of Scottish and Irish Economic and Social
       History 1600-1900 (Edinburgh, 1977). See also Ravina G.: Burguesia extranjera y comercio
       Atlantico. La empresa comercial Irlandesa en Canarias 1703-1771. (Tenerife, 1985).
37.    For a geographer's model, see Vance, J.: The  merchant's world: The geography of
       wholesaling (Englewood Cliffs, 1970).
38.    Crawford, W.H.: The political economy of linen: Ulster in the eighteenth century. In:
       Brady, C., O'Dowd, M. and Walker, B. (ed.): Ulster, an illustrated history (London,
       1989), pp.134-57. See also Crawford, W.H.: Domestic industry: the experience of the Irish
       linen industry (Dublin, 1972) and Crawford, W.H.: The evolution of the linen trade in
       Ulster before industrialization. In: Irish Economic and Social History XV (1988), pp.32-53.
39.    Crawford, W.H.: Ulster landlords and the linen industry. In: Ward, J. and Wilson, R.
       (ed.): Land and industry. The Landed Estate and the Industrial Revolution (Newton Abbot,
       1971), pp.118-23.
40.    Kriedte, P.: Peasants, landlords and merchant capitalists. Europe and the world economy
       1500 - 1800 (Leamington Spa, 1983).
41.    Lemon, J.: The best poor man's country: a   geographical study of early south eastern
       Pennsylvania (Baltimore, 1972).
42.    Truxes, op.cit. note 5, Ch. 10.
43.    Cullen, L.M.: Economic development 1750-1800. In: Moody and Vaughan (ed.), op. cit.
       note 6, pp.159-95.
44.    Cullen, L.M.: Ireland and France 1600-1900. In: Cullen, L. and Furet, F.(ed.): Ireland et
       France XVIIe-XXe siècles (Paris, 1980), pp.9-20.
45.    Botigheimer, K.: Kingdom and colony: Ireland in the westward enterprise 1536-1660. In:
       Andrews, K., Canny, N. and Hair, P. (ed.): The Westward enterprise. English activities in
       Ireland, the Atlantic and America 1480-1650 (Detroit, 1979), pp.45-65.

# 14

# Beef Cattle Production in the European Periphery: A Case Study from the Island of Zealand, Denmark

Karl-Erik Frandsen

## 1. Introduction

The population increase and urbanization in Western Europe, in particular in the Rhineland, Northern France and the Netherlands from the beginning of the 16th century, created a growing demand for victuals. The increasing demand for grain could partly be covered by import from the Baltic region (Poland, East Prussia, Livonia) and to a smaller degree by production in Denmark and North Germany.

The increasing demand for meat, however, had to be covered by two major regions: Hungary and Denmark, both of whom had certain advantages on the markets. Hungary had the advantage that for climatic reasons the cattle could stay out all winter, and within a production system that can be described as ranching.[1] In Denmark, however, it was necessary to stable cattle during the winter, and the associated stall-feeding required a higher degree of investment. But Denmark had the advantage of being closer to the Northern European markets than Hungary, which was important even if the product was able to transport itself almost directly to the consumers.

## 2. Production and Export of Oxen in Denmark

The systematic production and export of oxen in Denmark on a larger scale started about 1480. At that time it was predominantly lean steers ("Græsöksne") which, after the last summer on grass in Denmark, were exported in the autumn for direct consumption in the great cities of Northern Germany (for example Hamburg, Lübeck and Bremen). The herds came mainly from Jutland and were driven down the old "Oxenroads" past the customs-houses at Ribe or Kolding, where export-duty had to be paid. In addition they had to declare for transit-toll through the duchies of Schleswig and Holstein at Gottorp and Rendsburg respectively, and continued to the great cattlemarket at Wedel on the Elbe river (Fig. 1).

Shortly after 1510 the character of the export changed fundamentally, and from 1521[2] the export of lean steers was strictly forbidden from the kingdom (but not from the duchy of Schleswig). The main product was now stalloxen ("Staldöksne") which after a winter of heavy stockfeeding in the stables reached the "Magermarkt" at Wedel, which started on 25th of March and continued for the next month. Even if the oxen were well fed, when they left the warm stable, they still lost much weight during the long march, where not much grass could be found so early in the spring, and were considerably skinnier when they arrived at the market. Therefore they had to spend a nice summer in the saltmarshes along the Frisian coast, before they could reach their final destination: Namely the salt barrels in the big West German cities and the Netherlands to be consumed during the winter and next spring.[3]

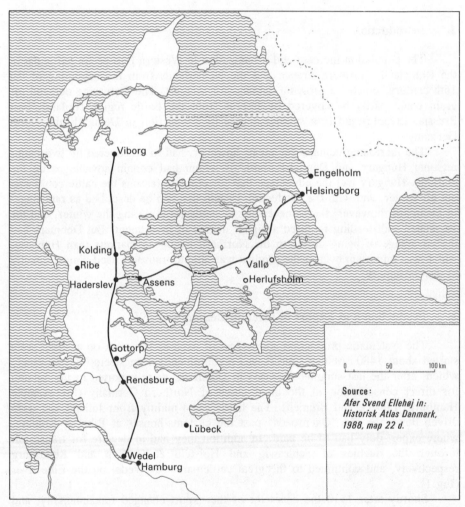

*Fig.1: Most important routes used for export of cattle*

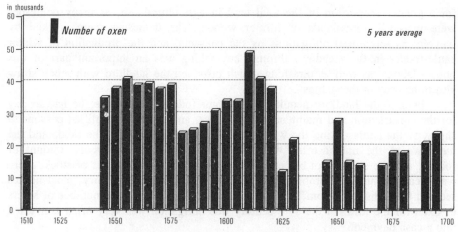

Fig.2: Oxen passing the customhouse at Gottorp

As can be seen from Fig. 2 this export of oxen increased rapidly during the first half of the 16th century and became very important for the economy and expansion of Danish rural society. By 1524[4] a royal decree had established that only noblemen and the king had the right to stall steers on fodder bought from other people. Peasants, townsmen and the clergy, by comparison, were only allowed to stall on their own produced fodder. This gave a *de facto* monopoly of larger scale production to the nobility and the Crown.

The production of oxen for the European market was organized in this way with a very clear and apparently rational division of labour between the estates in society. Nearly all the calves were born at the peasants' farms (of which 50 % belonged to the Crown after the reformation in 1536 and the rest to the nobility). The heifers were sold or bred to cows, while with a few exceptions all the bull calves were castrated. Post castration they were fed up as steers on the villages' grassland during the summer and in the peasants stables, on hay and straw, during the winter.

When the steer was 3 years old it could be used to draw a plough, and when it reached the age of 4 or 5 years, it was normally in October sold to a nobleman or the king's bailiff to be put into a large stable, where it was very well fed until next spring. At this point, now as a fully grown stalloxen, it began the long march to the soup tureens of the Netherlands. Normally the landlord would sell the stalloxen to a merchant, who hired a man to drive the herds to Wedel and sell them there at the market. At the customs-houses the man would show a paper containing documentation on where the oxen had been stalled. If this had been on a noble or royal manor, the oxen were exempt from duty, otherwise a customs duty had to be paid.

The Danish manorial system must be described as essentially a *Grundherrschaftssystem*, where the demesnes were small by European standards accounting for less than 9 % of the total arable area. However, they did account for a greater proportion of the total land in agricultural use, as many demesnes were placed in grasslands or former woods. The demesnes were worked by 'corvée' as seigniorial obligations by the peasants, and the production of winter cattle-fodder in the meadows through haymaking was an important part of this unpaid labour, as well as threshing which provided the landlord with much straw also to be used in the stables.

Even if the landlord acquired the winter fodder almost free, he had to pay for the construction and maintenance of stables, and for more qualified persons to supervise the cattle during the winter to prevent diseases. Since the fields and the meadows of the manor often were unable to supply the necessary amount of fodder, and the stables not big enough to hold the desired number of stock, there was great potential for problems. However, during the 16th century, most nobles secured the obligation from their tenants to stall or feed a fixed number of steers or oxen for the lord. In the 17th century this obligation was very often converted into a cash payment.[5]

The export of oxen reached a level of about 40,000 per annum in the years 1550-1575. As a consequence of the revolt in the Netherlands the demand rapidly diminished, and roads became insecure. Consequently, export over the next two decades was only 50 % of the previous years. For the many noblemen who had made great investments in stocks and new stables this was a heavy blow and forced many of the smaller landlords into the dangerous path of credit on the Kiel money market ("Kieler Umschlag"). This development certainly also effected the peasants, as the lords stopped buying the steers or paid lesser prices. In addition, they attempted to compensate for their export losses by strictly demanding the payment of manorial rents and increasing the claim for villeinage.

The situation gradually improved again, and in the years 1610-15 the export reached a new peak of about 50,000 animals per annum.

The involvement of Christian IV in the 30-Years War caused a rapid decline in export. When Jutland was occupied by Wallenstein's army in the years 1627-29, export stopped completely, and it never really recovered. It stayed at the lower level of about 20,000 heads, except in 1640 when it rose to 38,000, but again suffered heavily during the wars against Sweden (1644-45 and 1657-60).

The main research has been done on the production and trade with oxen from the Jutland peninsula, so the following section will concentrate on the development of production on two manors in Eastern Denmark: the estates of Vallö and Herlufsholm both on the island of Zealand.

## 3.    Production of Steers at the Manor of Vallö.

The manor of Vallö still has a very beautiful main building that originally was erected in 1586. It is situated on the eastern part of Zealand, near the small city of Köge, approximately 40 km south of Copenhagen. Its history goes back to the beginning of the 14th century. In 1554 the manor, as well as the demesnes and the dependent tenants, was divided into two parts, and the part under study here is the eastern part only: "Öster Vallö". From 1616 it belonged to the Marsvin family, represented by the powerful Ellen Marsvin, who in 1620 donated it to her daughter Kirstine Munk. In 1615 this beautiful young lady was married "by the left hand" to King Christian IV, and he gradually and from 1626 effectively incorporated Vallö into the Crown lands.

In 1630 the marriage was dissolved, and regardless that Kirsten Munk still held the title of property to Vallö, the king kept it in his possession on behalf of their common children. As such Vallö, together with the smaller Lellingegård, was administered by the governor of the fief of Tryggevælde. For this reason the very detailed annual accounts of the bailiffs at Vallö were sent for auditing at the Exchequer (*Rentekammeret*) in Copenhagen. It is in this archive that they still are kept.[6]

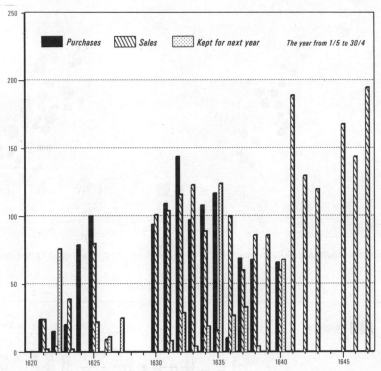

*Fig.3: Purchase and sale of steers and oxen at Vallø*

There is a marked difference between the 1660s and the next decade with regard to the management of the oxen production at Vallö. As can be seen from Fig. 3 the purchase of cattle as well as the sales were very irregular in the earlier period, the reason being that Vallö was used as an element in the royal management policy applied on all Crown estates on Zealand. The king was very keen to encourage the economic management of his domains not least with regard to the production of oxen,[7] and he very often ordered the transfer of steers from other manors to Vallö to be fed there. For example on 23 November 1621

Christian IV ordered 47 steers to be transferred from the castle of Frederiksborg to Vallö to be fed during the winter, and again in the next spring on 7 April the bailiff was told to receive 56 steers from the manor of Esrum to be kept on grass at Vallö during the summer. During the same year (1st of May 1621 to 30st of April 1622), 100 oxen were sent to merchants in Funen in return for grain, and without any payments to the bailiff. This, of course, makes it very difficult to calculate the profits from this production.

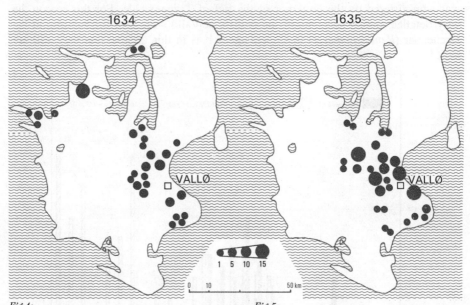

*Fig.4:*
*Steers bought for stalling at Vallø autumn 1634*

*Fig.5:*
*Steers bought for stalling at Vallø autumn 1635*

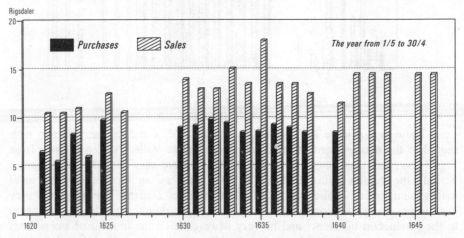

*Fig.6: Prices of steers and oxen at Vallø*

The stables at Vallö were not big enough to accomodate all the desired steers. In the winter 1624-25 16 young steers from the manor stock as well as 22 heifers were sent to the 'corvée' peasants to be stalled as a part of their seigneurial/manorial obligations. In most other years this obligation was expressed as a cash payment. Some years it seems to have been decided to keep rather many stall-fed oxen for an extra year. As Fig. 3 indicates, this may have been the case in 1622-23, when 76 stalloxen were kept for the next spring.

In normal years, however, the steers were bought from the peasants. Many steers were bought from the peasants belonging to the manor of Vallö and living in the vicinity, but some years the bailiff had to go far away to get the necessary stock. In the autumn of 1624 he bought 63 steers from peasants around Vallö, mostly a pair from each peasant, and 16 steers on the market of Slagelse, the point to which the peasants had brought them.

From then 22nd to the 25th August 1625 the bailiff went to Engelholm in Scania to buy 48 steers, and the king had to pay him 8 Rigsdaler (Rdl.) for travel costs including the transport of the steers over the Sound by Elsinore. The same autumn the bailiff bought 100 steers from the peasants in the eastern part of Zealand. Besides the normal cash payment, that year he also paid 2 to 6 carloads of firewood for each steer, which of course makes the calculations of prices (Fig. 6) rather difficult.

The unfortunate involvement of Christian IV in the 30-Years War and the occupation of Jutland by the Emperor's forces made the cattle trade impossible. In the spring of 1626, it was still possible to sell 80 oxen to a merchant from Copenhagen for 25 Rdl. the pair, but the councillor Frederik Reedtz appointed by the king to supervise Vallö in his absence, decided not to buy new steers in the autumn of 1626. Only 22 oxen from the previous year were kept in Vallö, and of those 9 were sold to butchers in Copenhagen.

The management of the manor had an alternative to the production namely dairy cattle. There was always a small stock of milch cows (and young animals) at Vallö, but in the spring of 1626 a stock of 125 milch cows is recorded, and this number remained steady during the next year. The manor could in this way use the stables, the grass and the hay to produce a substantial amount of milk and butter as well as many calves. In the spring of 1628 there were still 94 cows on Vallö, but during the next summer, everything seems to have gone wrong: Many of the cows were sold, some died and some were slaughtered and only 54 remained. In October 1630 all the cows, with the exception of 9 were transferred to Copenhagen, and the stables filled up with 94 newly bought steers and 48 young steers bred on the manor as well as 60 steer-calves. The steers born and reared at Vallö were called "opfödningsoxer" in contrast to the "köbeoxer", steers bought from the peasants, and the former category was sold at substantially higher prices than the latter. In 1634 the merchant Steffen Rode paid 23 Rdl. per head for "opfödningsoxer" but only 13 Rdl. for the normal stalloxen.

After this severe crisis for the cattle-trade during the war 1626-29, the new bailiff continued to buy and sell oxen very regularly. There seems for some years to have been a shortage of good steers for stalling around Vallö. As the map

(Fig. 4) indicates, in 1634 the bailiff in 1634 had to buy steers a long distance from Vallö: At Odsherred near the castle of Dragsholm in the north-western part of Zealand, approximately 60 km from Vallö, he bought 15 steers from the peasant Niels Jensen. The next year (Fig. 5) he could acquire all 117 steers in the vicinity of Vallö, and it was certainly important to acquire so many, because expecting great profits from the production of oxen, the king in 1635-36 had ordered the construction of two new big stables. The expenses to the craftsmen alone cost more than 2,000 Rdl., so it was important to get them filled up with steers. Unfortunately it was not possible to sell the oxen in the spring of 1636, so as shown by Fig. 4, 124 stalloxen had to be retained until the spring of 1637, when the Copenhagen merchant, Steffen Rhode, bought 100 oxen for the reasonable price of 27 Rdl. the pair. 13 oxen were slaughtered at the manor. Again in the spring of 1641, there seem to have been difficulties in selling the stall-fed cattle. 68 oxen had to stay at Vallö during the summer and next winter. The costs of having the stock for another year were not very large but the manor naturally lost revenue. The costs in wages for employed cattlemen in 1636-37 were only 18 Rdl. plus lodging and food. It is important to remember that there were hardly any winter-fodder production costs because it was cut by the peasants as villeinage.

The buyers of the oxen in the 1620s were various merchants from Copenhagen and Köge and often in small numbers. From 1631 onwards it was normally just one purchaser who took all the cattle. In 1631 and 1632 it was the Copenhagen merchants Niels Aagesen and Jacob Madsen respectively. From 1633 onwards the dominant business connection was the great merchant of Copenhagen, Steffen Rode, who had emigrated from Lübeck and had close family-relations with cattle-merchants in Flensburg. Thanks to Steffen Rode's famous account-book (also compiled by his widow, Marie and her sons after 1638) it is possible to follow his transactions during the years 1637 to 1650.[8] Normally Steffen Rode's firm bought all the oxen from Vallö as well as from most of the royal estates on Zealand. Only in 1639 did the king sell the oxen to Morten Michelsen, a merchant from Köge, who, in the same year, took a lease of the home farm of Lellingegård.[9]

Steffen Rode and his heirs were very much involved in the tottering state economy during the last years of Christian IV's reign. So, in 1645 the firm did not pay for the oxen, but instead the sum was debited against its account at the Exchequer, where its credit was large as the firm had made great loans to the state. In 1637, however, he did pay for the purchase at the Exchequer but only on 20th May, and since he must have received the oxen at the beginning of March, he got nearly three months credit.

1638 was the last good year for the production and trade in oxen. Vallö delivered 60 oxen to Steffen Rode at the price of 13.5 Rdl. per head. That year he bought 747 in Scania, 786 in Zealand and 520 in Funen and Jutland, altogether 2,053 oxen. He sold the oxen at the market in Wedel for 37,348 Rdl. (18.2 Rdl. per head), but as the costs of transport were rather high (Scania: 4.0 Rdl. per head, Zealand: 3.1, Jutland-Funen: 2.4), the firm only had a net surplus of 2,119 Rdl. (1.0 Rdl. pr. head).

1639 was worse, with a deficit of 2,662 Rdl. despite 216 oxen being shipped directly to Holland. Denmark and her merchants did play a more dominant role in the trade in oxen during the next years, which certainly was not to benefit the firm of Steffen Rode's heirs, the account of which showed a deficit every year.

The detailed accounts of Vallö stop in the spring of 1641. Thanks to the accounts of Steffen Rode it is possible to follow the development a few years more as shown in Fig. 3 and 6.

Another important source are the accounts of the customs officer of Assens on the west coast of Funen, where the toll was paid before shipping oxen over the Lillebælt to Haderslev in the duchy of Schleswig. From these records it is possible, for a few years, to trace the oxen from Vallö on their march from the periphery to the European core.[10]

On the 5th of April 1640 828 oxen belonging to Steffen Rode's heirs arrived at Assens driven by a man called Morten Pletz. 86 of the oxen were from Vallö as confirmed by a letter from the Governor of Copenhagen, Corfitz Ulfeldt, dated 17 March.

In 1643 Marie Rode exported 1,213 oxen of which 124 were from Vallö, but the war 1644-45 interrupted the trade completely. On 12th of April 1645 Marie Rode bought 160 oxen from the governor of Vallö, Sivert Urne, which had been stalled on the manor since the 25th of November 1644, at a price of 14.5 Rdl. per head. Due to the war she could not export them, but instead kept them on different manors until the spring of 1646. Her men arrived with a herd of 1,045 oxen at Assens from the 24th to 28th March, and in this herd were also 144 oxen from Vallö, stalled there between the 11th of November 1645 and the 14th of March 1646.

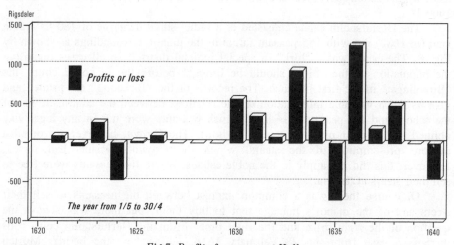

Fig.7: Profits from oxen at Vallø

The exporters' profit seems not to have been big, but at least the stalling at Vallö did bring some profit to the owner, as shown in Fig. 7. After the introduction of the regular and planned production of oxen in 1630, most years showed a clear profit, even if the costs of production are excluded from this calculation. However, as mentioned above, the costs of wages were negligible. More costly was the construction of stables and losses incurred by dead animals particularly when a cattle-disease ravaged as in 1635-36 when 12 oxen died. The years 1635 and 1640 must have been most awkward when it was not possible to sell the oxen, and, as a result, the income from the sale was lacking.

## 4.    Production of Steers at Herlufsholm

To examine whether those described structures were unique to Vallö, another manor from the same region will be briefly discussed.

The Herlufsholm estate is located near the city of Nästved on Zealand, 90 km from Copenhagen. At this place a monastery was founded in 1135, the Skovkloster, belonging to the order of the Benedictines. After the reformation in 1536 the monastery was secularized as a royal fief, but in 1560 by an exchange of real estate it was acquired by Admiral Herluf Trolle and his wife Birgitte Göye, together with 131 peasant farms. In 1565 this noble couple erected Herlufsholm as a foundation for a school for children of noble or honest parents, which today it still is. The Council of the Realm appointed a director of the school, and he was aided by a bailiff, who managed the agricultural production of the demesnes fields as well as the big estate of peasants farms. The director rendered an account of the estate's economy for revision by the Council of the Realm every year, which was made by the Exchequer. Accounts are preserved for most years between 1611 and 1660.[11]

The Herlufsholm estate consisted of a rather small demesne of 180 ha arable land (in 1682) but with 155 peasant farms in the manor surroundings as shown by Fig. 8. The estate was a "birk", a special judicial district, so all cases regarding the population of the "birk" should be brought before the manorial court, the "Birketinget" in the first instance. The records of the "Birketing" are printed and give us much valuable information on the relations between the administration of the school and the peasantry.[12] The Danish peasants were not in any legal way obliged to sell their products to the landlord. The Crown, however, enforced a dubious pre-emption of the steers, which the Crown's tenants had raised. However, this did not apply to the noble estates, where the peasants were free to sell their steers at the market.

Of course there was a common interest between the peasants to sell their steers and of the manor's management to buy them, but there was always the question of the right price and the right time. From the Herlufsholm court rolls there is a case from the 23 February 1618, telling how the bailiff, Morten Andersen, came to a peasant in the village of Appenæs, Mads Pedersen, to buy a pair of steers for the school. Mads Pedersen replied that he would not sell them before he had done his own ploughing and his corvée on the demesne. The bailiff

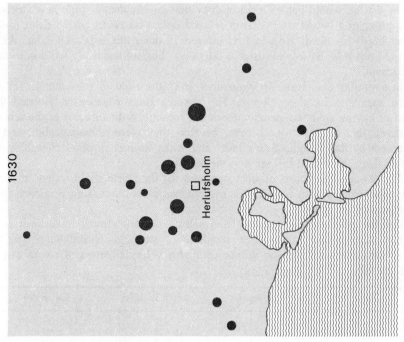

1630

Herlufsholm

*Fig.9:Steers bought for stalling at Herlufsholm autumn 1630*

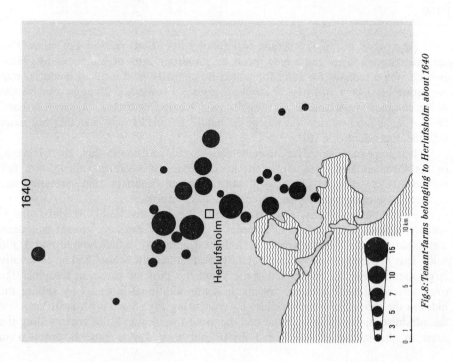

1640

Herlufsholm

*Fig.8:Tenant-farms belonging to Herlufsholm about 1640*

told him that he would get so much money that he could buy some new steers to use for ploughing but Mads Pedersen refused unless he could get 18 daler (=7.2 Rdl. per head) for them. He was first offered 13 daler but at last 18 daler. As he still would not sell, he was accused at the court, but unfortunately we don't know the judgment.

In a similar case from 26 November 1632 the right of pre-emption for the landlord seems to have developed. The peasant Hans Madsen in Holsted was accused of having sold two steers without the consent of the director of the school. He claimed that he had to sell them, because they were unmanageable, and as replacement he had bought a new pair, and when he had finished ploughing the land with them, he would sell them to the school.

Fig. 9 maps the points of steer purchase for the Herlufsholm estate in 1630, and by comparing it with Fig. 8, it is evident that most purchasing was done from the peasants of the school.

For the peasants the breeding and selling of steers must have been quite important. One of the most stable suppliers of steers to Herlufsholm was the peasant Hans Iversen from the village of Ladby. His deliveries of steers can be seen in the table below:

| Year | Number | Price in daler | Sum in Rdl. |
|------|--------|----------------|-------------|
| 1640 | 5 | 10 | 41.5 Rdl. |
| 1641 | 7 | 10 | 58.1 Rdl. |
| 1642 | 2 | 11.5 | 19.1 Rdl. |
| 1643 | 1 | 10.2 | 8.5 Rdl. |

As a barrel of rye at this time cost about 2 Rdl. Hans Iversen's incomes from the breeding of steers must have been an important part of his economy. Hans Iversen was a tenant of a farm for which he annually paid a token manorial rent consisting of: 1 barrel of oats, 2 lambs, 1 goose, 4 fowls and 20 eggs. For his use of the common wasteland he additionally paid 1 barrel of barley, ½ barrel of oats, and for some land outside the village he paid 3 barrels of oats, together not more than the equivalent of 6 Rdl.

It is important, however, to remember that in addition to this, Hans Iversen, or his farmhands, had to work on the demesne, as soon as called, with full equipment of plough or wagon with draught animals and without any compensation except the consumption of food and drink.

Fig.10 shows many similarities with Fig.3 from Vallö. Herlufsholm is unique because during a number of years many more steers were bought than oxen were sold: In total for the whole period the figures are 2,055 bought and 1,306 sold. This occurrence can be explained by the fact that the school had an especially large requirement for meat, and some years the director chose to slaughter the oxen instead of selling them, and made some additional business by selling the hides to the shoemakers in Roskilde. By comparing Fig. 3 and 10 it is striking that the administration of Herlufsholm had the same troubles as Vallö when selling the oxen in the spring of 1635 and 1640, but this was a general problem in

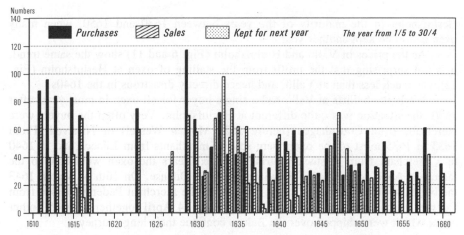

*Fig.10: Purchase and sale of steers and oxen at Herlufsholm*

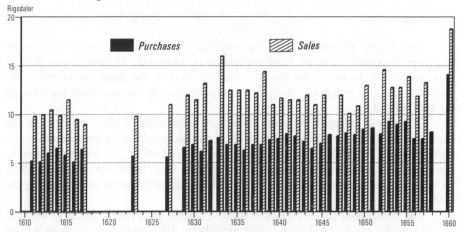

*Fig.11: Prices of steers and oxen at Herlufsholm*

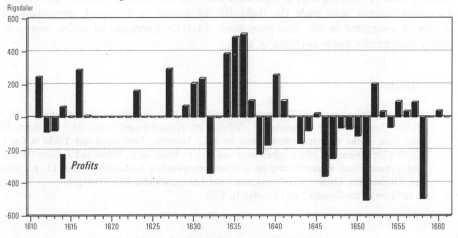

*Fig.12: Profits from oxen at Herlufsholm*

Herlufsholm in the majority of the years between 1630 and 1650, when many oxen had to be kept another year.

As the prices of Vallö and Herlufsholm (Fig. 6 and 11) show the same trend, it is not surprising that the profit from the stalling of oxen at Herlufsholm (Fig. 12) was much less than at Vallö, and hence directly disastrous in the 1640s.

While the bailiffs at Vallö had a single buyer for all oxen in most years after 1630, the situation was quite different at Herlufsholm. Very often the buyers were merchants from Copenhagen (before 1620 Morten Wessel) or Næstved. From 1633 to 1635 most of the oxen were sold to merchants from Lübeck and in 1640 to a merchant from Rostock. During the next years the oxen were sold to the merchant Anne, widow after Knud Lauritsen in Odense. On 16th of March 1643 the bailiff sold 40 oxen for 23 Rdl. the pair to the merchant Peder Pedersen from Odense. He arrived with the herd at Assens on 24 April together with 450 other oxen, which were shipped over the Belt to continue their long journey.

## 5.   Conclusion

The supply of meat to the core regions of Europe developed into a very important occupation for the rural population of Denmark. In many years it must have been very lucrative for the peasants, as well as the large landowners, the merchants and, in particular, the state, which derived its income from the trade in the form of tolls.

The whole business, however, was very vulnerable to slight changes in the European political, economic and military scene and could bring great losses to the participants in the trade. It is evident that, in times of crisis, the noblemen tried to compensate for the loss of income by enlarging their demesnes and demanding more labour from their peasants. The Crown reacted to the same fall by raising the taxes.

The creation of the European economic system with consumption in the core of meat and cereals produced in the periphery did certainly contribute to the development of both regions in a positive way, but the strong connections with the European market also made the daily life of people in Denmark dependant on events in countries in the European core, literally hundreds of miles away, of which they hardly knew anything at all.

## Notes

1.    The best publication on the European production and trade in cattle is: Westermann, E. (ed.) (1979): Internationaler Ochsenhandel. Beiträge zur Wirtschaftsgeschichte Bd.9, Stuttgart. In this book is the very important article by E.L. Petersen, Production and Trade in Oxen 1450-1750: Denmark. Much information can also be found in H. Wiese and J. Bölts (1966): Rinderhandel und Rinderhaltung im nordwesteuropäischen Küstengebiet vom 15. bis zum 19. Jahrhundert. Quellen und Forschungen zur Agrargeschichte XIV, Stuttgart.

2.    The "Law of the Country" og Christian II. § 93 .

3.  Enemark, P. (1971): Studier i toldregnskabsmateriale i begyndelsen af det 16. århundrede, I-II, Arhus.
    idem (1983): Oksehandelens historie ca 1300-1700. In "Sortbroget kvaeg".

4.  Decree of July 15th 1524.

5.  Frandsen, K.-E. in: Det danske landbrugs historie II 1536-1810. Odense, 1988, pp.23-27.

6.  Rigsarkivet. Godsregnskaber för 1660: Vallö og Lellingegard 1621-1640 (except the years 1628-29 and 1639).

7.  Frandsen, K.-E. (1988): Christians IV og bönderne. In: Ellehöj, S. (ed.): Christian VIs Verden, pp. 170-192. For a fully developed plan of the stalling of oxen on royal manors in the winter 1622/23 see royal orders of March 12th 1622 in: Laursen, L. (1922): Kancelliets Brevböger 1612-1623, pp.276.

8.  Det kongelige Bibliotek: Gl. kongl. Saml. 992 fol.: Steffen Rodes regnskabsbog. Olsen,A. (1919): Steffen Rodes Regnskabsbog over Studehandel 1637-1650, in: Historisk Tidsskrift 9. raekke bd. 1.

9.  Engqvist, H. H. in : Köge Bys Historie bd. 1. 1288-1859. 1985, pp.180-187.

10. Rigsarkivet: Registrant nr.108 B, 171 a-d: Fortegnelse pa de fra Assens udförte Öxne.

11. Rigsarkivet: Godsregnskaber för 1660. Herlufsholm.

12. Landbohistorisk Selskab (ed.) (1954, 1957, 1979): Herlufsholm Birks Tingbog 1616-1619 and 1630-1633.

# 15

## The Hungarian Economy within the Modern World-System

Vera Zimányi

### 1. Introduction

In Europe, not only the differing natural resources of the individual regions, but also differing levels of development, population density, urbanization, time of emergence and quality of the state organization, for example were already clearly marked in early medieval times. The existing variations in turn became specialized to an extent as never known before; these various regions increasingly established contact with each other, each exploiting and developing further its own advantages, improving its own specialization. Naturally they also influenced each other in the process. It was only in the course of the 16th century that this cooperation constituted what one could describe as a modern world-system. At the same time, the antecedents go back much further in history.

The flourishing Middle Ages witnessed the emergence of two economic centres in Europe: Northern and Central Italy on the one hand, and the region of the lower Rhine, the Low Countries, on the other. These two major centres were linked by a string of important German towns, the whole together constituting the axis of economic development in Europe as early as around 1300. The most densely populated and urbanized centres had already used up all their internal reserves and were therefore incapable of further internal expansion - thereby creating a relatively "overpopulated world" ("un monde plein"). This eventually led to a crisis in the 14th century. When the plague of 1347-48 hit these societies, which had starved and biologically weakened over two or three generations, it caused the death of one third, thereby putting an end to relative overpopulation. However, slow recovery and reorganization now took off from a higher level than at the dawn of the Middle Ages, also drawing on the resources of more distant regions to meet the requirements.

This was how Hungary, too, started to join mainstream economic development in Europe, primarily with the exploitation of its precious metal reserves. Hungarian gold and silver, and later copper, were important at an international level, prior to the arrival of precious metals from America. Also significant was the exporting of livestock, which began in the second half of the 14th century. Early cattle and small horse deliveries went to Northern Italy (primarily Venice), the southernmost point of the European economic axis.

## 2.  The Spatial and Temporal Evolution of the Hungarian Economy

### 2.1.  The Ranch-Type of Livestock Breeding on the *Pusztas*

The Hungarian economy was not alone in utilizing its relatively small population density in extensive livestock breeding over comparatively large areas. Cattle were especially suited for driving great distances to the major town centres. The constantly growing tradition of exporting livestock dated from medieval times in other countries also. The Hungarian Great Plain

> "belonged to the external zone of Northern and Eastern Europe specialized in livestock breeding. Certain coastal regions of North Germany, Denmark, Southern Sweden, Eastern Poland, Volhinia, Podolia, the Ukraine, Moldavia, the Danubian Romanian principalities and Hungary belonged to this belt, which drove livestock to the northern and southern German towns and to Northern Italy."[1]

But the Iberian Peninsula was also part of this network. Here, ranch-type livestock breeding had already evolved in the Middle Ages. Recent research has shown that the organizational forms of the ranch system, its technologies and terminology, as well as cattle suited to extensive keeping, were taken to the American colonies at a developed stage.[2]

Tamás Hofer gives the following definition of this system of breeding:

> "Ranch farming is livestock breeding based on grazing and has a commercial purpose. It is characterized by extensive utilization of land and human labour, and, accordingly, by large proportions: large herds and large grazing areas. It is independent of land cultivation, the livestock are not fed with grown fodder, their labour and manure does not serve land cultivation, and they are bred in numbers that far exceed the requirements of the local population. The livestock is raised exclusively on natural grassland. Climatic and vegetational conditions determine the possibility of ensuring grass for the full year, whether the cold and snowy winter, or perhaps the dry summer should be weathered by moving on to more distant pastures, or by reaping some grass for winter, as nomadic groups also do under certain circumstances."[3]

He adds, however, that whilst in the case of the nomadic groups the livestock form the principal basis of sustaining life, yielding the necessary food, wool and skin, under the ranch system the purpose of the breeding is the large-scale marketing of either the animal itself or one of its main products (skin, wool, etc.).

On the Iberian Peninsula and in Hungary alike, a special kind of large, hardy cattle evolved, which was suitable for breeding on ranches, and which at the same time differed from the livestock kept in the stables of the peasant farms. Only a small proportion of livestock exported from Hungary came from stock raised in peasant households (this was sold on the domestic market), or from the manors of the landowners. In contemporary Europe the Hungarian cattle regarded as being of the best quality was the longhorned, grey-and-white, heavy cattle bred on the Great Plain. This was referred to as a separate breed as early as the 15th century.[4]

The first bone finds of this corpulent animal date from the 14th century, and it was around this time that the *magni boves* appeared in the Esztergom customs list of 1365 - a definition which later gained general currency.[5]

It had earlier been presupposed that the migrating Magyars had brought with themselves from the steppes the taller, heavier *Primigenius* breed of cattle, which differed from the smaller *Brachyceros* type widespread in Europe. More recently, however, the general opinion is that it was the Cumans (Cumani) who brought it. They were a fragmentary nomadic people from the steppes and had settled in Hungary in the 13th century.[6] Having themselves become impoverished, however, the land and livestock allotted them gradually came into the possession of the affluent Hungarian peasant burghers ("*civis* farmers", a well-to-do stratum of peasantry, somewhat similar to the yeomanry) of the neighbouring market towns. The affluent *civis* deliberately opted for the large scale breeding of the new, corpulent breed of livestock. Indicative of this is their aim, from the second half of the 14th century, to annex as much pastureland to the outskirts of their town as possible. They also made use of the depopulation affecting some minor communities; during the 14th century the town of Debrecen added to its 2,000 hectare outlying area a further 46,000 hectares, largely comprising pastures.[7]

The breeding places of export cattle could be mapped on the basis of customs records which contained information about the place of origin of the livestock. According to this, the overwhelming majority of the livestock registered at the Vác customs office in 1563/64, was raised in the area between the rivers Danube and Tisza (where the Jazygians and Cumans had originally been settled), in the belt of confluence of the Tisza and its estuaries, and on the outskirts of Debrecen, a town that acquired a leading role at an early date.[8]

The number of exported livestock continuously increased, especially in the course of the 16th century. Indeed, this is not surprising; in the slowly recovering western world in the wake of the crisis of the 14th century, a similar marked population growth may be observed at the end of the 15th century, and particularly in the course of the 16th century. This, however, did not culminate in another crisis. Areas again relatively overpopulated on the one hand could procure additional food from the outside (the technique of shipment had to be improved for this), and, on the other, could more efficiently use their own resources. In some areas, primarily the country that was to become Holland, but also elsewhere, agricultural technology was improved and more intensive cultivation commenced. All of this was only possible, however, through the acquisition of additional products from more distant regions (grain, cattle), the production of which called for large areas and rather extensive methods. At any rate, demographical tension modified the given price structures, and the prices of basic foodstuffs of vital importance rose by a greater extent than those of manufactured articles. For this reason, the explanation that the price revolution of the 16th century was a simple consequence of the flooding of Europe by precious metals from America, is insufficient. Had this been the case, this would not have restructured the international price situation to the extent that it could be described as a price revolution.

In the meantime, the terms of international trade became so favourable for agricultural produce and so unfavourable for manufactured articles as to bring about an international boom in agriculture, which lasted from the mid16th century to the first half of the 17th century. The precise times varied slightly from region to region.[9] Thus, as early as at the end of the Middle Ages, the peripheries surrounding these areas started to respond to the food needs of an overcrowded Western World. It then becomes understandable that the zones (core, semiperiphery, periphery) became further specialized over the centuries and thus increasingly interrelated and interdependent more so than ever before.

The framework of production that had already evolved in the Middle Ages enabled Hungarian agriculture also to respond to the challenge of the great 16th century agricultural boom by substantially expanding its production and thereby becoming capable of stepping up its exports. This was indeed accomplished in an economically rather favourable 16th century, a century that had tragic political and military consequences for development in Hungary: In 1541 the Turks occupied the central third of the country, this being precisely the Great Plain, where livestock was bred.[10]

The significance of Hungarian livestock exports in the total foreign trade turnover of the 16th century is borne out by the customs records of foreign trade to the northwest for the year 1542. According to these, cattle accounted for 87.4% (oxen 85.1%) of exports to these areas. At his time wine exports accounted for a mere 2 % (calculating at customs value). Examining both exports and imports, one finds that cattle accounted for 60.78% of total trade turnover. This figure does not however include the livestock driven southwest, to Italy, primarily to Venice. It is only from 1574-76 that we have synchronous data recording customs revenues from exports to both the northwest and southwest.[12]

Carefully assessing other data from Hungarian and Turkish customs records, which unfortunately vary in time and location, László Makkai came to the conclusion that in 1587 beef exports to the west, northwest and southwest exceeded 120,000 animals.[13] In an exceptional year, the figure could be higher still. This was a huge quantity if one takes into consideration that at around 1580 the Danes exported approximately 45 000 cattles and the Poles some 50 000. The latter figure was largely the result of herds being brought into Poland from Moldavia, that is to say, it was largely a transit commodity.[14]

What method was used to produce this huge number of livestock in Hungary? Slight similarity with nomadic livestock keeping had already been pointed out. Some studies describe sales-oriented cattle breeding on the Great Plain as "monoculture". The same indeed holds for wine production in Hungary's famous wine regions. However, in this instance the term 'monoculture' is not being used in the sense in which it is applied in the colonies, where it implies the exclusivity of a certain product.

Differing in type from draught cattle used on peasant farms and from milch cows, livestock raised for slaughter cattle was not kept by the peasants or on the landowners manors. Neither were they grazed on pastures on the outskirts of the village. "A belt-like utilization system was established on the borders, to which the *pusztas* were added. The vineyards as well as the tiny plots of the vegetable gardens and hemp plots remained on the inside as did the grazing zone of the draught and milking animals which were driven out daily. The belt of arable land, also on the inside, was sometimes inadequate even for subsistence. Outside were to be found the 'summering places' of the extensively kept herds, their summer pastures and winter grounds, sometimes with hayfields."[15] This great abundance of land could only come about after the medieval village network had become so sparse in the 14th and 15th centuries that the inhabitants of the remaining settlements moved to the big market towns or into a few "giant" villages. All of this took place long before the appearance, in the 16th century, of the Turks.[16] The Turks' presence and the fleeing of the noble landlords as a consequence, only served to intensify this process in the 16th century.

The market towns strove to acquire large areas that had become *pusztas*. These, however, they did not divide up properly amongst their burghers; certain parts of the *pusztas* could only be rented. Who were the leaseholders, namely, who owned the herds? Naturally, they were the well-to-do peasants of the market towns, who ran their farms with a permanent staff of farmhands and periodically employed journeymen. The livestock was guarded by herdsmen who spent most of the year out in the *puszta*, initially just residing in crude huts. In winter they drove the cattle to more sheltered places, nonetheless keeping them in the open all the time. At most, they would build them a simple thatched shelter. Alongside the summer pastures, however, they cut hay, piling it up for winter. When in the course of time the herdsmen were joined by their families, they even started sowing some grain for their local needs. This was how homesteads emerged in the 18th century.[17]

Recent research in this area has also changed some of the earlier ideas on both "semi-nomadic" livestock keeping and on "monoculture type" animal husbandry. It soon became evident that there was largescale production and trade of fodder, precisely for the purpose of providing the large number of livestock with better winter supplies. The market towns did not rent out every part of the *pusztas* to individual big farmers; thus small producers were also able to reap hay on this land. In that way the town municipality profited from the possession of these pusztas.[18] The important discovery in these studies was, however, that in the 17th century this kind of largescale farming was not necessarily of the "monoculture" type; it also extended to grain production and grape growing. Extensive livestock breeding and substantial grain production were combined on the farm.

> "They did not produce a substantial quantity of grain just because their livestock breeding was dependent upon it, but because the scale of livestock raising made arable farming possible. The ploughing of the more fertile lands of the rented *pusztas*, as well as reaping hay, keeping wild and domesticated livestock and growing grapes also called for a permanent staff of farmhands, and, at the same time, a larger force of periodically employed wage-labourers or sharecroppers. It was thus inevitable to produce a relatively large quantity

of bread grain, but also spring grains, the straw of which, besides the hay, was used to feed the livestock. The sowing of spring grains provided both feed and fodder straw. However, livestock keeping nonetheless remained extensive, albeit in some areas and on some farms there were qualitative differences".[19]

In all probability extremely rich, "nabob" type peasant farmers were among those who lived on the livestock keeping Great Plain, some of whom could possibly have owned several hundred animals, rather than thousands as some scholars assume.

> "The number of livestock kept only by grazing on the vast *pusztas* may seem just as unlimited as the size of the *pusztas* themselves. If, however, we do not regard the extensive livestock keeping pursued on them as nomadic but as animal husbandry also producing hay and grain and with a definable farm organization, then the quantitative limitations of affluent peasant animal husbandry become apparent."[20]

The study of records dating from 1687-1690 from Nagykörös, a large market town on the Great Plain, leads one to a similar conclusion. In the best documented 660 households, one tenth, that is, 66 farms owned 40-80 % of the registered livestock, as well as at least 40% of the grain production. The majority of the recorded gardens were also in their hands. Only the ownership of vineyards reveals a more even distribution. However, half of the households did not own any kind of land, let alone vineyards.[21] The same study illustrates that farming by the inhabitants of the market town began to transform from the mid-17th century onwards. Presumably also due to marketing problems, arable farming and grape growing, as well as more intensive animal husbandry based on hay, gradually superseded extensive livestock breeding. The sales-oriented livestock breeding of the first half of the 17th century was gradually transformed and supplemented by more labour-intensive activities.[22]

## 2.2. The Cattle Trade to Central Europe and Venice

From the period of early transformation, it would now seem suitable to return to the most prosperous period of the large-scale livestock economy, and to review the organization of the cattle trade. Apart from the demand of the domestic market (which was not necessarily met from this cattle breed) the major buyers were the Austrian and South German towns, to some extent Moravia, as well as Venice and her hinterland. According to estimates, in the most prosperous period some 120,000 cattle were exported annually,[23] of which 19,000 went to Venice. At around the turn of the 1520s and 1530s, the town needed an annual 14,000 beeves, 13 000 calves and 70 000 small livestock, sheep, goats, pigs - according to Marino Sanudo.[24] Subsequently, this quantity only increased: The exact number of cattle purchased by Venetian butchers is known: 18-19,000 in the 1620s, 14-18,000 in the 1630s, 13-16,000 in the 1640s. Over 24 years, Venice purchased on average some 15,000 livestock annually.[25]

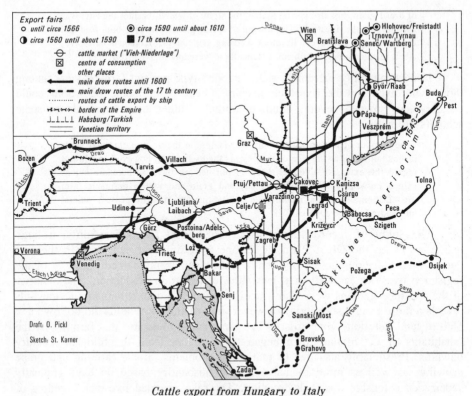

*Cattle export from Hungary to Italy*

Source: O. Pickl, Der Viehhandel von Ungarn nach Oberitalien vom 14. bis zum 17. Jahrhundert.
In: Internationaler Ochsenhandel (1350–1750). 7th International Economic History Congress,
Edinburgh 1978, ed. by E. Westermann, Stuttgart 1979, p. 47.

How was this enormous quantity collected? We have already seen that each
farmer, even the most well-to-do, never possessed more than a few hundred cattle,
of which he sold one or two dozen annually. At the cattle fair professional
Hungarian cattle merchants bought up a few hundred (seldom more than three
hundred) livestock. Their assistants drove these cattle to one of the cattle markets
on the country's borders, where either Italians purchased the livestock, or
Hungarian traders took them to Vienna or Moravia. There were, however, also
merchants with smaller capital, and they were the majority, those who drove only
several dozen livestock across annually, on a single occasion.[26] Most evidence
exists for the cattle driven towards Italy. To supply Venice, Italian merchants
settled in Laibach (Ljubljana) and Pettau (Ptuj), and handled the driving of
livestock to Görz (Gorizia) and from there on to Venice (see map). They became
enormously rich, with some merchants even buying themselves into the nobility.

Pettau even developed the Forced Marketing Right (Niederlagsrecht) for its own interests. Possessing greater capital, the Italian merchants gradually pushed out Austrian merchants, and in exchange for large cash advances to the Habsburg archduke, they practically monopolized cattle driving, which was suspended only briefly on several occasions, so great was the dependency of the archdukes on cash advances.[27]

The distance the livestock had to cover to reach Venice was at least 500 kilometers. When in 1566 Szigetvár, the great cattle fair, fell into Turkish hands, the major fair centres had to be moved further north. Győr and Pápa became the centres for buying cattle and this meant an additional 200 to 300 kilometers for the animals (see map). An organization was required for the transit of such large numbers of livestock, or for their temporary keep of certain duration. This is reflected in the settlement system of Győr: some of the town's citizens arranged gardens situated outside the town to store hay and straw, and these became important additions to the fairs. The cattle drivers and cattle merchants could keep their livestock here and also spend the night; they were not allowed to enter the towns. This was also where their business deals took place. From spring to late autumn there were always people in these compounds. Being outside the town precincts however, they were, of course, vulnerable to pillaging and attack by the Turks. In addition to this, however, some of the town's burghers began to rent more distant *pusztas*, where they grazed the cattle arriving from the Great Plain after their long journey and before marketing.[28] On average, herds were able to cover 20 kilometers daily, and therefore it required efficient organization to prevent any hitches in the course of a journey of several hundred kilometers.

A new situation arose in 1597, when Emperor Rudolph II prohibited members of Venice's Compagnia del Partito della Beccaria from buying at the northern Hungarian markets because this had caused meat shortages in Vienna. Possessing considerable capital, the Italian merchants did not, however, leave Venice empty-handed, but paid the prices that the Hungarians asked. The cattle-driving routes to Italy had to be changed, however (see map). One route passed from Turkish occupied Hungary through Turkish occupied Balkan land to Zadar (Zára), where the cattle were transferred to ships. Another route was from Turkish occupied territories, from which cattle were driven along the river Drava to Légrád and Csáktornya: the latter was the estate of the rich Zrinyi Counts. From here the cattle reached Buccari, a small port on the Adriatic Sea, across a whole chain of Zrinyi estates in Croatia. Buccari also belonged to the Zrinyi family. From here the livestock was shipped to Venice.[29]

Nevertheless, it often happened that the cattle were also driven on the mainland, that is to say across Austrian territory, accompanied by the armed escorts of the Zrinyi Counts. The Counts, however, referring to their rights as nobles, refused to pay the higher Austrian customs which were several times those demanded in Hungary. A veritable trade war was waged between the Counts and the Austrian authorities, between the Graz and Hungarian treasury.[30] Large landowners in Hungary benefited from the huge profits of cattle exports not by participating in production, but by becoming involved in the trading itself. In addition to the Zrinyis, it is also known that the Batthyánys were actively engaged in mediating (via their stewards and foreign correspondence), between the cattle drivers of the Great Plain and Italian and Austrian cattle merchants and butchers. They signed an agreement with the cattle drivers, organized into companies of three to five members, that in exchange for protection and organizing from the landowners, the latter would receive a share of the profit as one of the parties, in addition to one Forint (Hungarian currency) for each ox sold. However, should the journey fail to make a profit, the landowner would not be affected by the loss.[31]

Only rarely could the *apaldo*, the Italians' exclusive purchasing privilege, be suspended; it always had to be restored. Other Italian and Austrian merchants outside the *apaldo* eventually lost the habit of trading. It required a rather complicated organization, as well as the building up of the necessary contacts, and the provision of pastures and other requisites. Their organization had to begin well in advance of the actual driving process.[32] In the last quarter of the 17th century, sharp conflicts, a veritable trade war in fact, erupted between the Austrian treasury administration, aiming at a mercantilistic economic policy, and Hungarian nobles with a vested interest in foreign trade. Complicated further still by political conspiracy, this state of affairs led eventually to the execution of Peter Zrinyi, the confiscation of his estates, and the subsequent suppression of trade at the port of Buccari.[33] Incidentally, from the second quarter of the 17th century not only did Venetian meat consumption show a marked downward trend, but demand in Austria and the German towns also flagged. The great agricultural boom of the previous period came to an end, profit on cattle driving dropped, and many Hungarian cattle merchants went bankrupt.[34] Those, however, who survived the slump, tried to consolidate their situation in the second half of the 17th century by buying themselves into the nobility. Not only the vinegrowing, but also the cattle-trading market town elite strove toward this end. Of the 660 tax-payers registered in the records of 1690 there were 32 nobles. These were not, however, nobles of the older stock who had moved into the town, but had, almost without exception, paid for their nobility in the second half of the 17th century[35] and arose from the local wealthy farmers.

It is well worth considering how the cattle which in medieval times were bred rather extensively in a semi-nomadic way, changed into an increasingly differentiated product, a product made within the framework of a production apparatus. It was shipped, in an increasingly organized manner, by international commercial capital, and became a much sought-after commodity. It is also of interest to see how the semi-nomadic Cumans, who had imported the new breed of livestock, were first suppressed by the well-to-do peasants of the market towns who controlled livestock breeding, and how finally the market town elite in full control of agricultural activities and in possession of most of the wealth obtained nobility in the second half of the 17th century. The way of life of the small producer participating in the process also changed radically in the course of the 18th century, when the emergence of homesteads at the locations of the former primitive dwelling places in the peripheral pasture zone of extensively bred livestock commenced.

It should be noted at this point, that not only beef cattle, but also sheep were raised in large numbers on the Great Plain, mainly for Muslim consumers. However, sheep breeding did not develop such spectacular and special forms of the kind produced by cattle breeding.

## 3.    Viticulture and Wine Export

In the 16th-17th centuries, wine was the second-most important export commodity of Hungary, a country now in the mainstream of the modern world-system. Viticulture was a far more peaceful and less "exotic" occupation than cattle breeding. Already in Roman times there had been wine making in Hungary, a land made suitable for vine cultivation by its natural resources. In the early Middle Ages the best wine was made in the southern part of the country, in the region of the Szerémség (Sirmium), situated between the Drava and Száva rivers. As early as the 13th century, they shipped it to places several hundred kilometers away; this was an area where considerable urbanization, development into market towns, had already begun in the 14th century. After the Ottoman occupation of the land, viticulture in the Szerémség declined, though it did not cease entirely.[36]

In the 16th century, the leading role in wine making was taken over by the Tokaj Hegyalja hill country in north-eastern Hungary. The shift was brought about by important technical innovation.

> "The soil was well prepared by terracing and new methods of dressing were introduced, but it was mainly through the postponement of the vintage to the end of October that the grapes were able to dry and undergo the 'noble rot' on the vine. The addition of the essence prepared from these grapes to the average wine led to the famous *aszu* of Tokaj."[37]

Naturally, even without *aszu* making, Tokaj wine was of excellent quality. Specialist literature refers to a wine monoculture in the Tokaj-Hegyalja region since the first half of the 16th century, with some reservation interpreting the expression that in both the work of the producers and in the value of agricultural production vine cultivation became predominant.[38]

Here, too, viticulture initiated a powerful market town development: over a small area, twelve market towns close to each other accounted for approximately 90% of the yield. One third of the approximately 2,300 vineyards at the end of the 1570s was owned by absentee proprietors (*extranei*), that is, people living elsewhere, with the proportion of feudal landowners' allodial vineyards accounting for a meagre 3-5%.[39] Nonetheless, a substantial quantity of wine came into the landlords' possession, as in addition to their own modest allodial yield, they were entitled to collect "mountain tariff" (*ius montanum*, *Bergrecht*), or a wine tithe of vineyards on hills of the estates. Apart from this, one tenth of the wine produced went to the Church - this revenue had, however, been rented by the landowners from the Church since the 16th century at a very low fee.[40]

Within the country, too, wine had an almost unlimited market. For most large estates, revenue from the seigniorial tavern brought in the greatest cash income. The best quality wines were, however, exported, at about double the tavern price. The produce of the northern wine region was shipped primarily by Polish merchants to the tables of Polish nobles and burghers. In 1610-1611, Tokaj Hegyalja exported 40-50,000 hectoliters of wine annually, whilst in the years between 1637 and 1641 the figure fluctuated between 32,000 and 42,000 hectoliters. The export value of this was identical with the price of approximately 32,000-50,000 fattened oxen.[41]

In the 16th and 17th centuries Sopron and the vicinity were regarded as the country's second most famous wine region. Vine had been grown here since the 14th century. At around the year 1600, it peaked - with vine being planted in all suitable areas. In the broader region vine was cultivated on some 40,000 yokes (a yoke held 0,575 ha). Nearly 83% of these wines were exported to towns in Silesia, 4,7% to Bohemia, 4,4% to Moravia, 7,8% to Poland, and 0,3% to Saxonia.[42]

North of this area, in the vicinity of Pozsony (Pressburg, Bratislava), too, there emerged an important wine region. In 1581, 752 heads of households from amongst the town's burghers produced a total of 17,636 hectoliters of wine. The most wealthy 56 burghers harvested an average 60 hectoliters. In 1578, the three wealthiest families produced 120, 162 and 196 hectoliters of wine. But the burghers of the neighbouring market towns also boasted an average 22/40 hectoliters of wine in some years.[43] There was also a flourishing viticulture in Tolna, a market town of the Turkish occupied zone in Transdanubia: almost one third of the registered 941 families possessed a vineyard in 1557. The combined wine production of Tolna and Szekszárd and surrounding area is estimated at 25-30,000 hectoliters in 1565-66, but some data for 153-54 seem to suggest double this figure. All of this is indicative of long-distance wine trade extending far beyond the region. This is, in any case, also borne out by other data.[44]

Within the different branches of agriculture, the so-called "monoculture" of wine resulted in a large-scale division of labour. Thus, for example, the wine-growing region of Tokaj was supplied with bread-grain mostly by the neighbouring grain-growing villages along the rivers Hernád and Tisza. The specialization was so intense that in many of these villages, even in an average year, considerable grain was produced for sale. Just like the great wine-growers, the wealthy peasants producing grain for sale had their fields cultivated by hired labour. Similarly, we find more intensive grain production around the major cattle-raising market towns,[45] despite the fact that some cattle-breeders themselves sowed grain. Although the Hungarian economy joined mainstream international agriculture primarily via the exports of cattle and wine, intensifying specialization nonetheless stimulated the internal market as well, exerting a favourable impact on grain production even in minor districts. Grain was, however, seldom exported and even then only in small quantities (mainly in the Csallóköz, near to Austria, along the river Danube). This was because Hungary's rivers flow from west to east, and from north to south, thus away from the direction of potential grain markets and not to them. It was not profitable to transport grain by towed ships or carts over great distances. These natural conditions saved Hungary from the impoverishment brought onto the Polish and Baltic peasantry by large-scale grain exports. Cattle breeding and wine growing was not pursued either on the serfs' tenures or the manors of the landlords. In consequence, broad strata of the peasantry were able to participate in it, and under freer conditions. Moreover, they themselves benefited from the international agricultural boom, which, in the early stages of a modern world-system, was the result precisely of the functioning of this world-system.

In Hungary, however, this development pointing towards market-oriented production, had its limits. Those peasants who had benefited from the agrarian boom, tried to acquire a coat of arms, mostly from the middle of the 17th century onwards when the agrarian boom declined. In this respect, the vine-growing elite was more successful than the elite of the cattle breeding market towns.

## Notes

1.  Hofer, T.: Manuscript p.4; W. von Stromer: Wildwest in Europa. Der transkontinentale Ochsenhandel in der frühen Neuzeit. Kultur und Technik, Zeitschrift des Deutschen Museums München 1979. Heft 2, pp.36-43; H.Kellenbenz: Südosteuropa im Rahmen der europäischen Gesamtwirtschaft. In: Die wirtschaftlichen Auswirkungen der Türkenkriege. Hrsg. O.Pickl, Graz 1972.

2.  Hofer, T.: Manuscript p.2; C.J. Bishko: The peninsular background of Latin American cattle ranching. In: Hispanic American Historical Review 1952, pp.492-515; T.G. Jordan, An Iberian lowland/highland model for Latin American cattle ranching. In: Journal of Historical Geography 15, 1989, pp.111-125.

3.  Hofer, T.: Manuscript, p.3; Grigg, D.B.: The Agricultural Systems of the World. An Evolutionary Approach. Cambridge 1974, p.112.

4.  Makkai, L.: Der ungarische Viehhandel 1550-1650. In: Der Außenhandel Ostmitteleuropas 1450-1650. Hrsg. I.Bog, Wien 1971, pp.486-487.

5.  Makkai, L.: Der Weg der ungarischen Mastviehzucht vom Nomadismus zum Kapitalismus. In: J. Schneider (ed.): Wirtschaftskräfte und Wirtschaftswege, Bd.5. (Festschrift für H.Kellenbenz), Stuttgart 1978, p.62.

6.  Makkai, L.: op. cit. note 5, p.69; Bökönyi, S.: Die Haustiere in Ungarn im Mittelalter auf Grund der Knochenfunde. In: Viehzucht und Hirtenleben in Ostmitteleuropa. Budapest 1961, pp.83-111; Matolcsi, J.: Historische Erforschung der Körpergröße des Rindes auf Grund von ungarischem Knochenmaterial. In: Zeitschrift für Tierzucht und Züchtungsbiologie 1970, pp.89-137.

7.  Makkai, L.: op. cit. note 5, pp.68-69; Balogh, I.: Adatok az alföldi mezővárosok határhasználatához a XIV-XV. században (Some data concerning the landuse system of the market towns on the Great Plain). In: A Hajdu-Bihar-megyei levéltár évkönyve III. 1976. pp. 5-21.

8.  Makkai, L.: op. cit note 5, p.64.

9.  Zimányi, V.: Mouvement des prix hongrois et l'évolution européenne (XVIe-XVIIe siècles). Acta Historica Academiae Scientiarum Hungaricae 19, 1973. pp.305-332.

10.  Zimányi, V.: Economy and Society in Sixteenth and Seventeenth Century Hungary (1526-1650). Studia Historica Academiae Scientiarum Hungaricae 188. Budapest 1987.

11. Ember, G.: Magyarország külkereskedelme a XVI. század közepén (The foreign trade of Hungary in the mid-16th Century), Budapest 1988, pp.169-170.

12. Zimányi, V./Prickler, H.: Konjunktura és depresszió a XVI-XVII. századi Magyarországon az ártörténet és a harmincadbevételek tanuságai alapján; kitekintés a XVIII. századra (Agrarian boom and depression in Hungary during the XVIth-XVIIth centuries, as illustrated by the customs duties and the price history; some observations on the XVIIIth century). In: Agrártörténeti Szemle 1974, pp.1-2.

13. Makkai, L.: op. cit note 5, pp.65-66.

14. Wiese, H./Bölts, J.: Rinderhandel und Rinderhaltung im nordwesteuropäischen Küstengebiet vom 15. bis zum 19.Jh., Quellen und Forschungen zur Agrargeschichte XIV, Stuttgart 1966, p.271. Wyczanski, A.: Le revenu national en Pologne au XVI siècle. Annales E.S.C. 1971, p.105. See also the contribution of A. Dunin-Wasowiczowa in this volume.

15. Hofer, T.: Manuscript, p.13, with slight stylistic modifications.

16. Balog, I.: op. cit. note 7; Makkai, L.: Pest megye története 1848-ig (The history of the county Pest up to 1848). Budapest 1958, pp.82-85.

17. Wellmann, I.: A magyar mezögazdaság története a XVIII. században (The history of the Hungarian agriculture during the 18th century, Budapest 1979, p.40-41.

18. Buza, J.: A hódoltság gazdagparaszti állattartásának néhány kérdése. Szénaárak 1660-ban (Some problems of animal keeping of well-to-do peasants in the Turkish occupied territories. Hay prices in 1660). In: Ethnographia 1980, 2., pp.242-243. The same author: Die großbäuerliche Viehzucht auf der ungarischen Tiefebene im 17.Jh. In: Zeitschrift für Agrargeschichte und Agrarsoziologie 32, 1984, pp.165-209.

19. Buza, J.: 1980, p.244; Orosz, I.: Mezögazdasági termelés és agrártársadalom. In: Hajdunánás története (Agricultural production and agrarian society. In: The history of Hadjunánás). Red. Orosz, I., Hajdunánás 1973, p.102.

20. Ibidem, p.247 (translated by the present author).

21. Sz. Sin, A.: Gazdaság és társadalom a 17. század végi Nagykörösön (Economy and society in Nagykörös at the end of the 17. century). Egyetemi doktori disszertáció, Budapest 1986, pp.39-41.

22. Ibidem, p.97.

23. Makkai, L.: op. cit. note 5, p.66.

24. Marino Sanudo, Diarii, Tomus 50, p.65.

25. Zimányi, V.: Velence szarvasmarhaimportja az 1624-1647-es években (Importation of beef in Venice between 1624-1647) In: Agrártörténeti Szemle 1972, 3-4. pp.389-397.

26. Szakály, F.: A Dél-Dunántul kereskedelmi utvonalai a XVI. század derekán (Trade routes of southern Transdanubia in the middle of the 16th century). In: Somogy megye multjából. Levéltári évkönyv 4. Red. by Kanyar, J. Kaposvár 1973, pp.81-82.

27. Pickl, O.: Der Viehhandel von Ungarn nach Oberitalien vom 14. bis zum 17.Jh. In: E. Westermann (ed.): Internationaler Ochsenhandel (1350-1750). 7th International Economic History Congress, Edinburgh 1978. Beiträge zur Wirtschaftsgeschichte 9, 1979,pp.45, 49, 53.

28. Gecsényi, L.: Györ határa a XVI-XVII. században. Kertek a város körül (The borders of Györ in the 16th-17th centuries. Stock-yards around the town) In: A Dunántul településtörténete VII. Red. by B. Somfai, Veszprém 1989, pp.165-166, 168

29. Pickl, O.: op. cit. note 27, pp.60-61.

30. Zimányi, V.: Documenta Zrinyiana, Pars Oeconomica, Tomus I. Budapest 1991.

31. Zimányi, V.: Adatok a Batthyányak XVII. századi marhakereskedésének történetéhez (Documents concerning the cattle-trading activity of the Counts Batthyány during the 17th century). In: Agrártörténeti Szemle 1961, 1, pp.60-84.

32. Zimányi, V.: Rauch Dániel nedelicei föharmincados jelentései és levelei (Reports and letters of Daniel Rauch, customs officer of Nedelice) In: Agrártörténeti Szemle 1979, 1-2, pp.229-230.
33. Zimányi, V.: La politique adriatique des Habsbourgs au 17e siècle. In: Rapporti Genova-Mediterraneo-Atlantico nell'e stá moderna. Atti del IVo Congresso Internazionale di Studi Storici, Genova, 4-7 dicembre 1989 a cura di Raffaele Belvederi. Genova 1990, pp.197.
34. Takáts, S.: A magyar tözsérek és kereskedök pusztulása. In: Szegény magyarok (The decline of the Hungarian cattle-traders and merchants. In: Poor Hungarians). Budapest, s.d.
35. Sz. Sin, A.: op. cit. note 21, p.56.
36. Szakály, F.: A Közép-Duna menti bortermelés fénykora (The great age of vine growing of the Middle-Danubean region). In: Dunatáj 1979, 2, pp.13-15; Buza, J.: Beitrag zur Geschichte des Weinhandels von Syrmien im XVIII. Jahrhundert. In: Acta Universitatis Debreceniensis de Ludovico Kossuth nominatae, Series Historica XLI. Red. by I. Orosz, Debrecen 1988, pp.145-154.
37. Makkai, L.: Agrarian landscapes of historical Hungary in feudal times. Studia Historica Adacemiae Scientiarum Hungaricae 140, Budapest 1980, p.13; Balassa, I.: A szölömüvelés és borkezelés változása a XVI-XVII. században Tokajhegyalyán (Changes in viticulture and treatment of wine in the 16th-17th centuries at Tokajhegyalja). In: Agrártörténeti Szemle 1973.
38. Kiss, I.N.: Szölö-monokultura a Hegyalján, XVI-XVIII. század (Vine-growing monoculture in Hegyalja, 16th-18th centuries). In: Agrártörténeti Szemle 1973. 3-4, p.383.
39. Ibidem, p.383.
40. Zimányi, V.: A rohonc-szalanaki uradalom és jobbágysága a XVI-XVII. században (The domain Rohonc-Szalonak and its peasantry during the 16th-17th centuries) Budapest 1968, p.39.
41. Kiss, I.N.: op. cit. note 38, p.85.
42. Prickler, H.: Zur Geschichte des burgenländisch-westungarischen Weinhandels in die Oberländer Böhmen, Mähren, Schlesien und Polen. In: Zeitschrift für Ostforschung 14, p.300., pp.514-515.
43. Zimányi, V.: op. cit. note 10, p. 41.
44. Szakály, F.: op. cit. note 36, pp.20-21.
45. Káldy-Nagy, G.: Harács-szedök és ráják. Török világ a 16. századi Magyarországon (Requisitioners and Rajas. Turkish World in Sixteenth Century Hungary), Budapest, 1970. passim.

# 16

# Wine in the Early Modern World-System: Profit, Production and Exchange

## Tim Unwin

"Up, and having set my neighbour Mr. Hudson, wine cooper, at work
drawing out a tierce of wine for the sending of some of it to my wife - I
abroad, only taking notice to what a condition it hath pleased God to bring
me, that at this time I have two tierces of claret - two quarter-cask of canary,
and a smaller vessel of sack - a vessel of tent, another of Malaga, and another
of white wine, all in my wine cellar together - which I believe none of my
friends...now alive ever had of his own at one time" (The Diary of Samuel
Pepys, 7th July 1665; Pepys, 1972, p.151).

In the 17th century entirely new wines began to appear on the market in
Europe. These were the product both of the cultivation of vines in lands newly
conquered by Europeans, and also of the investment of capital in the production of
new types of wine in traditional areas of European viticulture. For those who
could afford them, these developments offered the opportunity to establish wine
cellars of previously undreamt of diversity. Samuel Pepys, writing in the extract
from his diary for 1665 noted above, was thus able to claim that the six different
types of wine that he had collected together in his cellar represented a unique
achievement among his friends. However, this claim was also representative of a
much wider change in the nature of the world in which Pepys lived; a change
characterized by the emergence of an increasingly dominant and all-embracing set
of economic imperatives.

Viticulture and the wine trade are of particularly interest in any analysis of
the development of this early modern world-system for four main reasons. First,
the wine trade had been of central importance to the medieval economy of Europe.
Thus the demand for wine by the urban elite of northern Europe provided a ready
outlet for supplies produced in the more southerly vineyards of France, Germany
and the Mediterranean world (Dion, 1959; Lachiver, 1988). However, secondly,
it is of interest because, in addition to its economic importance, wine had clear
political, ideological and social significance. It therefore provides insights into the
whole restructuring of European society and the moves towards global integration
that took place in the 16th and 17th centuries (Unwin, 1991). Thirdly, the
environmental adaptability of the vine meant that it could successfully be
cultivated in most of the regions over which Europeans imposed their political
authority during this period, and fourthly the 17th century also saw the
introduction of a range of new types of wine production in the traditional
European heartland of viticulture. Viticulture and wine therefore provide a useful
focus for an examination of key elements of Wallerstein's (1974, 1979, 1980,

1983, 1989) conceptualization of the emergence of a modern world-system, and in particular of labour relations, production systems and exchange mechanisms.

## 1. Wallerstein: Elements of a Critique

Wallerstein's comprehensive analysis of the emergence of a modern world-system has been both highly influential (Butlin, 1987), and also subject to a number of substantial criticisms (Brenner, 1977, 1982; Laclau, 1979; Wolf, 1982; Corbridge, 1986; Dodgshon, 1987). It is not the purpose of this paper directly to develop a further theoretical critique of Wallerstein's corpus of work, but rather to examine how a consideration of his ideas can assist in developing an understanding of the structural changes that took place in the wine industry, focusing in particular on the 16th and 17th centuries. Three main problems with Wallerstein's approach, however, require some initial exposition: its Eurocentric focus, its insistence on the existence of several different types of capitalism, and its association of particular kinds of labour control with specific types of production.

Much of Wallerstein's thesis is based around the central assertion that "It was in the sixteenth century that there came to be a European world-economy based upon the capitalist mode of production" (Wallerstein, 1974, p.67). In developing his argument he therefore concentrates heavily on alterations taking place within Europe itself, and he pays only limited attention to the structural changes occurring in other societies that were eventually to lead to their domination by an economic system originating in Europe. In the 15th century most of Europe was a side-show on the global stage, and the centres of cultural, economic and technological efflorescence lay to the east, in what are now China, India and the western Asian Ottoman empire. Any understanding of the way in which Europe came to conquer and dominate these other cultural spheres must therefore examine in detail their internal contradictions and constraints, and the precise mechanisms at the interface of conflict: war, repression, and legitimation.

Wallerstein's European world-economy was forged from the reorganization of the earlier separate economic systems focused on Flanders and northern Italy. He claims that three processes were essential for the establishment of the new world-system: an expansion of the geographical size of the world in question; the development of variegated methods of labour control for differing products in different zones of the world economy; and the creation of relatively strong state machineries in the emerging core states of the world. However, in developing his arguments, he suggests that a number of different types of capitalism existed within Europe during the 16th century, each with its own particular form of labour organization. While Wallerstein (1974, p. 77) in general accepts Marx's overall view that capitalism developed as "the only mode" of production "in the sense that, once established, other 'modes of production' survived in function of how they fitted into a politico-social framework deriving from capitalism", he nevertheless also accepts Braudel's argument that at least during this period there were in effect several different types of European capitalism. Wallerstein (1974, p.77) thus goes on to claim that "it is precisely this existence of several capitalisms which gave importance to the increased stock of bullion, for the velocity of its circulation was precisely less in the beginning in northwest Europe than in Mediterranean Europe". While it is widely accepted that industrial capitalism can be seen as having developed out of an earlier stage of merchant capitalism, it is by no means clear that Wallerstein's different capitalisms in the 16th century had sufficient in common to enable them to be grouped together under the single general conceptual heading of 'capitalism'. In seeking to clarify this problem, it is evident that much depends on the definitions of capitalism that are chosen, and it may well be that the time has come for historical geographers to move beyond the use of general over-defined terms such as 'feudalism' and 'capitalism', and instead begin to focus much more precisely on the dynamics of particular attributes of systems of surplus extraction existing at any given place and time. The mere existence of profit and a desire for the accumulation of money are not by themselves satisfactory defining characteristics of capitalism.

A third difficulty with Wallerstein's views is his argument that the various different capitalisms emerged because "each mode of labor control is best suited for particular types of production", and the division of labour in Europe assured "the kind of flow of surplus which enabled the capitalist system to come into existence" (Wallerstein, 1974, p.87). Wallerstein thus argues that slaves were used on large-scale enterprises such as those required for sugar cultivation, and that "a slightly less onerous form of labour control" was required for grain production, cattle-raising and mining (Wallerstein, 1974, p.90). The evidence he adduces for this assertion is only partial, and one purpose of this paper is therefore to examine the systems of labour used for viticulture and wine production in different parts of the world during the 16th and 17th centuries in order to assess Wallerstein's assertions concerning the links between types of production and systems of labour control.

In examining the emergence of capitalism, Braudel (1982) has usefully drawn attention to the importance of credit and the use of profit in the 16th and 17th centuries. For Braudel (1982), much of the gross capital of the later medieval period melted away, and was never invested in the expectation of increased profits. Furthermore, he argues that rather than being central to the production process, "Capital was most at home in the sphere of circulation" (Braudel, 1982, p.374). It was thus in the new banking systems and credit arrangements that capitalism found its earliest true home. Braudel (1982) identifies three key phases in the development of credit: the late-13th and early-14th century banking arrangements of the Florentine families such as the Lucchese, Frescobaldi, Bardi and Peruzzi; the increased use of paper bills of exchange by the Genoese in the 16th century; and the unprecedented use of paper credit in 18th century Amsterdam. The importance of these credit and banking arrangements was that they underlay and facilitated changes in the organization of trade. Quite simply, long-distance trade could not operate successfully without the presence of a complex network of credit arrangements. Moreover, as Braudel (1982, p.400-401) argues, "the great merchants, although few in number, had acquired the keys to long-distance trade, the strategic position par excellence;...they had the inestimable advantage of a good communications network at a time when news travelled very slowly and at great cost;...they normally benefitted from the acquiescence of state and society, and were thus able, quite naturally and without any qualms, to bend the rules of the market economy".

It was only when this system began to break down in the 18th century that attention turned to the use of capital in the processes of production.

## 2.    Wine in the "Age of Discovery"

The Portuguese and Spanish voyages of discovery in the 15th and 16th centuries opened the door to the cultivation of European varieties of *Vitis vinifera* throughout the world. Although there is evidence of viticulture and wine production in China and many other parts of Asia well before this period (Unwin, 1991), there is no indication that vines were cultivated and wine made from their grapes in America or the southern hemisphere prior to the European conquest and settlement of these lands.

The explosion of energy expressed in the Iberian voyages of the 15th century can be explained by consideration of a wide range of economic, social, political and ideological factors (Bell, 1974). In economic terms, the Spanish and Portuguese were eager to gain control over the trade routes to the east from whence came luxury goods such as spices and silks, and they were also keen to acquire the resources of African gold and American silver. In political terms, the voyages provided an opportunity for the kings of both countries to pacify a restive nobility, and socially they detracted attention from increasingly difficult internal problems at home. Furthermore, as Marques (1972, p.140) has suggested, while these explanations provide a reasonable basis for action, "they omit the colorful wrapper that each man requires to rationalize his own actions and to convince others of a noble and idealistic undertaking: the fight against the infidel and the salvation of souls". These ideological factors, stemming in part from the history of Iberian reconqest, are central to any understanding of the voyages of discovery.

Moreover, it is religion and ideology that have most often been used to explain the widespread subsequent expansion of viticulture in Latin America. Hyams (1987, p.255) thus argues that "One of the problems which faced the *conquistadores* in their conquests and colonizations in America was that of providing a supply of wine for the Mass. It should not be thought of as a minor problem; to the Spaniard of the sixteenth century it was of the very first importance". However, by that date it is evident that the amount of wine consumed in the Mass or Eucharist formed a negligible part of total wine production in the world. Indeed, the practice of the Catholic Church in the 16th century generally prohibited the partaking of wine by the laity at Mass; it was only the monks and priests who celebrated in both kinds (Unwin, 1991).

## 2.1. Viticulture and Wine Production in Latin America

Prior to the Spanish conquests of the Aztecs in Mexico (1519-21) and of the Incas to the south (1531-34) there is no clear evidence of vines being cultivated for wine in the Americas. However, vines are indigenous to Mexico and north America, and sources as diverse as the Vinland Sagas and the writings of Spanish missionaries confirm the identification of vines there by the earliest European visitors. The main traditional alcoholic beverage of central America was *pulque*, made from the agave, but other drinks such as *tesgüino*, made from sprouted maize kernals, and *balché*, a honey mead were also common (Bruman, 1967), and to the south, in Peru, *chicha*, produced from fermented maize, was the most widely available form of alcohol. However, the reasons why the Aztecs and Incas appear not to have made wine from grapes are unknown. It may have been because it was easier to make alcohol out of other plants, but it seems also that it reflects a different cultural and ideological symbolism between Eurasia and the Americas (Unwin, 1991).

In his "Historia de los Indios de Nueva España", written in 1536, Motolinía (1950, p.217) noted that

> "In many places in the mountains there are big wild grapevines, and no one knows who planted them. They grow out very long shoots and bear many clusters and produce grapes which are eaten green. Some Spaniards make vinegar out of them, and some have made wine, but only in small quantities".

Such wine apparently did not suit Spanish tastes, and consequently considerable effort was expended on the importation and planting of Spanish rootstocks. Cortés in particular appears to have placed specific emphasis on vine cultivation. Thus in the municipal ordinances of the new Mexico City dating from 1524, it is recorded that proprietors of new estates had to plant vines of the best quality. The main vine type imported seems to have been the Criolla, or Mission vine. Unfortunately, many of these vines suffered from frost damage, and it was only gradually that the best sites for viticulture were identified. Nevertheless, viticulture and wine production spread extremely rapidly in the wake of the Spanish conquests. It is likely that vines were introduced into Peru in the 1530s soon after Pizzaro's conquest of the Incas, and by the mid-1550s they were also to be found cultivated in Chile and Argentina.

Most vineyards at this time appear to have been found on secular estates, and production soon became based primarily on slave labour. At first, attempts were made to cultivate vineyards based on the *encomienda* system which involved the allocation of both land and Indian labour to the Spanish *encomenderos*. However, while this system was reasonably successful for the expropriation of indigenous Indian products, it proved much less suitable for the cultivation of imported Spanish crops. Furthermore, by the middle of the 16th century, growing royal opposition to the *encomienda* system, and the dramatic indigenous population decline, led to its replacement by *haciendas*. Initially, the labour for these was again usually provided by Indian slaves, but as Cushner (1980) reports for Peru, legislation was passed in 1601 which prohibited Indians from working in vineyards unless they were paid a daily wage. The outcome was therefore that the owners of vineyards increasingly turned to slaves from Africa, in order to provide the necessary labour for their cultivation. This frequently involved considerable expense, and as Cushner (1980, p.81) once again notes, "Within the context of the Jesuit *haciendas* of coastal Peru, black slave labor represented a considerable capital investment, frequently over fifty percent of total variable costs".

In explaining this expansion of American viticulture, it is essential to look outside the purely religious motives that have usually been alluded to. In economic terms wine was bulky and thus expensive to import from Iberia. Moreover, the quality of such wine was usually low, having suffered from the long sea voyage across the Atlantic in high temperatures. Of most importance, though, was probably the social prestige and cultural significance of wine to the *conquistadores*. The creators of *Nueva Espana* were eager to create a veritable new Spain across the Atlantic, and in this "New World" they sought to introduce as many European crops as possible. Wine consumption played a central part in the lifestyles of the Spanish nobility, who did not wish to go without it in their new home. As Motolinía (1950, p.148) noted, guests at a wedding of eight Spaniards in Telzoco in 1526 brought the couples gifts of "good jewels and also much wine - the jewel which gave most joy to all of them".

By the end of the 16th century this expansion of wine production in the Americas was beginning to have severe repercussions for those in Iberia who had previously produced and exported wine to the new colonies. In a remarkable parallel to Domitian's edict of A.D. 92, Philip II therefore promulgated a decree in 1595 forbidding further planting of vines in Spain's American possessions. Despite its attempted enforcement by subsequent viceroys, this legislation was only of limited success. Its main influence seems to have been to lead to a shift in wine production from secular to monastic estates, since religious orders were exempt from the decree. Once again, though, it is probable that the expansion of monastic vineyards was not simply a result of the need for wine for religious observances, but rather because of the recognition that wine production for secular consumption was an easy way of making a financial profit.

In summary, therefore, while the development of viticulture in Latin America did indeed become essentially based on slave labour, according well with Wallerstein's arguments concerning peripheral areas of his modern world-system, such wine production was destined primarily for consumption in New Spain, and there is negligible evidence of wine from Latin America being exported back across the Atlantic to Europe. The vineyards of the New World did not serve as cheap production sites for wines consumed in the European core.

## 2.2  Viticulture in the Habsburg Semi-Periphery: Spain and Hungary

The Spanish conquest of Latin America in the first half of the 16th century coincided with the incorporation of Spain into the Habsburg Empire. The subsequent shifting balances of European power had important implications for wine production and trade. In particular the establishment of Spanish suzerainty

over the Low Countries provided a considerable boost to Spanish wine production
and exports. Moreover, the expansion of Ottoman power in the eastern
Mediterranean, accompanied by the reconquest of Iberia from the Moors, led to a
westward shift in sweet wine production in the Mediterranean. Wines such as
rumney and malmsey, which had previously been produced for the north European
market in the eastern Mediterranean, now began to be made in Spain. The
significance of such wine was that it could travel better and last longer than the
light wines of France and Germany, because of its higher alcohol level and
sweetness. Moreover, this was also the period when sack, the forerunner of
modern sherry, began to be exported in large quantities from southern Spain. The
expansion of Spanish wine exports, particularly to England, was further enhanced
by the final capture of Bordeaux by the French in 1453, and the increased efforts
of English merchants to import wines from countries other than France.

More important than the increased overseas demand for Spanish wines was
the growth of the domestic market, influenced both by the demands of the Court
and also the rapidly rising urban demand for wine from all classes. Both of these
sources of demand coalesced in Castile, both at the Court in Valladolid, and in the
burgeoning industrial towns of Segovia, Salamanca and Burgos. High on the
continental plateau to the south-west of Valladolid there thus emerged a specialist
area of wine production known as the Tierra del vino (Huetz de Lemps, 1968). In
this relatively dry area, with cold winters, yields were low and the resultant wines
were of good quality. Furthermore, during the 16th century there seems to have
been a widespread increase in urban alcohol consumption and drunkenness
throughout Europe (Braudel, 1981), and this acted as yet another incentive for the
expansion of wine production. In Valladolid, for example, it is estimated that wine
consumption had reached 100 litres per person per year by the middle of the 16th
century (Braudel, 1981, p.236). Wine production and marketing were closely
regulated, but a range of different labour practices existed for the cultivation of
vines, and it is extremely difficult to make any broad generalizations that would
validate or contradict Wallerstein's assertions concerning labour in his semi-
periphery. The demands by the Court for quality wine did, however, lead to
attempts to produce wine of higher quality during the latter part of the 16th
century, and there is some evidence of capital being invested in the ageing of wine
(Unwin, 1991).

At the eastern end of the Habsburg Empire, in Hungary, Zimányi (1987) has
examined a different set of processes transforming the viticultural landscape. Once
again, the economic alterations taking place in Hungary during the 16th and 17th
centuries closely reflected wider political changes at the European scale. Despite
the political fragmentation production increased considerably. This was associated
with an increase in demesne farming, and a decline in the number of small peasant
landholdings that had been characteristic of the 15th century. New laws in 1514
stipulated the imposition of weekly one day labour services, and it was this *robot*
system that enabled the expansion of demesne farming to take place. This increase
in serfdom associated with grain production accords well with Wallerstein's (1974,
1979) model of an emerging capitalist world-system. However, whether or not the

serf mode of production should be seen as feudal or capitalist remains a matter for considerable debate. If the expansion of serfdom in 16th century Hungary is considered as being essentially a reimposition of feudalism, then capitalist expansion in the core of Europe can be seen as being enabled by the enforcement of a different mode of production at the periphery. On the other hand, if serfdom is considered in this instance to be part of a capitalist mode of production, then Wallerstein is correct in arguing that different types of labour control existed within the expanding capitalist world.

The position of viticulture in Hungary was however substantially different from that of grain production (Zimányi, Makkai, Kirilly and Kiss, 1968; Szakçly, 1979). Prior to the 16th century, many Hungarian peasants sold their wine to regions which lacked their own production. However, thereafter the enforcement of the right by which landlords could purchase the wine produced by their serfs before any other buyer, meant that the landlords obtained greater control of the wine trade. However, unlike arable land, vineyards lay outside the system of servile tenure, and their owners only had to pay a vine-tax to the landlords and a tithe to the Church (Zimányi, 1987). The peasants, by concentrating on wine production, could thus avoid some of the excesses of the *robot* system. Moreover, vine cultivation is more labour intensive than arable farming, and thus required the use of additional paid labour by the owners of vineyards. Consequently, viticulture and wine production in 16th century Hungary was associated with a very different system of labour control than that which pertained with grain production, and can be seen to have incorporated several elements of an essentially capitalist system of surplus extraction from the early 1500s.

Coincident with these changes in labour organization, a new style of wine was also beginning to be made in Hungary. From the last third of the 16th century, producers in Tokaj began to add a small amount of juice from late harvested grapes to their wines. This created a particular style of sweet dessert wine, which under the name *Tokay azsu* became famous throughout Europe. Production continued to expand into the 17th century, and Zimányi (1987) notes that throughout most of the viticultural regions peasant commodity production of such wines successfully resisted the competition of demesne dominated wine production. Despite the pressures on them from landlords and urban owners of vineyards, the peasant producers were still able to sell their wines at a reasonable profit. Much of this wine was consumed within Hungary, but there was also considerable export trade, particularly of the sweet Tokaj wines to Poland (Carter, 1987) and elsewhere in northern Europe.

## 2.3   French and German Viticulture in the 15th and 16th Centuries

The vineyard regions of France and Germany lay nearest to the core of Wallerstein's emerging modern world-system. With increasing population levels, and the development of a thriving commercial and industrial economy, the Netherlands, England and northern Germany provided a ready market for wine. At the beginning of the 16th century, the patterns of trade and commercial activity, however, continued largely as they had done in the medieval period. Within Germany, for example, the merchants of Cologne and Frankfurt continued to dominate the trade in the wines of the Rhine, and vineyards continued to be cultivated in the far north of the country at places such as Stettin and Königsberg (Schröder, 1978). However, the increasing dominance of Dutch mercantile activity, and the opening up of more long distance trading networks, led to a substantial reorganization of the wine trade (Enjalbert and Enjalbert, 1987). In particular, the greater ease with which wines from Iberia could be imported into northern Europe, and the unreliability of the yields of northern vineyards, meant that many vineyards in areas such as Flanders were abandoned. While this may in part have been associated with the onset of the so-called Little Ice Age, it is evident that vineyards could still be cultivated successfully in northern Europe throughout the 16th and 17th centuries, and that the restructuring of vine cultivation probably therefore owed more to changing economic rather than environmental factors.

Within France, the dominant factor influencing the wine trade during this period appears to have been the growth in urban population, and in particular that of Paris. The wines of the west coast, with easy access to the sea continued to supply the bulk of the country's exports, as they had done in the medieval period (Lachiver, 1988), but the growing urban demand for wine led to an expansion in production in the Ile-de-France and in Champagne. Orléans, on the Loire, was in a particularly fortunate position able to supply both markets. Moreover, this expansion was also associated with increasing specialization in the types of vine cultivated in different regions. Thus, while Burgundy had been renowned for its Pinot Noir wines from at least the 14th century, the 16th century saw producers around Anjou concentrating on the Chenin Blanc, those of Orléans turning to specialization on the Cabernet Franc, and the Muscat rising to prominence in Alsace. The great diversity of types of labour control used to produce wine in both France and Germany at this period makes any generalizations on the subject extremely hazardous. Nevertheless, it is clear, particularly in France that during the 17th century peasant viticulture continued alongside that of the landlords, and sharecropping continued beside the use of paid labour. Lachiver (1988), also, for example, argues that the period between 1300 and 1600 saw the French nobility abandoning direct exploitation of vineyards in favour of leases.

## 3.    Capital in the Sphere of Production

Viticultural and vinification practices between 1400 and 1600 owed much to traditions built up over previous centuries. However, they also experienced a structural reorganization which laid the foundation for yet further changes in the 17th and 18th centuries. Prior to about 1600 profits from wine production appear only rarely to have been reinvested in the production process. Profits were certainly to be made from the wine trade, but these were largely generated by the demand for wine in areas where it was not produced. As commercial activity increased, the vineyards of northern Europe became less economically viable, and the sweeter, heavier wines of the Mediterranean began to provide growing competition for the light dry wines of France. The dawn of the 17th century, however, saw the emergence of a new system of wine production, in which capital began to be invested both in the profit-oriented cultivation of vineyards in new lands and also in the production of new types of wine.

## 3.1 Vintage and Fashion: the Creation of New Wines

Since early medieval times wine had served as one of the essential symbols of the nobility and urban elite of northern Europe. With the increasing availability of wine throughout Europe in the 16th and 17th centuries, and the spread of wine drinking to other elements of urban society, it became important for the ruling classes to identify a new type of beverage with which to celebrate their rituals. During the 17th century this was one of the factors leading to a differentiation in wine production, between makers of high quality wines for the elite, and low quality wines for the mass urban market.

The key feature of the new high quality wines was that they had greater levels of capital invested in their production than had previously been the case. In particular it had been discovered that ageing certain wines in bottles gave them a distinctive taste, full of nuances which had previously been unknown. The essence of ageing wines was that the return on financial investment in their production was delayed, and consequently the ageing and maturing of wines was an option only available to those with sufficient capital to invest over a long period. If wine is kept for more than a year in barrel, it is necessary for the producer to obtain more barrels in which to put the succeeding year's vintage. Such investment was usually beyond the means of most small producers, who needed to obtain a rapid return on their labour and investment soon after the vintage in order to pay off debts incurred throughout the previous year. Indeed, many small producers were usually in constant debt to merchants who provided them with the wherewithal to sustain themselves on the guarantee of their annual grape harvest. It was not just in the ageing of wine, though, that increased levels of capital investment were required. The new wines of the 17th century were based on improved methods of cultivation, the use of specific varieties of vines, the selection of particular grapes, and improved methods of vinification.

Such increases in capital investment, however, were only viable if it was possible for higher prices to be obtained for the wines, and thus for increased profits to accrue to the merchants or large producers. This need to realize profits lay at the heart of the creation of particular wine fashions in the 17th century, and was closely related to the flamboyance of the French and post-Restoration English courts. The activities of the French soldier, philosopher and courtier St.-Evremond in popularizing champagne in 17th century London are thus widely commented upon (Simon, 1962), as is the popularity of champagne at the French court in the 18th century among such personages as Madame de Pompadour, the poet Voltaire and the Duc de Richelieu.

## 3.2 The Origins of Champagne

Wines from the region around Reims and Epernay had been well known in Paris during the medieval period, and had been widely purchased by the nobility of neighbouring countries, such as the Counts of Hainault (Sivéry, 1969). However, this wine was still, and very different from the sparkling wines that began to be made in Champagne during the 17th century. From about the 1650s a number of wine producers began experimenting with new techniques of viticulture and vinification. The most widely reported such experiments were those of Dom Pierre Pérignon, cellar master of the Abbey of Hautvillers (Gandilhon, 1968), who is reputed to have reorganized the abbey's vineyards, and to have selected particular grape varieties, such as the Pinot Noir for its wines. Moreover, at the vintage he is said to have chosen individual panniers of grapes to blend together to make the particular wines that he wanted. These grapes were then pressed rapidly to minimize contact with the skins, and having fermented they produced an almost white still wine, known as *vin gris*. The result was a high quality wine which could sell at premium prices.

Following the Restoration in England in 1660, these wines became increasingly popular. The practice of keeping wine in bottles had also emerged in England at about this time, and it was soon discovered that if the wines of Champagne were so stored they developed an attractive effervescence. It seems that the natural fermentation of the wines made in northern France was not usually completed before the cold winters had set in, and the drop in temperature caused the fermentation to cease while there was still a substantial amount of sugar in the wine. Consequently, in the spring when the temperature began to rise again, the fermentation would recommence within the bottle, and the carbon dioxide given off during fermentation would be absorbed within the wine. However the additional pressure that this caused meant that many bottles broke, and it was therefore only when strengthened glass, produced through the addition of lead oxide in its manufacture (Charleston, 1984), became available, and when more sophisticated techniques of fitting and extracting corks were used, that sparkling champagne production became reliable and successful.

## 3.3 The New French Clarets

A second region of France where the investment of capital in quality wine production took place was in the south-west around Bordeaux. In 1663 Pepys (1971, p.100) recorded that he drank a new sort of French wine, known as "Ho Bryan", for the first time. This was none other than that produced on the property of Haut-Brion in the Graves by Arnaud de Pontac, the first President of the Bordeaux Parlement. Following the collapse of the wine trade with England in the 15th century, the merchants of Bordeaux had turned increasingly to the Netherlands as a market for their wines during the 16th and 17th centuries. However, after the outbreak of war between France and the Netherlands in 1672, the Dutch looked elsewhere for their wines, and began importing considerable amounts from Spain and Portugal. This served to encourage producers in the vicinity of Bordeaux to concentrate increasingly on quality wine production, a process which they had embarked on somewhat hesitantly during the previous decades.

During medieval times most of the light wines exported from Bordeaux had been produced to the south and east of the city, but from the middle of the 17th century the relatively infertile sands and gravels to the north-west saw a veritable explosion of vineyard development, as money was invested in the production of high quality wines (Pijassou, 1980; Enjalbert, 1953). Many of the owners of these properties were lawyers and magistrates who had acquired their status and wealth through their membership of the Bordeaux Parlement, and Forster (1961) has noted that by the 18th century more than 70 per cent of the aggregate wealth of sixty-eight families of this *noblesse de robe* came from the sale of wine. Gradually, improved methods of viticulture and vinification were introduced, largely as a result of trial and error, rather than any theoretical understanding of the processes of wine making. During the first half of the 18th century particular types of grapes were being selected prior to vinification, oak barrels were introduced for ageing the wines, racking was widely practised, and day labour had been introduced on the best quality vineyards.

The market for such wines was primarily northern Europe, and once peace had been restored between France and the Netherlands, Amsterdam became an important focus for their redistribution. Moreover, despite more or less constant conflicts between Britain and France at this period, the so-called New French Clarets became particularly popular amongst the English nobility. This can be seen very well, for example, in the Book of Expenses belonging to John Hervey, first Earl of Bristol, which records that he purchased increasing amounts of such wines from the properties of Haut-Brion, Margaux, Latour and Lafite during the first few decades of the 18th century (Hervey, 1894). Not only were these wines expensive in England as a result of their costs of production, but war with France meant that they were in very scarce supply, and often only came onto the market as prizes captured from French fleets, or as a result of smuggling acivities.

## 3.4 Wine from New Places

The examples of champagne and claret are the best known instances of the increased capital investment in wine production that took place in France during the 17th century. However, experimentation in improving the quality of wines was equally taking place in Germany, and in northern Portugal the period also saw the beginnings of port wine production. Elsewhere in Europe, other producers were concentrating on the supply of bulk low quality wine for the growing mass market of urban consumers, and the Dutch introduction of distillation to western France saw the beginnings of brandy production in Cognac and Armagnac. The essential characteristic of brandy was that for an equivalent amount of alcohol it was cheaper than wine to transport, and it could readily be diluted to whatever level was required at its final destination.

In addition to these developments taking place in the heartland of viticultural activity, the 17th century also saw vineyards being planted for the first time in other areas of the world. Typical of these investments in new vineyards was the rapid expansion and then subsequent demise of wine production in the Canaries (Steckley, 1980). By the beginning of the 17th century Tenerife had started to benefit from the increased demand in northern Europe for the sweet wines that had traditionally been produced in the eastern Mediterranean. Earlier, the population of Tenerife, like that of Madeira, had responded to the boom in European sugar consumption, and much land had been put down to sugar cane. However, when Caribbean production began to undercut the price of sugar from the Atlantic islands, the owners of land sought to diversify into another crop, and the one which they turned to was the vine.

Tenerife was also well situated to an expansion of viticulture as a result of its location with respect to the American trade, and this provided an added boost to demand. From the 1640s both arable and pasture on Tenerife was increasingly being converted into vineyards, and by "1681 London customs officers taxed enough Canary wine to fill roughly 4.5 million bottles" (Steckley, 1980, p.343). Although a small English Factory was established on Tenerife, there is little evidence that the merchants were investing directly in production, and what investment that took place seems largely to have been in the hands of the Spanish landowners. As demand for Canary wine increased, though, it became increasingly difficult for the English merchants to finance their activities; it was only possible to export a limited amount of woollen goods from England to the Canaries where a much more equitable climate prevailed. Consequently, as Steckley (1980, p.345) has illustrated English merchants "regularly handled Canarian tax payments to Madrid, sending bills of exchange drawn upon English partners in the peninsula and retaining the unique Canarian coinage for wine purchases". This, once again, reinforces the importance of credit arrangements in enabling the changes that took place in 17th century viticulture and wine production to have occurred. The wine boom in the Canaries, however, was short lived. By the beginning of the 18th century with rising customs duties in London and the outbreak of the Spanish War of Succession, English merchants were forced to leave Tenerife, and the expansion of Portuguese wine production provided them subsequently with a nearer and cheaper source of supply.

A very different example of further direct investment in the production of wine in the 17th century is afforded by the colonial policies adopted by the British and Dutch. The Virginia Company established in north America by King James I, for example, was designed specifically to exploit the economic potential of the New World. As part of its activities the Company was therefore actively involved in attempts to produce wine which could be sent back to the growing market at home, and thus reduce the amount of wine imported from other countries. Initially the settlers produced wine from the indigenous grapes, but this proved to be unsuccessful (Adams, 1984; Pinney, 1989). The Company then introduced French varieties, and even French vine growers, but the vines failed, largely as a result of the then unknown problems of phylloxera. Eventually, by the late 1620s, with the rapid success of tobacco production, and the continued failure of viticulture, attention gradually shifted away from vine growing in Virginia.

Dutch fortunes in their Cape Colony in southern Africa, however, were very different. Here, from the 1650s onwards, vine cultivation and wine production proceeded successfully. Initially, the quality of the wine was not particularly good, and by the 1670s much of it was distilled into brandy (Leipoldt, 1952), which was in any case cheaper to transport back to northern Europe per equivalent unit of alcohol. These early experiments were reinforced by the arrival of a large number of French Huguenots following the Revocation of the Edict of Nantes in 1685, and by the beginning of the 18th century wine production had become firmly established in the colony. Although anyone could make wine, the government retained a monopoly on its sale. Because of the low prices that they were prepared to pay there was little incentive for producers to improve the quality of Cape wine, and with the exception of the wines of Constantia they were generally inferior to those available to Dutch merchants elsewhere. Consequently, although the Dutch East India Company was seeking to supply their home market with wines from the Cape, they met considerable opposition from merchants in Holland. Nevertheless, these early investments provided the basis of an industry which subsequently successfully survived the annexation of the Cape by the British at the beginning of the 19th century.

## 4.    Profit, Production and Exchange

The developments of viticulture, both within Europe and more widely at a global scale, provide good evidence of the complex set of processes giving rise to the restructuring of the European economy during the 15th and 16th centuries, and its subsequent expansion to the global arena in the 17th century. During the medieval period merchants had invested profitably in the infrastructure of trade, exemplified by the highly successful activities of wine merchants throughout the continent. Profit was possible as a result of market knowledge and the ability of merchants to transport goods over relatively long distances. However, there is very little evidence that these merchants invested directly in the production of the new sorts of wine that began to be made in the 17th century. The profit motive certainly existed amongst the successful medieval wine merchants, but it was a motive that found its expression primarily in exchange relations.

By the 17th century there had been a subtle change in emphasis, and it seems to have been the landowners of Europe who were able to take advantage of the new social and economic conditions prevailing. Central to any understanding of the reorganization of the wine industry was the emergence of a differentiated market. With the rising urban consumption of all kinds of alcohol in the 16th and 17th centuries, it became possible for wine producers to concentrate on different sections of the market. Some, probably most, turned their attention to increasing their output, regardless of the quality, in order to fuel the rising trend of alcoholism among the urban poor. Others, though, were able to invest in new methods of viticulture and vinification, and to make wines destined to serve as social symbols of the elite of late 17th century Europe. These vintage wines were able to command prices several times greater than those of the common wines, and by investing directly in their production landowners and merchants were able to reap greater profits than had previously been possible.

In conclusion, Wallerstein's arguments concerning the origins of the modern world-system do provide some insights into processes that were of relevance in explaining the changes that took place in vine cultivation and wine production. However, it is evident that in the core and semi-periphery of Europe many different types of labour control were used to produce the wines that were consumed throughout the continent. Owners of land seeking to benefit from the export market in wine, often tended to concentrate on paid wage labour since it was more skilled than that available through other forms of labour control. However, sharecropping arrangements continued to be widespread, as did vine growing and wine making among the peasantry. Rather than speaking of different kinds of European capitalism, we need to focus attention much more directly on specific methods of profit appropriation, on particular types of labour organization and production, and on the commercial relationships involved in the exchange of particular commodities. Above all, though, this paper has stressed the need to incorporate social, political and ideological factors alongside economic ones in any explanation of the emergence of a modern world-system

## Acknowledgements

This paper was first presented at a conference on the early modern world-system in geographical perspective held in Göttingen in April 1990. I am grateful to the comments made then by participants at the conference and in particular to Hans-Jürgen Nitz. I have also learnt much from John Dickenson and members of the London Group of Historical Geographers who have subsequently commented on a revised version of the paper.

## REFERENCES

Adams, L.D. (1984): The wines of America, 3rd ed., London: Sidgwick and Jackson

Bell, C. (1974): Portugal and the quest for the Indies, London: Constable

Braudel, F. (1981): Civilization and capitalism 15th-18th century, volume I: The structures of everyday life, London: Collins Braudel, F. (1982): Civilization and capitalism 15th-18th century, volume II: The wheels of commerce, London: Collins

Brenner, R. (1977): The origins of capitalist development: a critique of Neo-Smithian Marxism. In: New Left Review, 104, pp. 25-93

Brenner, R. (1982): Agrarian class structure and economic development in pre-industrial Europe: the agrarian roots of European capitalism. In: Past and Present 97, pp. 16-113

Bruman, H.J. (1967): Man and nature in Mesoamerica. In: B. Bell (ed.), Indian Mexico: past and present, Los Angeles: UCLA Latin America Centre, pp. 13-23

Butlin, R.A. (1987): European rural transformations: some reflections on the context of agrarian capitalism. In H.-J. Nitz (ed.), The medieval and early-modern rural landscape of Europe under the impact of the commercial economy. Göttingen: Department of Geography, University of Göttingen, pp.87-104

Charleston, R.J. (1984): English glass and glass used in England, London: George Allen and Unwin

Corbridge, S. (1986): Capitalist world development: a critique of radical development geography, Basingstoke: Macmillan

Carter, F. (1987): Cracow's wine trade (fourteenth to eighteenth centuries). In: The Slavonic and East European Review 65(4), 537-78

Dion, R. (1959): Histoire de la vigne et du vin en France des origines au XIXe siâcle, Paris: privately published

Dodgshon, R.A. (1987): The European past: social evolution and spatial order, Basingstoke: Macmillan

Enjalbert, H. (1953): Comment naissent les grands crus. In: Annales, âconomies, Sociétés Civilisations 8, pp.315-28; 457-74 Enjalbert, H. and Enjalbert, B. (1987): L'histoire de la vigne et du vin, avec une nouvelle hierarchie des terroirs du Bordelais, Paris: Bordas

Forster, R. (1961): The noble wine producers of the Bordelais in the eighteenth century. In: Economic History Review, 2nd series 14(1), pp.18-33. Gandilhon, R. (1968): Naissance du Champagne: Dom Pierre Pérignon, Paris: Hachette

Hervey, J. (1894): The diary of John Hervey, first Earl of Bristol. With extracts from his book of expenses 1688-1742. Wells: Ernest Jackson

Huetz de Lemps, A. (ed.): (1968): Apogeo y decadencia de un viñedo de calidad: el de Ribadavia. In: Anuario de Historia Economica y Social 1, pp.207-25

Lachiver, M. (1988): Vins, vignes et vignerons: histoire des vignobles franìais, Paris: Fayard

Laclau, E. (1979): Politics and ideology in Marxist thought. London: Verso

Leipoldt, C.L. (1952): 300 years of Cape wine. Cape Town: Stewart

Marques, A.H. de Oliveira (1972): History of Portugal. New York: Columbia University Press

Pepys, S. (1971): The diary of Samuel Pepys, volume IV, 1663, ed. by R. Latham and W. Matthews. London: G. Bell & Sons

Pepys, S. (1971): The diary of Samuel Pepys, volume VI, 1665, ed. by R. Latham and W. Matthews. London: G. Bell & Sons

Pijassou, R. (1980): Un grand vignoble de qualité: le Médoc, Paris: Librairie Jule Tallandier

Pinney, T. (1989): A history of wine in America : from the beginnings to prohibition. Berkeley: University of California Press

Schröder, K.H. (1978): L'ancien extension de la viticulture dans le nord-est de l'Europe centrale. In A. Huetz de Lemps (ed.), Géographie historique des vignobles, tome II. Bordeaux: CNRS

Simon, A. (1962): The history of champagne. London: Ebury Press

Sivéry, G (1969): Les Comtes de Hainault et le commerce du vin du XIVe siécle et au début du XVe siécle. Lille: Université de Lille

Steckley, G.F. (1980): The wine economy of Tenerife in the seventeenth century. In: Economic History Review, 2nd series, 33, 335-50.Szakçly, F. (1979): A közép-Duna menti bortermelés fénykora, Dunataj 2, pp.12-24

Unwin, P.T.H. (1991): Wine and the vine: a historical geography of viticulture and the wine trade. London: Routledge

Wallerstein, I. (1974): The Modern World-System: Capitalist Agriculture and the Origins of the European World-Economy in the Sixteenth Century. London: Academic Press

Wallerstein, I. (1979): The Capitalist World-Economy. Cambridge: Cambridge University Press

Wallerstein, I. (1980): The Modern World-System, volume II: Mercantilism and the Consolidation of the European World-Economy, 1600-1750. London: Academic Press

Wallerstein, I. (1983): Historical Capitalism. London: Verso

Wallerstein, I. (1989): The Modern World-System, volume III: The Second Era of Great Expansion of the Capitalist World-Economy. London: Academic Press

Wolf, E. (1982): Europe and the people without history. London: Faber

Zimányi, V. (1987): Economy and society in sixteenth and seventeenth century Hungary (1526-1650). Budapest Akadémiai Kiadó

Zimányi, V., Makkai, L., Kirilly, Z. and Kiss, I.N. (1968): Mezögazdasági termelés és termelékenység Magyarországon a késöi feudalizmus korában (1550-1850). Agrartorteneti Szemle 10, pp.39-93

# 17

# The Timber and Naval Stores Supply Regions of Northern Europe During the Early Modern World-System

Ian Layton

## 1. Introduction and Hypothesis

The boreal forests of Scandinavia and Finland have played an important role in European commerce since at least medieval times, when the pelts of their furry denizens and charcoal-smelted iron and copper were their principal export commodities. The more direct utilization of these extensive softwood resources for the production of tar, pitch and timber for foreign consumption (especially for ship-building and housing construction) gathered impetus during the early modern period, when the increasingly populous core regions began to experience shortages in their domestic supplies. During these formative centuries western Europe's maritime exploration, conquest, and trade began to weave the complex and dynamic network of contact patterns that characterize the modern world-system[1].

Few historical geographers have concerned themselves with the international commercial and economic development of the countries of northern Europe over the long term, and even more seldom has their attention been focused on the early modern period[2]. On a highly generalized map (Pounds 1979, p. 279), it appears that by about 1800 the export trade from the Scandinavian countries to the major markets of northwestern Europe took place along relatively few routes and was of less importance than the trade from countries bordering the southern and eastern Baltic. Nevertheless, the period prior to 1800 involved the early development and expansion of trade in forest commodities from the northern periphery of Europe and it is of interest to analyse this trade in more detail, in order to try to understand the causes and processes of these changes in resource utilization.

Starting from the simple hypothesis that the form of forest usage in a region is primarily a function of distance from the market, it can be an amusing exercise to consider a hypothetical mental map for London's tar and softwood merchants during the early modern period, based on sailing distances from the ports of export (Fig. 1). This topological map distorts the actual configuration of coastlines by relocating the various tar and timber ports so as to allow them straight-line contact with London.

Naturally, true compass direction and shape have to be sacrificed in this effort to map the relative proximity of these ports to the London market. Archangel, for example, is portrayed twice since it could be reached by two quite different routes. The purely maritime route involved an extra distance of about 100 nautical miles but, even during the summer season, this was a hazardous voyage for the small vessels of the early period. The alternative route via rivers

and lakes to St. Petersburg and thence through the Danish-controlled Sound
involved much extra cost and time, with timber taking two to three years on the
continental leg - exemplifying contemporary difficulties of overland transport.
Both routes were inaccessible during the frozen grip of winter and, indeed, the
Baltic was generally closed to shipping for six to seven months - not, necessarily,
due to the physical presence of ice but more to the psychological barrier and risks
ice posed for sailing-ship owners. Winter storms in the North Sea may also have
contributed further to the concentration of shipping activity to the summer months.

It would be interesting to construct a similar map based on freight costs
rather than physical distances; for Sound tolls, changes in freight rates, and
variations in different nations' import and export custom duties, also played
important roles in the prices paid in London. Cheaper transport costs could
'shrink' distances, so that more remote ports and their hinterland regions would
appear to be relatively nearer to the core of consumption than nearby rivals with
high freight costs and other economic barriers to free trade. Conversely, distances
were generally 'increased' through various national mercantile measures.

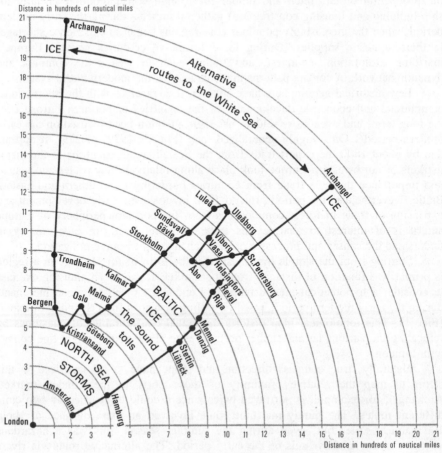

*Fig.1: A hypothetical mental map for London's tar and softwood merchants*
*(Based on sailing distances)*

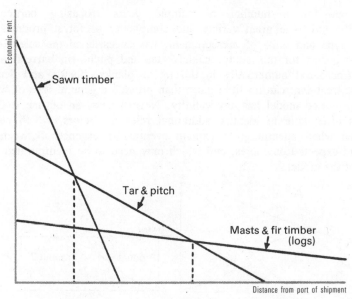

Fig.2: *A Von Thünen interpretation of coniferous forest utilization*

The notion of the importance of distance leads naturally to a von Thünen line of reasoning, and a zoning of coniferous forest utilization can therefore be postulated for this period (Fig. 2). The sawing of timber by water-driven sawmills was more capital intensive and had higher production costs than tar burning, or the selective felling of suitable masts, ship-timbers and logs for export. Sawmilling was also subject to more locational constraints than these other activities, and a suitable falls site near to the coast was of prime importance. During such a long period as three centuries, however, considerable changes took place both in relative prices and in transport and production costs. This model ought therefore to be considered as having a dynamic potential rather than being purely static.

Empirical studies of trade and economic change within regions in northern Sweden over the long term clearly reveal that, while the volume of forest exports has steadily increased, the composition of trade has changed, with the major types of product normally following similar sequences from ships' stores to tar, hewn timber, sawn timber, pulp and, most recently, paper.[3] Figure 3 transposes these 'production cycles' (3A); first into relative values (3B) and then into a kind of 'time-geography' diagram (3C) which serves (I) to illustrate how a single region experiences changes in product dominance and (II) to suggest how the production of one particular commodity changes over time within a number of regions located at increasing distances from the core region of consumption. This latter interpretation provides the main theme of this paper.

In order to test the validity of these normative generalizations at the macro-scale, a preliminary analysis of northern Europe's export trade data has been carried out for a number of sample years, focusing particularly on Sweden/Finland. The great variety and complexity of forest products, both in terms of type and units of measurement, has necessitated the selection of two major categories for this study; namely 'tar and pitch' (in barrels) and 'sawn timber' (measured numerically in dozens of planks, boards and deals). This pragmatic treatment can do little more than provide a general idea of whether or not the proposed model has any validity. Nevertheless, an outline study of this nature is often able to identify additional relevant factors, which need to be considered when attempting to explain eventual divergences between the real world and expected outcomes, and which may need to be incorporated in a later version of the model.

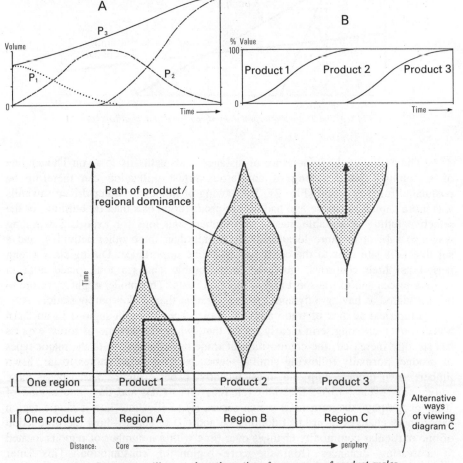

Fig.3:Diagrams to illustrate how the notion of a sequence of product cycles might be transposed into a dynamic spatial model
(see text for explanation)

Much of the Baltic trade was recorded by the Sound Toll Registers throughout this study period, and the records from the late 15th to the mid-19th centuries (1497-1856) have been preserved at the Danish Records Office in Copenhagen.[4] The modest amounts passing through the Great Belt in small vessels were probably no more than 15% of the total Sound traffic. The Little Belt was rarely used and the overland route between Lübeck and Hamburg was of far less importance, especially for bulky, low-value commodities.[5] Since tar barrels and bulky timber products could not easily be overlooked or smuggled, they are likely to have been fully recorded. Unfortunately for the researcher, Swedish/Finnish vessels were largely exempted from paying tolls, especially during the period 1651-1720, which greatly reduces the value of the Sound Toll records for analysing the total trade leaving the Baltic.[6] Another serious drawback is that 'port of departure' can refer to last port of call and not to the original port of lading; also, in the published material at least, the smaller ports are often grouped under national headings rather than being specified individually.

As it has therefore been impossible to obtain a reliable picture of forest commodities leaving the Baltic prior to 1720 using only the Sound Toll records, national trade archives have also to be employed - especially those of Sweden/Finland.[7] Even the Swedish records on the foreign export trade have their drawbacks since they only list the ports of export and, in a few cases (notably Stockholm), they do not reveal the real ports of origin for the various commodities. For Stockholm's entrepôt trade, however, the sources of her tar and timber imports can be identified for certain years through diligent searches in the Stockholm City Archives.[8] Thus, by combining Stockholm's domestic imports of tar and timber with the national foreign export records, it becomes possible to map the ports of origin for these commodities with considerable accuracy. When combined, these sources also permit the reconstruction of a simple statistical overview of the relative importance of different regions for selected years, thereby illustrating major locational changes in the commercial utilization of forest resources.

## 2.    The Tar Trade

Outside the Baltic pond, Norway and western Sweden played important rôles in the early trade in timber and naval stores with the markets of western Europe, especially Britain and the Netherlands. These regions had the comparative advantages of not having to pass through the Sound and pay dues on their way to market and of being located nearer to their customers in the core regions of consumption. Norway was an important source for balks, sawn deals and smaller mast dimensions, as well as for tar and pitch (used mainly for waterproofing ropes and rigging, and for caulking vessels). Although tar could be widely obtained throughout the coniferous forest belt, Sweden/Finland long dominated this trade with superior quality products and was less inclined to export timber. Instead, Sweden elected to safeguard charcoal supplies for the more important iron industry and therefore imposed severe restrictions on the felling, sawing and export of timber throughout most of the early modern period. Even the export of this high quality tar was subject to mercantile restraints and, between 1648 and 1714, a succession of Stockholm-based monopolistic tar companies maintained standards and controlled all production and trade from the regions north of the capital city.[9]

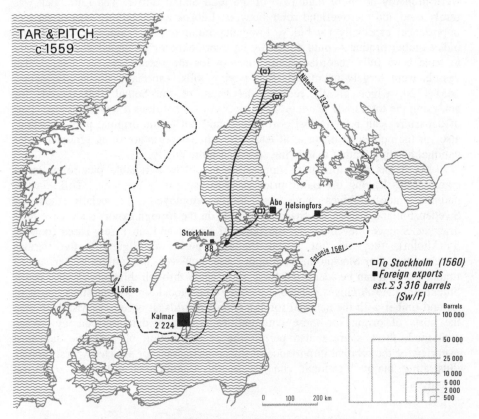

*Fig.4:  Shipments of tar and pitch from ports in Sweden/Finland c.1559  (Source: see Table 1)*

Sweden/Finland's export records are sadly incomplete for the 16th century, but some material has survived. Using excerpts from tables in Forssell (1875, pp. 24-26), a very approximate reconstruction can be made for the years around 1559 (see Tab. 1 & Fig. 4). As there appear to be gaps in the 1559 records, other customs houses have here been represented using corresponding data for the nearest available year. In addition, the origins of tar deliveries to the main export port of Stockholm in 1560 have been broadly identified using the Castle Toll records (slottstullen).[10] The main region for tar production in the mid-16th century was in the south of the kingdom, in Småland, and some 68% of Sweden's export trade was focused on Kalmar. The only other region of any significance was the south coast of Finland, which accounted for 29.4% of Sweden/Finland's exports.

At this time the south Finnish port of Viborg, later to be so dominant in the tar trade, was of little importance with exports only amounting to 0.9% of the total volume. According to the Sound Toll lists for 1562 (Tab. 2), there were no tar or pitch cargoes leaving the Baltic from Swedish or Finnish ports. Over 61% of the tar traffic passing through the Sound sailed from ports along the south coast of the Baltic and 38% from the east.[11]

By 1600 (Tab. 1), the external demand had already stimulated the rapid development of tar burning in the best pine forest regions of south-eastern Finland, so that the port of Viborg alone accounted for over a third of Sweden/Finland's exports (6,550 barrels, cf 30 in 1558). Stockholm also exported 6,550 barrels that year, but these were derived mainly from Österbotten rather than being produced locally. It is clear that Österbotten was already the chief region supplying Stockholm and that Norrland had only had a modest output (mainly in the northern province of Västerbotten). This pattern continued throughout the 17th and 18th centuries and was largely due to variations in environment and resources. Conditions for agriculture were better in the low-lying coastal plains of Österbotten, so that villages tended to be more numerous and larger. Natural resources such as fur-bearing animals, salmon and other fish, were less bountiful than in Västerbotten and there were fewer Saamis (Lapps) to trade with. The plentiful labour and need to supplement farming with other sources of income meant that maritime and forest resources came to be more intensively utilized, especially by organized seal hunting and tar burning.

Table 1. Regional changes in the origins of Sweden/Finland's tar exports
for selected years (%).

| Year | W.coast | S.Sweden | S.Finland | Österbotten | Norrland | Total[1] |
|---|---|---|---|---|---|---|
| c 1559[2] | 0.5 | **68.3** | 29.4 | 0.9 | 1.0 | 3,468 |
| c 1600[3] | - | 23.8 | **54.6** | 21.5 | 0.2 | 18,730 |
| c 1641[4] | 2.0 | 9.6 | **50.2** | 33.5 | 4.7 | 124,218 |
| 1686 | 6.5 | 13.2 | 34.9 | **43.2** | 2.2 | 112,524 |
| 1760 | 2.7 | 21.2 | 2.3[5] | **64.8** | 9.0 | 139,075 |
| 1805[6] | 5.5 | 6.9 | 0.2 | **77.5** | 9.9 | 177,610 |

1   Number of barrels exported (excluding Stockholm) plus deliveries to Stockholm; except for 1805, which is based solely on foreign export data.
2   Stockholm's import data and origins are from 1560.
3   Stockholm's imports (34.9%) came mostly from Österbotten (21.5%) and the remainder (13.4%) probably from South Finland (Norrland too?).
4   Stockholm's import data and origins are from 1648.
5   Viborg was lost to Russia in 1721.
6   Stockholm's exports amounted to 61.8% of the national total and, as in earlier years, were derived from the Gulf of Bothnia, ie Österbotten and Norrland. The percentages for these regions have been based on the direct exports plus the proportions derived by Stockholm from the north in 1760. Complete data on the origin of these has not yet been traced, but Luleå alone sent 11,954 barrels to the capital city and it is clear that the Norrlandic towns were beginning to rival those of Österbotten and that another locational shift in tar production was in the offing.

Sources: c 1559: Forssell (1875:Tables IIa-c, facing p 24). 1560: Friberg (1983: 290ff, Tables A & B). 1600: *Lokala räkenskaper, Tull och accis*, KA (Hallberg 1959:172) and *Slottstullängderna*, KA (Calonius 1936:145 & 153). 1641: Boëthius & Heckscher (1938:36-53) and 1648: Tirén (1937:224). 1686: *Handel och Sjöfart 12*, KA (Åström 1988:24, table 1:6). 1760: *Årsberättelser: Utrikeshandeln ser 2*, KmkA, RA, and *Borgerskapets äldsta arkiv F I:ser 7 nr 26*, SSA. 1805: *Årsberättelser: Utrikeshandeln ser 2*, KmkA.

Tab. 2: Tar and pitch passing through the Sound Toll for selected years.

| Year | Denmark | | S. Baltic | | E. Baltic | | Sweden/Finland | | Total |
|---|---|---|---|---|---|---|---|---|---|
| | barrels | % | barrels | % | barrels | % | barrels | % | barrels |
| 1562 | 588 | 0.7 | 50,934 | 61.3 | 31,602 | 38.0 | - | - | 83,124 |
| 1600 | 672 | 0.9 | 42,018 | 59.0 | 15,936 | 22.3 | 12,552 | 17.6 | 71,178 |
| 1641 | 3,732 | 3.2 | 9,636 | 8.2 | 3,072 | 2.6 | 101,616 | 86.0 | 118,056 |
| 1686 | 834 | 1.2 | 2,136 | 3.2 | 7,848 | 11.7 | 56,133 | 83.8 | 66,952 |
| 1724 | 708 | 1.9 | 1,009 | 2.6 | 491 | 1.3 | 35,794 | 94.2 | 38,002 |
| 1760 | 804 | 0.8 | 1,152 | 1.2 | 392 | 0.4 | 92,780 | 97.5 | 95,128 |

Sources: Sound Toll Tables for 1562 (Bang 1922, p. 5), 1600 (ibid, p. 199), 1641 (ibid, p. 467), 1686
(Bang & Korst 1939, pp. 309-314), 1724 (Bang & Korst 1945, pp. 59-65), and 1760 (ibid, pp. 752-761).

The Sound Toll Tables for 1600 (Tab. 2) record the westward movement of
12,552 barrels of tar (no pitch yet) from Sweden/Finland, which was 67% of the
nation's exports and 17.6% of the total passing through. The remainder came
mainly from the southern (59%) and eastern (22.3%) Baltic, showing that the
south was still the primary production region. Admittedly Sweden/Finland must
have shipped 6,178 barrels (33% of her total exports) to precisely this region, but
this addition to local supplies was less than one tenth of the volume shipped
through the Sound from other countries.

The 1641 situation (Tab. 1) reveals a sevenfold increase in the volume of
Sweden's tar and pitch exports since 1600.[12] Reasons for this remarkable upswing
include the rapid growth in demand in the core regions for tar, as a lubricant and
for shipbuilding (especially in relation to British and Dutch maritime rivalry).[13] In
1628 the Dutchman Jan van Swinderen (also recorded as Hans von Schwindern)
was given leave to travel round Sweden in order to find the best locations for tar
and pitch distillation. In the following year he and his partner David Grunau (or
Grunow) were given permission to use the Crown forests in Vadsbo, Värmland
and Tiveden (free of charge for six years), for the production of tar using a new,
less wasteful technique. In 1634, he was also granted the sole right to establish a
pitch distillery in Göteborg and a monopoly on the tar trade in the counties of
Älvsborg and Göteborg, on the understanding that he instructed the peasantry in
his new method of tar burning.[14] Here is the evidence for a technical innovation
which was presumably diffused rapidly throughout the country from the west coast
region.

The dominant export ports for tar and pitch were Stockholm and, during the
16th and 17th centuries, Viborg. Several other Finnish and Swedish towns outside
the Gulf of Bothnia (south of Åbo and Stockholm) also contributed minor
quantities. Stockholm, however, was a special case since the exported tar was not
produced in the city's natural hinterland. Instead, it was derived from 'overseas' -
ie from the shores of the Gulf of Bothnia (especially from the Finnish province of
Österbotten), where trade was subject to the so-called Bothnian Trade Restraint.

Originating in the 14th century (and only fully repealed in 1812), these regulations prohibited direct foreign trade and forced the Bothnian merchants to deliver their merchandise to the staple towns of either Åbo or Stockholm. Since Stockholm was the capital city and major port for Sweden/Finland, it was both the focal point for the collection of taxes in kind and the main market for many of the products from the Bothnian and Lappmark provinces. Åbo thus played only a minor role, and Stockholm functioned as the nation's principal entrepôt. The city's merchants were thus in a powerful position, and mercantile policies also resulted in the formation of monopolistic companies to control the trade in salt and tar. The Crown was keenly interested in controlling commerce, both as a source of direct profit (through the sale of various taxes in kind, tolls and other duties) and as a means of steering the utilization of the nation's resources and of maintaining high prices abroad by restricting production.

An important measure to ensure the high quality of tar and pitch exports was the appointment in 1641 of tar 'wrackers', or inspectors, in all of the staple towns from which the products were shipped. They tested the contents of each barrel to ensure the correct quality and quantity, and this gave Swedish tar a reputation for excellence that was reflected in the high prices paid. Van Swinderen attempted to obtain a personal monopoly on Sweden's entire tar trade, but had to be content with a compromise, in which other wealthy Dutch merchants played the leading role in establishing the Norrlandic Tar Trade Company (Norrländska tjäruhandelskompaniet). Viborg and Stockholm merchants later increased their influence (Fyhrvall 1880, pp. 293-5). This was the first in a succession of monopolistic tar trading companies which completely dominated production north of Stockholm and Nyen, especially in the Gulf of Bothnia, and which led to the city's primacy in terms of Swedish/Finnish tar and pitch exports (Figs. 5 & 6).

Viborg continued to be the largest single direct exporter of tar (37,706 barrels) in 1641, followed by Åbo (14,616) and Helsingfors (6,984). Stockholm's entrepôt trade accounted for 44% of the total tar and pitch exports but, assuming that conditions in 1641 were the same as in 1648, all of this tar was most likely derived from the eastern shore of the Gulf of Bothnia.[15] As in 1559 and 1600, Kalmar (still Sweden's southernmost town) continued to serve the largest producing area in mainland Sweden (ie Småland), but only shipped 6,439 barrels in 1641. The dramatic increase in tar burning had thus taken place in the two main regions of Österbotten and southern Finland, each accounting for about three-sevenths of the total exports, and another northward locational shift in production was in progress.

Together, Sweden and Finland shipped 101,616 barrels of tar and pitch through the Sound in 1641 (Tab. 2), which accounted for as much as 86% of all westward-moving tar shipments (cf 17.6% in 1600). Göteborg, the sole west coast outlet for Sweden, exported a further 2,455 barrels - so that 78.8% of the nation's total exports went to consumers on the Atlantic seaboard and only 21.2% to markets in the southern Baltic.[16] Clearly, the British and Dutch markets were responsible for both the increase in production and for the redirection of trade since 1600.

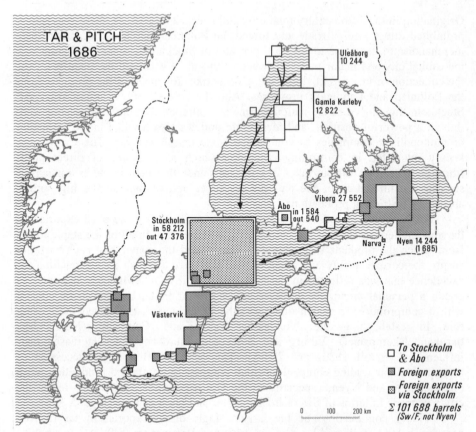

Fig.5: Shipments of tar and pitch from ports in Sweden/Finland in 1686 (Source: see Table 1)

It is possible to reconstruct the pattern of tar and pitch production and trade from Sweden/Finland even more accurately during the latter part of the seventeenth century. Figure 5, for example, shows the situation in 1686. The foreign export trade has been mapped by ports and the precise origins of the tar delivered to the major entrepôt of Stockholm have also been identified for the same year, using complementary data on the city's domestic tar imports. The main changes since 1641 were the marked decline in exports from Åbo and the south coast of Finland and the continued expansion of northern Österbotten. Viborg retained its pre-eminence, however, but Nyen had also emerged as a rival tar port. Territorial gains from Denmark, in the south of Sweden, contributed little to the tar trade - mainly because the area corresponded closely to the zone of broadleaf deciduous forest which was less suitable for tar burning and already much cleared for cultivation.

Tab. 3: Likely tar consumption in Bothnian towns for 1686 (tar barrel equivalents)

| Österbotten | Local tar deliveries | Shipments to Stockholm | | | Town balance |
|---|---|---|---|---|---|
| | | Tar | Pitch x 2 | Total | |
| Uleåborg | 8,186 | 10,244 | - | 10,244 | -2,058 |
| Brahestad | 8,173 | 6,814 | 978 | 7,792 | 381 |
| Gamla Karleby | 13,657 | 12,145 | 1,354 | 13,499 | 158 |
| Jacobstad | 7,580 | 5,092 | - | 5,092 | 2,488 |
| Ny Karleby | 9,863 | 4,348 | 5,432 | 9,780 | 83 |
| Vasa | 9,201 | 2,197 | 5,126 | 7,323 | 1,878 |
| Kristinestad | 2,231 | 1,296 | - | 1,296 | 935 |
| Total | 58,891 | 42,136 | 12,890 | 55,026 | 3,865 |
| Västerbotten & Ångermanland | | | | | |
| Torneå | 360 | - | - | - | 360 |
| Luleå | 728 | 1,392 | - | 1,392 | -664 |
| Piteå | 1,108 | 708 | - | 708 | 400 |
| Umeå | 409 | 433 | - | 433 | -24 |
| Härnösand | 173 | - | - | - | 173 |
| Total | 2,778 | 2,533 | - | 2,533 | 245 |
| Grand total | 61,669 | 44,669 | 12,890 | 57,559 | 4,110 |

Source: Handel och sjöfart 12, Kammararkiv, RA.

The records of the 'petty toll' or town toll (lilla tullen, 1622-1810), paid on the merchandise entering a staple town, have not been particularly well preserved but summary extracts remain for 1666-1686.[17] The quantities of tar delivered to Bothnian towns in 1686 can thus be compared with the amounts sent to Stockholm and Åbo in the same year.[18] The same source also contains summaries of the tar and pitch exports from Sweden's (including Finland) staple ports, enabling the construction of Figure 5. The differences between the amounts delivered to and shipped from the various Bothnian ports may provide some indication of local consumption - especially by shipbuilding - but the negative values clearly suggest that stocks must also have been retained at the towns from one year to the next. The apparently large discrepancy between deliveries and shipments diminishes to more realistic proportions when the numbers of pitch barrels are converted back to their tar equivalents (two barrels of tar were needed to distil one barrel of pitch). In 1686 Stockholm's imports consisted of 46,751 barrels of tar and 11,461 barrels of pitch (equivalent to another 22,922 barrels of tar), or circa 69,673 tar units. Her exports were 31,692 barrels of tar and 15,684 of pitch (ie 31,368 in tar equivalents), representing 63,060 tar units. The number of actual barrels of tar and pitch imported was 58,212, and 47,376 were exported - a difference of 10,836 barrels. The city apparently retained 15,059 barrels of tar and shipped 4,223 more barrels of pitch than she received during the year. Since the extra pitch was equivalent to 8,446 barrels of tar, the real difference between the imports and exports was in fact only 6,613 barrels! Clearly, like the Bothnian towns, Stockholm distilled pitch, built and repaired ships, and also maintained reserve stocks between one shipping season and the next. Högberg (1969, pp. 161-162) found that between 1802 and 1806 Stockholm apparently consumed about half of her tar and pitch surplus (the difference between imports and exports). However, he did not attempt to calculate how much of her tar import was converted into pitch, a process which reduced the number of actual barrels exported and thereby accounted for a sizeable amount of the city's apparent consumption.

Table 3 shows the deliveries of locally produced tar to Bothnian towns, together with a simple calculation of the balance between deliveries and shipments to Stockholm in 1686. The production of pitch at the town-ports was probably larger than the amounts sent to the capital, for its presence is indicative of local shipbuilding, which was particularly important in Österbotten. Details of vessels built or under construction at the individual towns in 1686 are not easily obtainable, but some idea of the regional significance of this skilled and high value-added processing of local raw materials can be given from secondary sources.

An analysis of all known vessels of over 100 lasts (svåra läster), built during the period 1680-1814, was carried out in 1956 by the National Museum of Maritime History (Sjöhistoriska museet).[19] During these 135 years 1,817 ships were listed, totalling 247,360 lasts. Apart from Stockholm (ranked 2nd with 113 ships, 24,520 lasts), the coastal towns and villages of Österbotten completely dominated merchant shipbuilding and accounted for 953 ships of 121,330 lasts - virtually half of the total numbers (52.4%) and tonnage (49%) covered by the survey. Thus, the Gamla Karleby district took 1st place, with some 300 ships (ca. 38,000 lasts); Uleåborg was 3rd with 22,400, Jacobstad 4th (21,000), Vasa 5th (12,500) and Kristinestad 12th (5,200). In the south of Finland, only Åbo (11th, 5,200) was of any importance. On the Swedish side of the Bothnian Gulf the coastal settlements produced 318 ships of 44,610 lasts, the main Västerbotten districts being Piteå (8th, 7,300 lasts) and Kalix (13th, 5,200), while shipbuilding in central and southern Norrland was focused respectively on Sundsvall (9th, 7,000) and Gävle (6th, 8,200). South of Stockholm, Västervik took 10th place with ca. 6,500 lasts and on the German coast Stralsund was 7th with 8,100. These 13 districts alone accounted for some 171,100 lasts (68.4% of the total volume), and in most cases they were important centres for the production and export of tar and pitch. The larger vessels were usually sold to Stockholm merchants and the smaller ones used locally, since the Bothnian Trade Restraint hindered the town merchants from making long-distance voyages. This was to change after the relaxation of restrictions in 1765, and Bothnian traders began to venture outside the Baltic and become involved in direct international commerce, selling both staple goods and finished sailing ships of a high quality.

In 1686 Sweden/Finland accounted for 56,133 barrels of tar and pitch passing through the Sound, which was equivalent to almost 84% of the total volume (Tab. 2). This was much the same proportion as in 1641 (86%), but it is probably an underestimate since 197 Swedish vessels left ports in Sweden/Finland and passed free through the Sound without their cargoes being registered.[20] Not only had the states of the eastern and southern Baltic lost their significance in the tar trade but Norway, too, was no longer a very important supplier, producing between 718,000 and 24,000 barrels per year around the end of the century (Sweden/Finland exported five or six times as much).[21]

Before 1712 the Swedish tar companies had been able to restrict production and exports in order to force up market prices. Naturally these measures were extremely unpopular, both among the suppliers (who were underpaid and yet forced to keep outputs well below local capacities) and among the maritime consumers. Dutch maritime and commercial power had waned towards the end of the 17th century and Britain took over the lead. Her response to Sweden's monopolistic policies was to seek alternative sources of supply, even though they might be more distant and involve greater freight costs. Archangel was thus drawn temporarily into the web of British commercial contacts and supplied tar and unprocessed timber. Although even more remote from the British core of consumption, the colonies in North America had the attraction of being under British control, politically and economically. This led to strategic rather than economically-viable investments in the development of American forest resources to supply naval stores and timber. The extra distances naturally involved high transport costs, which resulted in impossible c.i.f. prices for London merchants. The British government was therefore compelled to pay high subsidies on North American tar and timber in order to ensure supplies and also imposed prohibitive customs duties on tar imports from the continent (Hautala 1963:19). Colonial tar was never considered to be as good as the Swedish, however; for it was thicker and became too hot when used for treating cordage. Furthermore, the century was punctuated by wars - especially between Britain and her new rival, France. Each time, tar and pitch were at a premium and when cut off from her supplies in the American colonies and former colonies Britain again turned principally to Sweden, but also to Russia (Högberg 1969, pp. 158-159).

Mercantile measures were adopted by Sweden in order to attempt to maintain control of her export trade. In 1645 Sweden's shipping regulations gave armed merchant vessels 'total exemption' (helfrihet) from customs duties, although in reality the reduction was only one-third. Other Swedish vessels had 'half exemption' (halvfrihet), which reduced customs by one-sixth. Foreign vessels, however, were required to pay full dues, which was an effective barrier to foreign contact and influence. These regulations were later modified and in 1782 all Swedish vessels had only to pay 50% export duties. The 1724 Production Edict (Produktplakatet, cf Britain's Navigation Acts) reinforced these privileges, by excluding foreign vessels from carrying Swedish commodities unless they arrived with cargoes of their own nationality or in ballast.[22] These measures had the undesirable effect of raising foreign freight rates and those of Swedish shipowners, who seized the opportunity to follow suit. Increasing the freight costs to Swedish ports thus pushed them relatively 'farther away' from the main markets and let more distant regions move 'closer' and enter the competition, especially in terms of local production costs. Herein lies part of the explanation as to why von Thünen's land use rings are less easily identified on a macro-scale, and why the observed distortions cannot be accounted for solely by variations in physical geography and land communications.

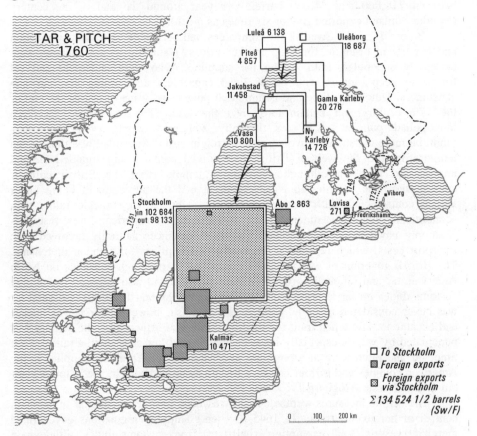

Fig.6:Shipments of tar and pitch from ports in Sweden/Finland in 1760 (Source:see Table 1)

The map for 1760 (Fig. 6) reveals the dramatic change in Sweden's tar trade brought about by territorial losses to Russia, particularly Viborg (1721) and its temporary replacement port of Frederikshavn (1743). Kaila (1931, p.36) portrayed Viborg's extensive tar production hinterland in the late 17th century, before the region's change-over to timber exports (see Fig. 7). Despite difficulties of trading during the Great Nordic War (1700-1721), when Viborg surrendered to the Russians in 1710, the town remained officially part of Sweden/Finland until the Peace of Nystad in 1721, when it formally became part of Russia.[23] Sweden thus lost access to her largest single tar-producing area and her total foreign exports fell by about one-third. The production of tar on the Russian side of the new border was abruptly terminated, however, by the Tsar who decided that it was a wasteful use of forest and encouraged the sawmill industry instead. This resulted in the spread of fine-blade saws and a rapid growth in the export of sawn deals from Viborg and other Russian ports in the Gulf of Finland.

In terms of production, the dominance of Österbotten was overwhelming (64.8%, see Tab. 1), but the Bothnian Trade Restraint continued to channel the region's trade through Stockholm. After the relaxation of Bothnian trade regulations in 1765, however, the town fleets of Österbotten became increasingly engaged in the direct foreign export of tar rather than shipping to the capital city, which came to rely increasingly on supplies from northern Norrland (Västerbotten).[24] The Sound Toll Tables for 1760 reveal no pronounced changes from the 1686 situation (Tab. 2). Sweden/Finland continued to dominate, accounting for 97.5% of the tar traffic. Although St Petersburg now appeared separately in the accounts it was of no importance whatsoever in the tar trade that year. However, the city had increased its share of the sawn timber trade and clearly the planned economic change-over mentioned above had taken place. During peacetime Russian tar and timber exports generally went through the new capital city rather than Archangel, but the reverse applied during the frequent wartime periods. In 1762, however, a discriminatory domestic toll was lifted from Archangel and the local, periodically Dutch-controlled, trading companies lost their tar monopoly - so that the northern route increased in importance during the latter decades of the century, particularly in relation to the British market (Högberg 1969, p. 154).

Table 1 only shows the percentage of direct exports from the various producing regions in 1805. At the time of writing, no details were to hand regarding Stockholm's supply areas (61.8% of the total tar exports), so that the city's imports have been allocated to the towns of Österbotten and Norrland in the proportions that applied in 1760. Österbotten undoubtedly increased its dominant position; but Västerbotten was also expanding its tar production as the region began a phase of economic development and colonization.[25] This regional growth was partly in response to Sweden's external losses of territory and resources that had begun in 1721 and ended with the loss of Finland in 1809.

No comparable data on trade through the Sound have yet been published for the years after 1795.[26] This is unfortunate, since the years of the French Revolution and the Napoleonic wars can be assumed to mark the end of a political, social and economic era, and the real beginning of continental industrialization as British innovations were diffused in the peacetime that followed.

This portrayal of regional changes in commercial tar production has so far been limited to the ports of shipment, implying that the tar-burning generally took place in their immediate hinterlands. Initially this was undoubtedly the case, but over time the activity was relocated farther and farther inland - thereby increasing the price of tar at the port. Reino Ajo (1946) has demonstrated the movement of the principal tar-burning areas across Finland during the 18th century (Fig. 7).

The major producing region in 1700 can clearly be identified as lying in the southeast (Savolaks & Karelia), where extensive areas of fine sand supported the pine
forests that supplied tar to Viborg and Fredrikshamn. It can also be seen that tar-
burning had virtually disappeared from there by 1750, since the Russians then
controlled 'Old Finland' and the former tar ports. Lumbering gradually took over
and the region formed the eastern limit for later tar production. Tar outputs
continued to expand, however, in a belt in Österbotten running parallel to the
Bothnian coast. By 1750 this belt had already moved inland, a process that
continued until about 1800 after which it contracted towards the Russian border.
Tar burning had originally replaced burn-beating (for rye cultivation) as the main
form of forest utilization. In its wake followed the felling of the better timber for
the later sawmill industry.

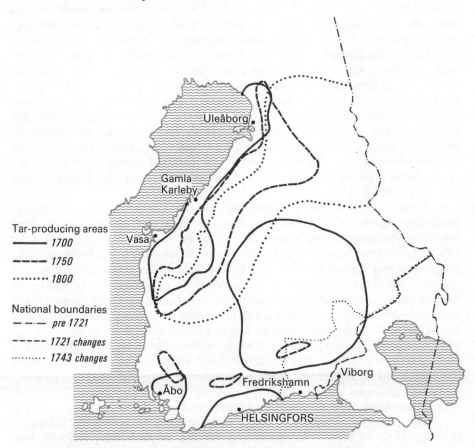

*Fig.7:Spatial changes in Finnish tar burning during the 18th century (after Ajo 1946:37,51 & 68)*

In Sweden, the extensive forests of Norrland were somewhat less accessible than in Finland due to greater variations in relief and the navigational hazards of the mountain and forest rivers.[27] Instead of simple zones of different forest usage running parallel to the coast, the pattern was more complicated with tongues penetrating inland along the river valleys.[28] A map of Norrbotten's tar production in 1822 (Hoppe 1945, p. 161, Fig. 23) shows that the centre of gravity of tar burning had already moved inland from the immediate coastal zone, along the major river valleys. In the decades that followed, this movement continued upstream (ie following the floatways) into Lappmark, where colonization gathered momentum in connection with the transport of iron ore and felling of timber for the export sawmill industry. Tar production provided an important subsidiary source of income for new farmers and labourers, especially since the products of hunting and fishing were no longer such desirable trade commodities.

Despite the numerous factors that have been introduced thus far to help explain the spatial patterns and changes in commercial tar production, the overall process does tend to follow the model proposed earlier in Figure 3C:II - with dominance shifting through a succession of regions, starting from those near to the core and moving gradually out towards the periphery of northern Europe. Each region thus experiences a cycle of tar production which can be illustrated, in both absolute and relative terms, by substituting regions A-C for products 1-3 in Figures 3A and 3B. To minimize the high costs of sea transport during this early modern period, resources near to the main markets were the first to be exploited. When a shortage of raw material arose and tar-burning was forced to move inland, away from the ports of shipment, local production costs and f.o.b. prices increased. This meant that other regions, more distant but with untouched resources, then had a comparative advantage in production costs which could offset higher sea freights and enable them to compete with and even take over from the earlier, more favourably located regions.

Clearly, the distance factor plays a key role in this dynamic process. Since increasing demand from the cores of consumption provides the prime motivation for the expansion and relocation of resource utilization, we might use the term 'demand wave' (as a less cumbersome alternative to 'resource utilization wave') to denote the movement of each form of forest utilization - outwards from the cores of consumption, like rings on water. European tar and pitch production fits this model remarkably well, but what about the case of sawn timber?

## 3.    The Sawn Timber Trade

The trade in softwood timber is more complicated than the tar trade and is more difficult to investigate, for the following reasons: Firstly, there are far more commodity categories and softwood cannot always be distinguished from hardwood. Secondly, most of the different categories and terms of measurement vary over time. Thirdly, even if sawn timber can be identified satisfactorily it is impossible to determine how it was produced; was it sawn by hand (or even hewn) or by water-powered sawmills, and did the mills have fine or coarse blades? The solution adopted here is simply to aggregate the different types of deals, planks and boards, and use a numerical measure (the dozen, ie 12 pieces) in order to try to follow the movement of the trade in sawn timber.[29] The timber trade of northern Europe as a whole appears to have experienced continual growth throughout the early modern period. Dutch influence predominated until the latter half of the 18th century, when the American Revolution and Napoleonic Wars finally ended their hegemony in the Baltic trade in timber and naval stores in favour of Britain, the new leading commercial nation. These two nations provided important markets as well capital and transport for the early timber trade. Norway was best situated to supply the rapidly growing demand and, in the absence of alternative means of livelihood, the country soon came to dominate in the supply of fuel and sawn timber to Holland and Britain, maintaining its lead throughout the period.

In the 16th and early 17th centuries the Baltic timber trade was focused mainly on the southern shores of the sea. Prussia and Pomerania were the main areas of supply, the logs being floated down-river in the form of rafts which often carried corn to the ports of export.[30] During the 17th century, however, the timber trade from the southern Baltic began to decline as accessible forests near the rivers were exhausted and overland transport distances increased between the remaining resource areas and the coast. Instead, the eastern and northern reaches of the Baltic Sea became the new focus of the trade in timber (Tab. 5). What, then, was the role of Sweden/Finland in this diffusion process?

Tab. 4: Regional changes in the origins of Sweden/Finland's sawn timber exports for selected years (%).

| Year | W.coast | S.Sweden | S.Finland | Osterbotten | Norrland | Total[1] |
|------|---------|----------|-----------|-------------|----------|---------|
| c 1559[2] | 55.2 | 26.0 | 15.7 | - | 3.1 | 22,808 |
| 1641[3] | 17.0 | 77.7 | 0.1 | - | - | 31,496 |
| 1685[4] | 30.5 | 53.3 | 6.0 | - | 0.8 | 56,732 |
| 1724[5] | 47.3 | 32.3 | 3.4 | 0.6 | 16.4 | 77,643 |
| 1760 | 22.5 | 32.7 | 6.5 | 1.6 | 36.7 | 178,860 |
| 1805[6] | 15.7 | 34.3 | 12.2 | 3.2 | 34.6 | 205,543 |

1  Number of dozens exported (excluding Stockholm) plus deliveries to Stockholm; except for 1641, 1685 and 1805, which are based solely on exports.
2  Data for Lödöse from 1546, and for Stockholm's import origins from 1560.
3  Exports only. Stockholm's share (5.2%) probably came from S Sweden, S Finland, and even Norrland.
4  Exports only. Stockholm's share (9.4%) probably came mainly from Norrland, but also from S Finland and S Sweden.
5  Data for Stockholm's import origins from 1723.
6  Exports only. Stockholm's share of the export trade (23.6%) has been apportioned to Österbotten (4.3%) and Norrland (95.7%), as in 1760. As Stockholm only re-exported c 30% of her imports, Norrland must really have accounted for over 50% of a domestic and foreign timber trade amounting to around 300,000 dozens in 1805.

Sources: c 1559 - Forssell (op. cit.);. 1641 - Boëthius & Heckscher (1936, pp. 36-53); 1685 - Boëthius & Heckscher (1936, pp. 139-339); 1724 - Boëthius & Heckscher (1936, pp. 437-457); 1723 - Handelskollegiet E:XI:b:2, SSA; 1760 - Årsberättelser: Utrikeshandeln ser 2, KmkA, and Borgerskapets äldsta arkiv F:I ser 7 n:r 26, SSA; 1805 - Årsberättelser: Utrikeshandeln ser 2, KmkA.

Tab. 5: Sawn deals, planks and boards passing through the Sound Toll for selected years.

| Year | Denmark | | S. Baltic | | E. Baltic | | Sweden/Finland | | Total |
|------|---------|------|----------|------|----------|------|----------------|------|--------|
| | dozens | % | dozens | % | dozens | % | dozens | % | dozens |
| 1562 | - | - | 3,569 | 90.2 | 58 | 1.5 | 330 | 8.3 | 3,957 |
| 1600 | - | - | 3,156 | 82.9 | 258 | 6.8 | 392 | 10.3 | 3,806 |
| 1641 | 251 | 2.6 | 4,998 | 52.5 | 815 | 8.6 | 3,450 | 36.3 | 9,514 |
| 1685 | 2,481 | 11.2 | 2,509 | 11.4 | 6,982 | 31.7 | 10,078 | 45.7 | 22,050 |
| 1724 | 357 | 1.1 | 7,094 | 21.2 | 4,821 | 14.4 | 21,221 | 63.3 | 33,493 |
| 1760 | 1,425 | 1.1 | 10,613 | 8.0 | 16,349 | 12.4 | 103,409 | 78.5 | 131,796 |

Sources: Sound Toll Tables for 1562 (Bang 1922, p. 4), 1600 (ibid, p. 198), 1641 (ibid, p. 466), 1685 (Bang & Korst 1939, pp. 294-300), 1724 (Bang & Korst 1945, pp. 59-65), and 1760 (ibid, pp. 752-761).

Table 4 shows the broad regional movement of sawing across Sweden and Finland during the study period. The pattern is not the same nor is it as clear as in the case of tar production (Tab. 1). The west coast region played a more complex role and Finland was of far less importance in terms of volume than is often implied. The focus of sawn timber production shifted abruptly from the west coast to Norrland between 1724 and 1760. This is an important observation, for previous evaluations of the significance of different regions have been based solely on direct export trade statistics. By including Norrland's domestic trade with Stockholm in the calculations, a more realistic picture is obtained and it can be seen that the region in fact experienced a take-off in sawmilling in the mid-18th century, ie a full century before it took the lead in the direct export trade through the application of steam technology.

The first record of a coarse-bladed, water-driven sawmill dates from 1447, when the monastery of Vadstena had a saw in Vadsbo (then in the western border region of Sweden). Älvsborg also had a saw by 1489 and others were already in use in Norway in the 15th century.[31]. In Sweden, the Church and the Crown had also established numerous saws by the 16th century, not only near the west coast but also in Norrland, for example Gävle in 1554 and others in 1573/4 (Carlgren 1926, p. 14). In turn, Finland received this innovation from Sweden in the 16th century, 1545 being the earliest record (Meinander 1945:311). The diffusion path clearly moved eastwards across these three countries (Åström 1988, p. 44). Nevertheless, it must be assumed that much of the deals, boards and planks were hewn or handsawn and that the export of 'sawn' timber does not necessarily indicate the use of water power.

Again, the 1559 picture can only be a composite of fragmentary data from several different years - nevertheless it illustrates the small scale of the trade in sawn timber and shows the importance of Sweden's only west coast outlet at Lödöse (Fig. 8). This region had locational advantages similar to those of Norway and was of interest to Dutch merchants. Within the Baltic 'pond' Sweden/Finland exported some 10,600 dozens; presumably most of this went to the ports of

Prussia, for the earliest usable records from the Sound Toll show that in 1562 only 330 dozens were listed as moving westwards from Swedish ports.[32]

The Sound Toll accounts for 1600 show that little change had occurred since 1562, with the west coast following the rapid expansion that was taking place in southeastern Norway (Boëthius 1929, p. 282). The situation was somewhat different in 1641, when Sweden/Finland can be seen to begin to rival the southern Baltic ports (Tab. 5). Within the Swedish realm the centre of gravity had shifted from the west coast to the east, where the ports of Småland and Östergötland dominated the export trade in deals and planks. Local production undoubtedly included oak timber for shipbuilding but, since most of this was reserved for the domestic usage, the export commodities were probably of the less valuable pine.

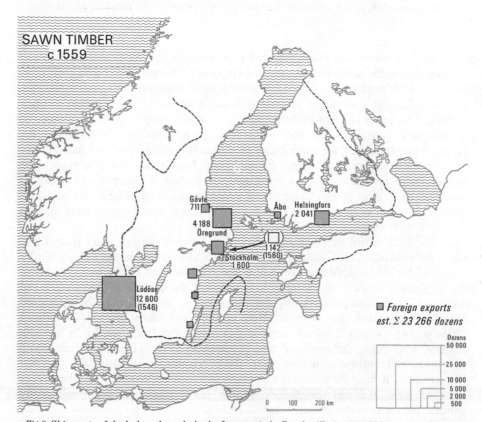

Fig.8: Shipments of deals, boards and planks from ports in Sweden/Finland c.1559 (Source: see Table 4)

Throughout most of the early modern period Sweden's exports of tar and sawn timber were insignificant when compared with the value of commodities derived from copper and iron mining. Even as late as 1750, the charcoal iron industry provided 75% of the value of Sweden's exports (Heckscher 1949, p. 652). The mining and refining processes required copious supplies of charcoal, which could only be transported for short distances overland before it became too broken to be of use.[33] This meant that the forests adjacent to mines and ironworks were soon depleted, transport costs increased and local shortages arose, which resulted in an acute fear of a shortage of forest during the 17th and 18th centuries (Wieslander 1936, p. 651). Powerful mining interests persuaded the state to therefore adopt protective policies which restrained the construction of sawmills, especially in the vicinity of ironworks and towns, and inhibited timber exports.

While Sweden concentrated mainly on the production of high quality iron and Finland on tar, timber constituted the major part of Norwegian exports (apart from the sale of rather poor quality tar). The Dutch were primarily interested in balks for pile-driving and dike construction, but the more important trade was in sawn timber, especially 'ordinary' deals. Britain became the leading consumer of Norwegian deals after the Great Fire of London (1666) and, indeed, throughout the 18th century "the fortunes of Norwegians were closely bound up with English fluctuations and developments" (Tveite 1961, p. 566). In the early decades of the seventeenth century, the Norwegian economy was expanding and trade in tar, masts, hewn and sawn timber was on the increase. Water-sawn deal exports may be used to indicate fluctuations and changes in the country's timber trade (ibid, p. 568): The customs records show that in 1668 deal exports peaked to around 720,000 dozens, but had apparently declined to 470,000 by 1699. Part of this decrease may have resulted from economic stagnation in continental Europe, which led to declining demands and falling prices, but it has been pointed out that Tveite's data are probably less reliable after 1688, when Norwegian export restrictions led to the under-recording of outgoing shipments by customs officials and merchants (Kjærheim (1964, pp. 91-95). Britain benefited from the economic situation, however, and imported increasing quantities of the now cheap Norwegian timber as the British economy expanded and her great naval and mercantile fleets came into being.

The growing shortage of English oak had widened the price gap between oak and Norwegian fir so that a substitution took place within the building trade, whereby fir replaced oak for structural purposes. London's Great Fire had undoubtedly accelerated this change-over and the economic growth of Britain was reflected in the architectural renewal and physical expansion of the capital city. The increased prosperity even penetrated country housing standards, as fir deals came to be used for flooring and panelling (Tveite 1961, p. 574). Thick Norwegian deals and planks were still either handsawn or roughly sawn at small water-powered mills using single, coarse blades - little different from those used in the 16th century. These products were usually reduced to thinner dimensions on arrival at Dutch and British ports, and Norwegian timber thus provided the semi-raw material for a consumer-based sawmill industry in the ports of the core regions. Such mills were naturally interested in obtaining logs, balks, etc, rather than fine-sawn, virtually finished boards, so that there must have been a conflict of interests between them and sawmilling in the resource-rich regions.

In 1685, Norway had around 1,700 sawmills, mainly still coarse-blade but possibly fine-blade in view of the long-standing influence of Dutch timber merchants. The country exported some 610,000 dozens, which was almost eleven times greater than Sweden/Finland's exports of 56,732 dozens.[34] Most of Norway's sawn deals came from the south-east (83.4%), where some 1200 saws were in operation. This illustrates well the heavy concentration of sawmilling in regions located near to the cores of consumption and outside the transport-economic constraint of the Sound bottle-neck and its tolls. Clearly, Norway was far ahead of Sweden with regard to sawmill technology and organization (Heckscher 1936, p. 431).

In the case of Sweden, the west coast region (centred on Göteborg and Uddevalla) frequently held the leading position regarding sawn timber exports for precisely the same reasons, trading mainly with Holland and Britain (see Tab. 4, 1559 & 1724; also Boëthius 1929, p. 286). Hostilities with Denmark/Norway had tended to inhibit the early trade through Lödöse and its successor Göteborg (founded 1621). But when Sweden gained the upper hand and took control of the whole of the west coast and Skåne (1658) she increased her timber exports from the west coast and, at the same time, obtained exemption for her ships from paying the 1% Sound dues on their cargoes until her fall from power in 1721. These privileges allowed Sweden's Baltic regions to develop Mediterranean and North Sea trade without the extra freight surcharge that the Sound system imposed on her competitors, temporarily bringing Sweden's Baltic ports a little nearer to the consumers in terms of cost-distance. The effect of the Sound as a permeable barrier to trade (ie reducing potential flows) can easily be exaggerated but, in association with changes in demand, freight rates, marine technology, mercantile policies and the international political scene, it undoubtedly played a significant role in the timing and character of forest utilization in the regions of northern Europe.

Along the coast south of Stockholm, the ports of Norrköping, Söderköping, Västervik and Kalmar, appear to have seized the opportunity to increase their timber exports and take over as the leading exporting region during much of the 17th century (Tab. 4). Calculations based on Sweden/Finland's total exports and the recorded amounts passing through the Sound (Tabs. 4 & 5), suggest that most of Sweden's sawn timber was shipped to the ports of the southern Baltic (72.1% in 1641 and 51.7% in 1685).[35] The situation changed radically during the 18th century, when Swedish timber exports took off and the proportions remaining in the Baltic further decreased to 25.3% in 1724 and 19.7% in 1760. Conditions were clearly favourable for increasing shipments to Holland and Britain and the west coast was experiencing an export boom in 1724 (Tab. 4 & Fig. 9) - possibly an effect of Sweden's loss of rights to toll-free passage through the Sound.

The map of 1724 reveals another distinctive feature of Stockholm's commercial role. Whereas almost all of the imported tar was re-exported, the city itself consumed about 70% of the boards received from Norrland - the remainder often being shipped in combination with iron cargoes.[35] The capital city was expanding rapidly and construction timber was therefore in great demand. This partly explains why Stockholm failed to dominate the export trade in sawn timber. There were no monopolistic trading companies as in the case of tar, but the

Bothnian Trade Restraint clearly served the city's interests by ensuring that domestic supplies were maintained rather than the timber being shipped directly abroad. Much of the timber imported by Stockholm from Norrland and Finland consisted of thinner board dimensions (enkla bräder), which were mostly consumed locally. Stockholm's exports, however, were mainly of thicker boards (halvbottenbräder) which, often in combination with cargoes of bar iron, were increasingly exported to southern Europe during the 18th century in exchange for salt (Högberg 1969, pp. 121-123).

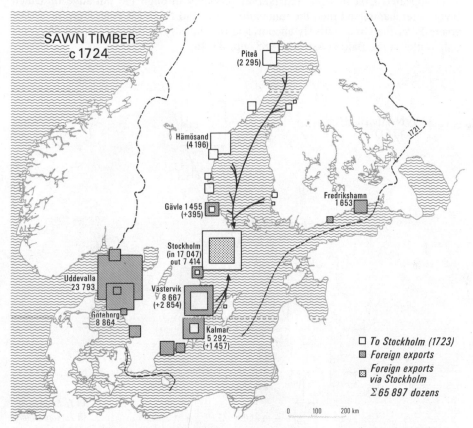

*Fig.9: Shipments of deals, boards and planks from ports in Sweden/Finland in 1724 (Source: see Table 4)*

After the Great Northern War (1700-1721), Sweden's period of greatness and control of the Baltic came to an end as Russia took control of Viborg and the Baltic states of Estonia and Livonia with their important timber ports of Riga and Narva. High freight costs meant that the more distant producing areas had to be satisfied with low f.o.b. prices if they were to compete with more favourably located supply regions. Since wage levels tended to be lower in the more peripheral areas, production and inland transport costs could be kept down;

however, it was clearly in the interest of entrepreneurs in Russia and the Gulf of
Finland to improve the quality and thereby the value of their export products. This
was undoubtedly an important factor encouraging the introduction of the Dutch-
type fine-bladed sawmill into the Baltic in the mid-17th century. These saws
usually had two or more fine blades set in frames and they were far more efficient
than the single coarse-blade mills. Even a single frame fine-blade mill could saw
three to four times as many logs per day and produce superior products with far
less waste than the older type of sawmill (Meinander 1945, p. 89). These new
saws appeared first in Riga and Narva, probably through the initiative of Dutch
timber merchants, and then the innovation spread to Nyen (St Petersburg in 1703),
where Peter the Great actively encouraged the industry in his new capital city - his
only outlet in the Baltic (Åström 1988, pp. 44-46).

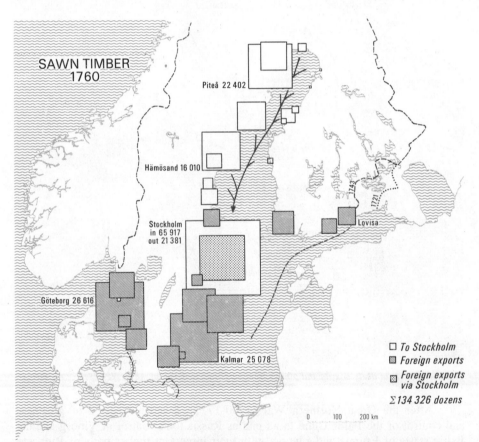

*Fig.10: Shipments of deals, boards and planks from ports in Sweden/Finland in 1760 (Source: see Table 4)*

The diffusion route then passed to Viborg where, after the town had submitted to Russian forces in 1710 and became Russian in 1721 (Peace of Nystad), the Tsar banned the "wasteful" export of tar and favoured the newly-introduced sawing technique.[37] Sweden's loss of Viborg, its major tar-exporting port apart from the entrepôt of Stockholm, led to the rapid expansion of Fredrikshamn for the export of tar and fine-sawn timber (for this port still had access by water to the remaining part of the rich forest areas of Karelia). The subsequent loss of Fredrikshamn together with the rest of the Karelian hinterland in 1743 forced the Swedes to attempt to repeat this exercise, by permitting the building of fine-bladed sawmills in Nyland, especially around Helsingfors. Borgå and the new border town of Lovisa (founded in 1745) were meant to take over the tar trade, but they lacked easy access to the former tar-producing tracts. Åström (1975, p. 3) shows how tar and pitch exports fell to nothing at Narva in the 1690s and Nyen in the 1710s, while the Viborg-Frederikshamn-Helsingfors regions also declined markedly at the turn of the century. Instead, in roughly the same order, the export of fine sawn deals began to dominate the timber trade. As had been the case for tar in the 17th century, the south of Finland again took the lead in the country's production of these new types of forest products for export.

Since this new trade appeared to be more profitable than the old coarse-blade sawmill products (and since the iron industry was less dominant), Sweden decided to give active support to the establishment of fine-bladed sawmills in Norrland by legislation in 1739, granting generous 'stockfångst' privileges (ie rights to fell a given number of trees on payment of a nominal fee per stump) in specific areas of Crown forest to the new entrepreneurs.[38] The first such mill in Sweden was that at Kramfors, Ångermanland - built by Christopher Kramm, who fled from Finland to Sweden during the Russian advance bringing his previous experience of this technique. From here the innovation spread rapidly northwards and southwards, with forest privileges being granted to new mills in most of the major river valleys, particularly in the coastal province of Västerbotten.[39] During the latter half of the 18th century a parallel movement also took place along the coast of Österbotten so that, after the relaxation of the Bothnian Trade Restraint in 1765, the new staple towns in the Gulf began the direct export abroad of sawn timber, as well as tar and sailing ships. Nevertheless, the traditional Stockholm market/entrepôt remained the dominant destination for Bothnian forest products until well into the 19th century .

This "rather peculiar route of diffusion" of Dutch-type sawmills (Åström 1988, p. 45) was undoubtedly the result of Sweden's negative attitude towards the timber trade. In effect, barriers to the free diffusion of new ideas and timber technology had been erected round Sweden, so that the Dutch timber merchants tended to invest their capital and know-how in the eastern Baltic, leaving Sweden to the iron and tar merchants. In more graphic terms one might say that the wave of technological innovation washed round the Swedish 'island'.

Table 4 reveals that by 1760 Norrland had taken over as the major producing region for sawn timber, a position it was to retain throughout the next century. Almost all of the region's output continued to go to Stockholm, however - even after the final removal of the Bothnian trade restriction in 1812. The map of the sawn timber trade in 1760 (Fig. 10) clearly shows the concentration of sawmilling on the western side of the Gulf of Bothnia, in striking contrast to the tar production on the Finnish side (Fig. 6). The Norrland forests had not been so heavily exploited as in Österbotten and, once cleared, the numerous mountain and forest rivers provided ideal floatways for the cheap transportation of logs to the sawmills and of rafts of sawn timber to the coastal timberyards and loading places (Layton 1981, pp. 38-42).

Apart from the continued increase in the importance of sawmilling in southern Finland and the decline of Sweden's west coast, the regional distribution of sawn timber exports was little altered between 1760 and 1805 (Tab. 4). As in the case of tar, the difficulties of obtaining naval timber during the frequent wars of the latter half of the 18th century had led Britain to invest in the development of the timber trade in North America.[40] Russian timber resources were also in great demand, so much so that the Tsar imposed a temporary ban on all timber exports in 1798 in response to domestic shortages for shipbuilding and fears among Russian iron manufacturers of a fuel shortage (Albion 1965, pp. 178-179).

Each country produced a wide variety of timber-based commodities, but the importance of each type was largely determined by relative prices. As tar prices fell in relation to sawn timber, for example, so did the tar-trade profits. It is not surprising therefore that regional and national production then changed over to the more lucrative method of utilizing forest resources. Local variations in the type, quality and dimensions of standing timber obviously played an important role in determining the available alternatives. When shortages pushed prices too high for comfort, the core markets often responded by finding a cheaper substitute. During the 17th century, for example, shipbuilding technology responded to the shortage of large masts by changing to composite masts, thus enabling Norway to continue to contribute to British naval stores requirements. Regional specialization was, however, the result of a variety of other factors - including suitable sites for water power, legal and physical accessibility of the forest, means and costs of inland transport, availability of labour, capital, etc (Tveite 1961, pp. 570-571).

## 4.   Conclusion

Despite the difficulties of classifying and measuring different types of sawn timber, this overview demonstrates the general north-easterly movement of production across the coniferous forest belt of northern Europe - a movement that was becoming increasingly evident in the 18th century and was continued in the 19th century. The relatively late entry of Sweden/Finland into the timber trade had advantages in the long term. The extensive stocks of mature forest in Norrland were largely untouched before the introduction of the fine-blade sawmill, and the impact of this new technology was relatively slight in comparison with what was to follow when steam power was introduced in the mid-19th century.

The production of tar normally preceded sawmilling in the countries and regions of northern Europe during the early modern period, but its relative importance was declining towards the end of the 18th century. The demand for tar was still increasing, since steam power and iron ships had not yet been developed. The market for sawn timber increased more rapidly, however, as the core regions of Europe began to experience dramatic growths in population. Urbanization and the beginnings of industrialization necessitated greater imports of fuel and building materials, which could not be satisfied from the depleted forests of nearby regions. The softwood resources of northern Europe were therefore of increasing importance, and their utilization generally provided the means and impetus for regional development in the periphery.

The evidence of direct and indirect trade links between the remote north and the European core supports Wallerstein's conception of a European world-system, with its origins in the beginning of the early modern period.[41] At first glance, the distance-based models proposed at the outset of this paper do appear to have some explanatory value regarding the process of spatial change in resource utilization. More detailed research will be needed at the regional level, in order to test their validity more rigorously and to better understand the effects and inter-action of the numerous physical, political and economic factors involved.

### Notes

1.   This paper can provide no more than a brief introduction to a recently started research project with the ambitious (and somewhat cumbersome) title "The geographical evolution of the demand for and the supply of coniferous forest products within Britain and northern Europe from c 1700 to the present day: economic expectation or knowledge diffusion?" The period to be investigated embraces the end of the early modern period and the dramatic socio-economic and technological changes of the past two centuries and thus follows on from the period under consideration at this present conference. Nevertheless, the theoretical lines of approach are similar and this paper serves as a prelude to the forthcoming study, which is financed by the Bank of Sweden Tercentenary Foundation.

2.   N.J.G. Pounds is one exception, however, and vol. 2 of his "An Historical Geography of Europe" covers this very period, 1500-1840 (Pounds 1979). In chapter 6 he takes up the pattern of trade and even produces a simple diagrammatic overview of the flow of trade from the Baltic about 1800 (ibid, p. 279, Fig. 6.1).

3.   Layton (1981, pp. 182-186), Layton (1988, pp. 205-206), & Layton (1989, p. 106).

4.   For 1497-1783, tables on the goods passing through the Sound (traffic, type, quantities, origins and destinations) have been compiled and published by Nina Ellinger Bang (1922) and Bang & Korst (1939-1953); while Hans Christian Johansen (1983) has continued this work by computerizing the records for 1784-95.

5.   Pounds (op cit). A link by water did in fact exist between Lübeck and Hamburg, the small Stecknitz Canal, built by Hansa interests in 1390, but this soon silted up and was too small for later vessels (Heckscher 1954;45). The Kiel canal proper was not opened until 1897.

6.   During the war years between 1710 and 1720 Swedish toll exemption was suspended, but finally rescinded at the Peace of Frederiksborg in 1720. Thereafter, Swedish/Finnish shipping was more fully recorded (Heckscher 1936:686, 1949:645). The great variety of timber products complicates the task of following the movements of the timber trade as a whole, so that in any eventual long-term analysis based on these sources it would really only be practical to attempt to follow changes in the major types and categories of commodities for ports belonging to nations other than Sweden/Finland.

7.    eg. Lokala tullräkenskaper, Kammararkiv (KA), and Årsberättelser: Utrikeshandeln ser 2,
      Kommerskollegii arkiv (KmkA), in Riksarkivet (RA), Stockholm.
8.    eg Borgerskapets äldsta arkiv and Handelskollegiet, in Stockholms stadsarkivet (SSA).
9.    The Swedish tar trading companies were based in Stockholm and operated from 1648 to
      1682 and from 1689 to 1714 (Fyhrvall 1880, Hallberg 1959). By such devices as setting
      prices and imposing quotas - first in 1663 (Hallberg 1959, p. 101) - for the various towns
      and producing areas, the companies were able to constrain supplies and thus keep foreign
      market prices up. Their monopolies were removed between 1683 and 1688, and after 1712,
      periods when production was virtually free from controls.
10.   Friberg (1983, appendix tables A & B).
11.   Bang (1922, p. 5). 1562 is the first year for which the flow of goods through the Sound was
      recorded (ibid, p. IV). Most of Sweden's exports were carried by vessels belonging to other
      nations, so that Sweden's early toll exemptions were of little statistical importance (ibid, p.
      IX).
12.   It has been estimated that it took 15 large trees to produce 1 barrel of tar, so that between
      1.5 and 1.8 million trees would have been required annually during the latter half of the
      17th century (Heckscher 1936, vol 1:2, p. 433).
13.   Improvements in marine technology enabled the construction of increasingly larger vessels
      with the capacities needed for carrying bulkier cargoes and heavier armament. Naval
      strength became necessary to safeguard merchant shipping from piracy and belligerent rival
      nations, and the naval and commercial power of Europe's maritime nations provided the key
      to economic and political expansion once the resources of the less technologically advanced
      New World, Africa and Asia became apparent and accessible by sea.
14.   Fyhrvall (1880, pp. 291-293) & Hallberg (1959, pp.162-164).
15.   1648 was the first year for which the records of the new tar company are available. The tar
      came principally from Österbotten where the new towns (particularly Vasa, Gamla Karleby,
      Ny Karleby and Uleåborg) had experienced a boom in their trade with the capital.
16.   Tar exports from the southern Baltic in 1641 were less than a quarter of the volume shipped
      in 1600 (9,636 cf 42,018 barrels, respectively), so that these four decades appear to mark
      the end of large-scale tar production in continental Europe.
17.   Handel och sjöfart 12, KA, RA.
18.   Åström (1988, pp. 24-25, Tabs. 1:6 & 1.7).
19.   See G. Zethelius (1963, pp. 203-212). In the 18th century the last measured a vessel's
      displacement and was equivalent to a weight of 2,448 kg or, in later terms, 1.8 net
      registered tons.
20.   A further 77 Swedish-registered vessels departed from other Baltic ports, together with
      another 128 toll-free ships that belonged to other Baltic ports and territories controlled by
      Sweden. Cargoes are therefore lacking for 402 vessels sailing westward through the Sound
      (Bang & Korst 1939, pp. 308-314).
21.   Hallberg (1959, pp. 89-92).
22.   The mercantile system only began to break down between 1825 and 1857, the year when
      both customs exemptions and the Production Edict were repealed (Layton 1981, pp. 72-73
      & Högberg 1969, pp. 28-30).
23.   Viborg in fact surrendered to Russian troops in 1710 (Åström 1988, p. 28) and the tar from
      southern Karelia was used by the Russian fleet and even exported as Russian.
24.   In 1765 six Bothnian towns were granted active staple rights, permitting them to sail direct
      to foreign destinations with their own vessels. Foreign shipping was still excluded from the
      Gulf of Bothnia until 1812, when the towns received passive staple rights and the Bothnian
      Trade Restraint finally came to an end (Layton 1981, p. 74).
25.   Trade in Bothnia and the Baltic stagnated until the 1815 peace treaty (Congress of Vienna),
      but not until after 1827 did Finland effectively ignore Stockholm and concentrate on her

own independent foreign trade. See, for example, Åström (1988, pp. 37-40) and Häggström (1967, pp. 43-47).

26. Hans Christian Johansen (1983) has published computerized data for 1784-95.

27. The rivers were not effectively cleared for timber floating until capital investments were made in the 19th century, first in association with the establishment of large fine-blade water-driven sawmills and later during the steam-powered sawmill era.

28. L-E Borgegård (1973, p. 35, Fig. 6, & ibid, p. 86, Fig. 15).

29. cf Högberg (1969, p. 107).

30. See also Peet (1972, pp. 5-8) and Mager (1960).

31. Benedictow (1977, p. 202). The monastery had foreign contacts and the technology may well have been derived from Hansa merchants in the late Middle Ages, before the Dutch took over as the leading commercial power.

32. Bang (1922, p. 4). Over 90% of the sawn deals and planks passing through the Sound were shipped from ports in the south Baltic, but the amount was small (3,569 dozens). Possibly, some of the Swedish timber may have been re-exported, but most must have been consumed locally.

33. Usually within a radius of c 25 km (Hoppe 1945, p. 156).

34. Mykland (1977, p. 235) and Tveite (1961, pp. 532-536, Tab. XXXII).

35. It should be remembered, however, that the cargoes of Swedish vessels were not accurately recorded during the period 1651-1720.

36. It was maintained by Stockholm merchants that sawn timber was needed as stowage for heavy iron cargoes. For further discussion see, for example, Högberg (1969, pp. 107-109).

37. In 1685 Viborg's foreign exports consisted of 31,968 barrels of tar and 4,248 of pitch, but only 6 dozen deals were exported beyond the borders of Sweden/Finland, showing that before the Tsar's initiative sawmilling was insignificant (Boethius & Heckscher 1938, p. 218).

38. Carlgren (1926, p. 44); Åström (1988, pp. 54-55).

39. The first fine-blade saw on the west coast appeared in 1740 and a rapid expansion took place. By 1825, there were 49 fine-blade saws in the old province of Västerbotten, out of a total of 68 in Norrland as a whole (Carlgren 1926, p. 154; Layton 1981, p. 205).

40. The complex effects of war on the British timber trade with the Baltic and North America have been taken up by many authors, notably Hautala (1963), Albion (1965), Kent (1973), & Lower (1973).

41. Wallerstein (1974 & 1980).


# References

Ajo, R. (1946): Liikennealueiden kehittyminen suomessa (French summary: 'Développement des espaces de circulation en Finlande' pp. 97-117). In: Fennia 69, pp. 1-123

Alanen, A.J. (1956): Stapelfrihet och de bottniska städerna 1766-1808. Skrifter utgivna av Svenska Litteratursällskapet i Finland: Historiska och Litteraturhistoriska Studier 31-32. Helsingfors

Albion, R.G. (1926, 1965 edn.): Forests and Sea Power: The timber problem of the Royal Navy 1652-1862. Cambridge, Mass

Åström, S-E. (1975): Technology and timber exports from the Gulf of Finland, 1661-1740. In: The Scandinavian Economic History Review XXIII, pp. 1-14

Åström, S-E. (1988): From Tar to Timber: Studies in Northeast European Forest Exploitation and Foreign Trade 1660-1860. Societas Scientarum Fennica: Commentationes Humanarum Litterarum 85. Helsinki

Bang, N.E. (1922): Tabeller oever skibsfart og varetransport gennem »resund 1497-1660. Vol. 1:2A: Tabeller oever varetransporten. Kobenhavn

Bang, N.E. & Korst, K. (1939, 1945, 1953): Tabeller oever skibsfart og varetransport gennem Öresund 1661-1783. Vol. 2:1: Tabeller oever varetransporten 1661-1720. Vols. 2:2:1-2: Tabeller oever varetransporten 1721-1783. Kobenhavn

Benedictow, O.J. (1977): Fra rike til provins 1448-1536. In: Knut Mykland (ed.): Norges Historia, vol. 5. Oslo

Boëthius, B. (1929): Trävaruexportens genombrott efter det stora nordiska kriget. In: Historisk Tidskrift 49

Boëthius, B. & Heckscher, E.F. (1938): Svensk handelsstatistik 1637-1737 (Swedish statistics of foreign trade 1637-1737. Introductory chapter in English, pp. xlv-lviii). Stockholm

Borgegård, L-E. (1973): Tjärhanteringen i Västerbottens län under 1800-talets senare hälft. Umeå

Bugge, A. (1925): Den norske trælasthandels historie, vol. I (pre 1544), vol. II:1-2 (1544 - end of the 17th century). Skien

Carlgren, W. (1926): De norrländska skogsindustrierna intill 1800-talets mitt. Uppsala & Stockholm

Calonius, I. (1936): Den Österbottniska sjöfarten och handeln under 1500-talet och böprjan av 1600-talet. In: Terra, pp. 136-161

Forssell, H. (1875):Anteckningar om Sveriges handel och om städernas förhållanden under de första femtio åren af Vasahusets regering. Stockholm

Friberg, N. & Friberg, I. (1983): Stockholm i bottniska farvatten: Stockholms bottniska handelsfält under senmedeltid och Gustav Vasa. En historisk-geografisk studie (English summary pp. 465-485). Uppsala

Fyhrvall, O. (1880): Bidrag till svenska handelslagstiftningens historia del I: Tjärhandelskompanierna. In: Historiskt Bibliotek 7, pp.287-347

Hallberg, A. (1959): Tjärexport och tjärhandelskompanier under stormaktstiden. In: Historiska och Litteraturhistoriska Studier 34, pp. 86-190

Hautala, K. (1963): European and American Tar in the English Market during the Eighteenth and Early Nineteenth Centuries. Annales Academiae Scientarum Fennicae B 130. Helsinki

Heckscher, E.F. (1935/6): Sveriges Ekonomiska Historia från Gustav Vasa: Medeltidshushållningens organisering 1520-1600 (del I:1); Hushållningen under internationella påverkan 1600-1720 (del II:2). Stockholm

Heckscher, E.F. (1949): Sveriges Ekonomiska Historia från Gustav Vasa: Det Moderna Sveriges Grundläggning 1720-1815 (del II:1-2). Stockholm

Heckscher, E.F. (1954): An Economic History of Sweden. Cambridge, Massachusetts

Hoppe, G. (1945): Vägarna inom Norrbottens län. Geographica 16. Uppsala

Häggström, N. (1967): Västerbottens and Österbottens utrikeshandel 1765-1880. In: Västerbotten 1967, pp. 36-58.Högberg, S. (1969): Utrikeshandel och sjöfart på 1700-talet: Stapelvaror i svensk export och import 1738-1808. Lund

Johansen, H.C. (1983): Shipping and Trade between the Baltic Area and Western Europe 1784-95. Odense

Kaila, E.E. (1931): Pohjanmaa ja meri 1600- ja 1700-luvuilla. Helsinki

Kent, H.S.K. (1973): War and Trade in Northern Seas: Anglo-Scandinavian economic relations in the mid-eighteenth century. Cambridge

Kjærheim, S. (1964): The Norwegian Timber Trade in the Seventeenth Century (review of Tveite 1961). In: The Scandinavian Economic History Review XII, pp. 91-95

Layton, I.G. (1981): The Evolution of Upper Norrland's Ports and Loading Places 1750-1976. Umeå

Layton, I.G. (1988): Industry and the local economy: Changing contact patterns within the Luleå district during the nineteenth century. In: Geografiska Annaler 70 B, pp. 205-218

Layton, I.G. (1989): Sundsvallsdistriktet i tid och rum: Regionala förändringar i ett internationellt perspektiv. In: Thule 2, pp. 105-113

Lower, A.R.M. (1973): Great Britain's Woodyard: British America and the timber trade, 1760-1867. Montreal & London

Mager, F. (1960): Der Wald in Altpreussen als Wirtschaftsraum (2 vols.). Köln

Meinander, N. (1945): En krönika om vattensågen (English summary: A story of the water power sawmill, pp 311-316). Helsingfors

Mykland, K. (1977): Gjennom noedsår og krig 1648-1720. In: Knut Mykland (ed.): Norges Historia, vol. 5. Oslo

Peet, R. (1972): Influences of the British market on agriculture and related economic development in Europe before 1860. In: Transactions of the Institute of British Geographers 56, pp. 1-20

Tirén, L (1937): Skogshistoriska studier i trakten av Degerfors i Västerbotten. In: Meddelanden från Statens Skogsförsöksanstalt 30, pp. 67-322

Tveite, S. (1961): Engelsk-Norsk Trelasthandel 1640-1710 (English summary). Bergen

Wallerstein, I.(1974 & 1980): The Modern World-System, vols I & II. New York and London

Wieslander, G. (1936): Skogsbristen i Sverige under 1600- och 1700-talen (English summary:The Shortage of Forest in Sweden during the 17th and 18th Centuries). In: Svenska Skogsvårdsföreningens Tidskrift 34, pp. 651-660

Zethelius, G. (1963): Handelsfartyg, typer och byggnad. In: Gustaf Halldin (ed.): Svenskt skeppsbyggeri. Malmö

## Part V    The American Colonial Periphery

## 18

## Economic Evaluation and Reconstruction under Spanish Dominance: Middle America and the Caribbean

Ursula Ewald

### 1.    Introduction

When viceroys of New Spain reflected on the world-system or rather on the relationship between Spain and her overseas provinces during the later decades of the 18th century, they then already arrived at some contorted interpretation. Their administration not only reflected Crown policies but also European attitudes towards colonies. During the past decades of the 20th century the intellectual's view of the world in many instances has been even more out of focus because of his faith in certain ideologies. To search for historical truth always will remain an evasive goal but since recent times at least quite a few scholars, when interpreting the past, have tried to give it an unbiased look. These scholars, however, do not appreciate how Immanuel Wallerstein fitted the Spanish Americas, particularly New Spain, into the framework of his interpretation of world economics or rather his world-system. He selected bits and pieces to support his conceptions and by-passed the rest. In order to support my view that on the eve of Independence Mexico, to a large degree, was not part of the world-system, three different topics will be discussed:

1) Hernán Cortés in his actions and in his writings;
2) the Relaciones Geográficas and
3) Alexander von Humboldt.

As regards the Spanish Americas there exists an enormous amount of published and, more often, of unpublished sources. But often they are difficult to find, difficult of access, difficult to read and difficult to interpret. Maps as well as written documents, as a rule, were compiled for entirely different purposes than for present day research interests. Since about the sixties of the 20th century the outpour of the printing presses has increased tremendously. However, inspite of great advances in research, a great deal of pioneer work still has to be done. So many conclusions, presented here, have to be taken as tentative assessments.

Considering the title of this paper, one could argue about the term "reconstruction". In the case of Mesoamerica with its advanced precolumbian civilizations, the conquest definitely was succeeded by an era of reconstruction. But in the case of the Caribbean Islands with their subsistence agriculture or of the new mining centres of mainland America "economic adjustment" and "economic development" perhaps might be better terms to use.

## 2.    The American Context

In the case of the Americas misconceptions abounded from the very beginning, - and be it only Christopher Columbus believing until his death that he had reached the eastern fringe of the Old World instead of having discovered the western fringe of a New World. Economic evaluation held priority when he persuaded the Spanish Crown to finance and to legalize his expedition westward to search for the treasures of the East. Since the first Spanish landfall in the Caribbean Christian missionary zeal barely veiled economic evaluation. Appropriation and exploitation of everything the Americas might offer only were the consequences. Any indigenous needs became marginal. As late as the end of the colonial period the Spanish government slighted the interests of its overseas subjects in many ways. But exploitation was not omnipresent.

Economic evaluation, as a rule, preceded all planning policies. The story of Christopher Columbus sometimes appears bizarre. Whatever he dreamed about, whatever he reported to the Crown, the Caribbean islands fell short of all expectations as long as they were under Spanish dominance. Neither their indigenous population nor their lands seemed to hold any promise. Only La Habana became an important harbour. First, French, British, and later American enterprise turned sugarcane growing temporarily into a profitable occupation. Still during the 19th century nobody visualized Jamaica's wealth in Bauxite or the area's potential for tourism or international banking. The same rapid drifting into political and economic backwater is true for the countries, today grouped together under the name of Central America. Surviving documentary material and later research reflect that already during the 16th century New Spain became the core of Spain's overseas provinces, the "jewel of the crown" . The Caribbean and the rest of Middle America still held some strategic importance but otherwise the rimland could be considered peripheral.

## 3.    The Case of New Spain

How did New Spain fare under Spanish dominance? Why did an economy come into existence which, to a certain degree, neither agreed with Spain's policies nor with 20th century ideologies of intellectuals?

### 3.1.  Hernán Cortés

If the discovery and the conquest of what was to become New Spain are to be condoned at all and if Hernan Cortés is not seen as some vile criminal, one might easily consider him not only as a highly able and gifted person, but as a man of vision. In his letters to Charles V, he reported the events before and after the *Conquista*. At the same time he tried to describe the newly discovered country and its peoples, both so different from the Caribbean Islands, known until then. The building of his own empire, the Marquesado del Valle, as it was to be called eventually, in a subtle way almost reveals better how Cortés assessed the situation. The Marquesado del Valle, the far-flung property and estates of Cortés, illustrates how Cortés evaluated the potential of the new country and how he claimed property as soon as he had the power to do so. But it also reveals how reconstruction succeeded destruction. Inspite of political defeat and the disastrous decline of the indigenous population during the 16th and 17th centuries, many Indian elements survived. Likewise the Spaniards in many instances adapted their traditions to the new country. So the emerging "*new economy*" might be better labelled as "*mestizo* economy" like the later population of Mexico. The new economic structure was a blend of many economies. Hernán Cortés must have believed in the future of the lavishly laid out new capital. His building lots in La Ciudad de México and in nearby Coyoacan already during the colonial period were worth a fortune.

Even considering that Hernán Cortés relied on his own observations, on tribute lists and on Indian informants, it is intriguing to speculate on what basis he amassed his agricultural estates. Beside some holdings in the Basin of Mexico they included choice areas in the Basin of Cuernavaca. Here, a lower altitude, sufficient water and fertile soils allowed the cultivation of sugar-cane. How did Hernán Cortés at this very early date foresee the pressure on land in central Mexico because of the introduction of Old World husbandry? His estate in faraway Jalapa, Oaxaca, bred the livestock for his highland haciendas[1]. Incidentally, his chain of agricultural holdings anticipated the policy of geographical and economic diversification which, at a much later date, the large Jesuit colleges of New Spain opted for in order to cut down on financial risks because of natural hazards and in order to be competitive in a competitive market. Likewise Cortés created his chain of agricultural holdings in the context of New Spain, not as a supplier for a cross-Atlantic trade.

Nothing indicates that Hernán Cortés would not have remained a loyal subject of the Crown. He definitely had glimpsed the wealth of the new land in precious metals. But his numerous references to them in his *Cartas* ought to be taken with slightly more than a grain of salt. In this he is in line with other explorers like Christopher Columbus or Sir Walter Raleigh who, probably often against better knowledge, reported to their sovereigns what these wanted to hear. Deliberate straying from the truth seemed a small price to pay for procuring legal sanction. What made Cortés realize that for centuries to come the new country's wealth would be derived from agriculture, not from mining?

Hernán Cortés and his followers lusted for gold and silver but at first they appropriated land, water and the people to cultivate the fields. After all, the Spaniards needed food for surviving. However, at the same time Hernán Cortés and later the Spanish Crown, to a certain degree, acknowledged Indian rights to property. Not only Spaniards, but also some worldly-wise Indians joined in the run on land. Likewise, having seen the appalling decrease of the indigenous population in the Caribbean, Cortés has to be credited with at least trying to protect the Indians in mainland America against excessive exploitation. That his failure, because of events out of his grasp, like disease and epidemics, took on apocalyptic proportions, Cortés probably realized. - So from the very beginning the history of New Spain on this point differs drastically from the British colonies and the later United States of America.

### 3.2. The *Relaciones Geográficas*

While events forced Hernán Cortés to evaluate and to act, that is to appropriate, Philip II tried to evaluate on a larger and on a more official scale several decades after the *Conquista*. In 1577 the Crown sent a questionnaire to her overseas administrators for evaluation of her provinces requesting the famous *Relaciones Geográficas*. Upon the background of renaissance, humanism and the first glimpses of the age of enlightenment the reasons for the *Relaciones Geográficas* were manyfold. They covered: 1) strategic and military purposes; 2) logistics of supply; 3) geographical, geomedical and economic assessment; 4) secular and ecclesiastical administration: 5) social and charitable considerations; 6) scientific evaluation and 7) history. The last two topics pass far beyond reporting the status quo because they explicitly ask for the changes due to the *Conquista*.

Considering the long tradition of requesting reports and of compiling questionnaires the very lack of preciseness is surprising. The questions obviously meant for different addressees, are repetitive, sometimes redundant, meandering and straying. In addition to the written reports the Crown asked for maps, plans and drawings. The highly diverse echo to the questionnaire reflects its loose structure. The evaluation of New Spain according to the questionnaire may be grouped together under the following topics: 1) the land; 2) surrounding waters; 3) natural hazards; 4) the people and 5) potential and economic development (see Table).

It is not surprising that faraway Spain paid great attention to all items concerning navigation. More intriguing is the Crown's interest in the mentality and in the values of her new subjects. If their potential as labour might have been the ultimate interest in these questions, this is definitely not true when asking e.g. for the architecture and for building materials of the settlements. Inspite of intermarriage, illegitimacy and the mixing of races the questionnaire still takes the division between Indians and Spaniards for granted[2].

### 3.3. Alexander von Humboldt

Hernán Cortés after the *Conquista* tried to reconstruct the shaken economy and the *Relaciones Geográficas* still dealt with some raw economy. Alexander von Humboldt was able to analyze New Spain at the end of the colonial period. He did not visit many places in this vast land. But he could rely on his own observations, conversations with the upper echelon of the Spanish bureaucracy and of society. The virtually free access to the viceregal achieves still surprises one today. His ability to compare New Spain with other parts of the Spanish Empire and with some European countries must be considered some special asset.

When one reads von Humboldt's description of New Spain, the influence of the intellectual climate of his times, his own background and experience become clearly visible.

Alexander von Humboldt might be regarded a liberal and a humanitarian. He was against colonialism, mercantilism, slavery and against the Catholic Church, all of which certainly existed in Mexico. Some of his statements taken out of context remind of leftwing intellectual thinking during the past decades[3]. It also cannot be denied that von Humboldt arrived at some surprisingly false conclusions. His view of the Church is distorted (e.g. v. Humboldt p. 318). Frequently he saw, what he wanted to see, like e.g. the badly administrated latifundia, where extended tracts of land lay barren[4]. Most likely he saw land under fallow, where for agricultural and ecological reasons moisture was to be conserved for future cultivation. His glowing picture of Mexico's potential for mining brought disaster to many investors later in the 19th century.

The scientist von Humboldt also remained astoundingly unreceptive towards the arts and architectural beauty. Cautioning with these examples against von Humboldt his more sound conclusions have to be delineated. It is intriguing that some statements contradict his own preconceived ideas. A. von Humboldt realized that, contrary to New Spain's de facto status as a colony and contrary to the ideas of mercantilism, an economy of its own had originated, in which agriculture by far, not mining, played the major role. In agriculture, cereal production for the home market, not cash crops for exports, prevailed, so that Mexico was rather independent of the vagaries of European politics (ibid. p.237 and p.320).

A. von Humboldt explained the large home production and home consumption by the distance between Europe and the Americas as well as by the vast expansion of Mexico. Transport facilities were poor. Shipping merchandise from Spain to any place in Mexico often exceeded 5 - 6 months. In von Humboldt's opinion smaller islands or coastal settlements and outposts were more easily administrated according to mercantilistic rules than a country with the history and extension of Mexico. Its highly gifted indigenous population, the abundance of labour and its large prosperous cities virtually forbade mere mercantilistic exploitation (ibid., p. 449). Inspite of his stress on the ubiquitous *latifundium* and its vices, he lists Indian agricultural commodities as being very important (ibid., p.326). So Indian agriculture must have existed after all - as it did! Even more revealing from the social point of view is his remark on the free status of Indian labour as compared with many parts of Europe[5].

As regards his judgment of the Indian whom, to present day opinion, he grossly underestimated, he cautioned that some transient visitor might not do him justice (ibid., p. 64).

His frequent dissatisfaction and exasperation with the state of affairs in Mexico are easily noticed. Coming from Prussia, knowing the wealth of German and English agricultural literature, it was small wonder that particularly in this field his fingers were itching for experiments and improvement. But inspite of these drawbacks, inspite of his at times biassed view, and his not always correct interpretation of what he saw, von Humboldt gave the best and most detailed survey of Mexico of the entire colonial period.

## 4.    Conclusions

On the eve of Independence certain regions of Latin America, New Spain being in the lead, had built up economies of their own which could hardly be labelled as colonial byproducts and dependencies. In spite of hundreds and thousands of Royal orders and of imposing bodies of legislation the Spanish overseas provinces generated, to a certain degree, economic changes and economic patterns of their own. There is no denying the Hispanization of vast parts of the Americas, there is also no denying that Spanish ships carried enormous amounts of silver and gold, of dyes and other commodities, across the Atlantic to Spain. But when the United States of America declared themselves independent in 1776, New Spain probably boasted a far better economic infrastructure than its northern neighbour. The architectural treasures of Mexico's colonial cities, to cite only one example, bear witness that not all wealth was drained off to Europe.

Which factors had shaped this economy? Spain demanded gold and silver, the dyes of cochinelle and indigo, cacao and vanilla. The population of New Spain and of neighbouring territories required food, clothing and other merchandise. As neither mining, nor manufacturing, but agriculture held priority, again it must be asked which factors determined this sector of the economy. Beside demand and the harsh constraints of the country the von Thünen principles are clearly to be delineated. But apart from physical geography and distance to the market the availability of labour has to be given far more attention than in Europe. All recent research at a macro- and microlevel confirm that village-, small holding- or rather *rancho* -, and estate-agriculture largely catered to Mexican, not to Spanish needs.[6] That the population in the mining districts and in the Far North had to be fed and clothed hardly has to be stressed. On the whole, village- and estate-agriculture supplemented each other. Friction was kept at a low level. Likewise, mining cannot be considered only in the context of colonialism and of dependence.

The paper argues that New Spain's economy and structure at the time of her Independence cannot be explained as colonial and as solely reliant on Spain and on other European countries. And yet Mexico was to experience until the forties of the 20th century more political turmoil then periods of peace. In spite of the potential of the country, Mexico still is an underdeveloped country. If one searches for reasons for dependency or underdevelopment it might be wiser to turn to the immaterial, not the material foundations of the country. The questions of the *Relaciones Geográficas* repeat themselves: what are the mentality, the disposition and the values of the population like? What are its traditions? One is afraid to tread on this ground because the terrain is shifting and full of quicksand. We only have glimpses of pre-*Conquest* values.

Octavio Paz in "The Labyrinth of Solitude" and other writings perhaps is only the most famous person to comment on the character of the Mexicans and on their problems. Spain, even Moslem traditions, and the Catholic Church bear much guilt, but cannot be held responsible for all of Mexico's blight. Spain might have been blamed for the almost pitiful lack of experience in administration after the War of Independence. But still continuing to blame everybody but oneself after almost two centuries of freedom appears ludicrous.

When the United States of America built up their trade, took over the Atlantic and Pacific runs, and expanded westwards and to the South, Mexico was torn by internal disputes. The "cosmic race", as its famous intellectual and educator José Vasconcelos named it, obviously had no "manifest destiny" like its northern neighbour. The racial and socioeconomic plurality of its population might account for the fact that the search for its identity still continues. It causes psychological instability and the feeling of inferiority. The closeness to the United States of America aggravates the problem. Mexicans highly esteem cultural values, they have an innate sense for beauty and some astounding creativity in the arts. But there is little faith in the benefit of steady work. They despise manual labour and they despise the people who do it.

The *bonanza* , the successful strike of the gold vein, is hoped for. Never ending toil definitely is seen as less rewarding. In spite of the almost fetish belief in the acquisition of land the cultivation is left to other people. If there is money, it is not being invested in business and industry, but in land. Correspondingly many Mexicans have only contempt for entrepreneurial skills and for business acumen. They scorn science and engineering as being inferior to the arts. Innovative, critical, open minds, and a sense for calculative risks are difficult to find. If risks are taken, often the *bonanza* spirit prevails. The interpretation of their history and the evaluation of the potential of the country is shrouded by myth and wishful thinking. Comparing Mexico with the old democracies of the western world the lack of civic virtues becomes painfully apparent.[7] Mexico's government might be seen as the very opposite to a liberal democratic one. The state still has pre-modern or archaic traits. The protection, the *habeas corpus*, of the individual is sadly neglected. Inspite of the omniscience of the state, no stable and just tax-system exists. The octopus-like bureaucracy is blemished by a deep-rooted inherent system of corruption which is only partly explained by *caudillismo*, nepotism and patronage. Exploition of the weaker party, be it man, woman, child, forests, soils or water, characterize many Mexicans. Few accept responsibility for people and tasks outside their sphere. Likewise instead of communication among various groups one looks for regulations from above - the list might be continued.

Do values leave marks on the land? In a subtle way they may often become more decisive than many government decrees. Certain values might hinder evaluation and prevent reconstruction. Psychological attitudes might keep Mexico at its present economic level in spite of being embedded in our world-system, by which not necessarily Wallerstein's world-system is meant.

## The Evaluation of New Spain
(Source: Relaciones Geográficas of 1577)

| The Land | Surrounding Waters | Natural Hazards | The People | Potential and Economic Development |
|---|---|---|---|---|
| Landforms (level, rugged or mountainous territory, character of the coast) | Patterns of tides; currents and winds; suitability for landfalls (rocks, reefs, shoals, natural harbours, beaches, prominent landmarks on the coast, character of the hinterland) | Shortage or excess of water<br>Disease-ridden countryside, volcanism | Ethnic groups (Indians)<br>Languages (existence of lingua franca)<br>Customs (heathen and religious beliefs, dress) | Conditions for navigation in the surrounding waters, character of coast, harbours and their hinterland<br>Trails, roads (easy or difficult), transportation and traffic |
| Climate and weather (the annual patterns of wind, rainfall and temperature; the seasons; humidity; hot or cool regions) | Islands and their main features | | History (warfare, feuds, alliances and enemies, the conquista by the Spaniards)<br>Mentality, intelligence, way of life, values | Land use (horticulture, agriculture and husbandry)<br>The introduction of European plants and animals and their performance in the New World, changes in land use since the conquista) |
| Soils (fertile or sterile) | | | Sustenance before and after the conquista | Use of water resources (irrigation) |
| Water resources (rivers, springs, lakes and lagoons, availability of water) | | | Health and common diseases<br>Economy and trade relations<br>Tribute obligations | Flora (the species, the use of fruits, of wood, dyes and medicinal plants); land (barren; under forest, vegetation cover or pasture) |
| Mineral resources: Salt | | | Secular and ecclesiastical administration | Wildlife (its use for food, dyes, etc.) |
| Vegetation | | | Charity and social welfare (hospitals, schools, other charitable institutions, their founders and endowments) | Subsoil mineral resources and salt |
| Animal life | | | Density and number of population (decrease or increase) | |
| Geographical position (latitude and altitude) | | | Settlements (site description, topography, healthy location, military aspects, legal status, layout, architecture and building materials, origin of building materials, outstanding building complexes like monasteries; native names of the settlements; if deserted, reasons of abandonment; density and distribution of settlements; access to water, transport, fields and pasture; distance to Church, justice, markets) | |
| Mountains and their altitude | | | | |
| Distances between settlements | | | | |
| Landmarks (volcanoes, caves or other scenic features) | | | The Clergy (orders, convents and monasteries; their number; buildings; means of support) | |

## Notes

1.  Cortes, ed. 1962; Garzia Martinez 1969 and Bazant 1969-70.
2.  For the *Relaciones Geográficas* see Wauchope 1972, vol. 12, pp.183 - 323.
3.  "Según las ideas que por desgracia se han adoptado hace siglos, estas regiones lejanas son considerados como tributarias de Europa..." (v. Humbold , ed. 1966, p.96).
4.  "El suelo de la Nueva España, así como el de la Vieja, en gran parte se halla en poder de algunas familias poderosas que han absorbido lentamente las propiedades particulares. Tanto en América como en Europa, hay grandes distritos que están condenados a servir de pasto para el ganado y a una perpetua esterilidad."; (ibid., p.318 sig.).
5.  "El cultivador indio es pobre, pero libre. Su estado es muy preferible al de los aldeanos de una gran parte de la Europa Septentrional. En la Nueva Espana no hay contribución de servicios corporales ni esclavitud,; el número de esclavos es casi ninguno; y la mayor parte del azúcar es fruto del trabajo de manos libres"; (ibid., p.237).
6.  For the macro-level consult e.g. the recent publications of Ouweneel (1988 and 1989). Garcia Martinez (1987) is an excellent example for the mirco-level; for the textile industry see Salvucci 1987.
7.  Mansilla (1989) gives an interesting survey on the lack of civic virtues.

## References

Bazant, J. (1969-70): Los bienes de la familia de Hernán Cortés y su venta por Lucas Alemán. In: Historia Mexicana 19, pp. 228-247
Cortes, H. (ed. 1962): Five Letters, 1519-1526. Trans. by J. Bayard Morris with an introduction. New York
Ewald, U. (1977): The von Thünen principles and agricultural zonation in colonial Mexico. In: Journal of Historical Geography 3, pp.123-133
Garcia Martinez, B. (1969): El Marquesado del Valle. Tres siglos de régimen colonial. México
Garcia Martinez, B. (1987): Los pueblos de la Sierra. El poder y el espacio entre los indios del Norte de Puebla hasta 1700. El Colegio de México. México
Humboldt, A. von (Ed. 1966): Ensayo político sobre el reino de la Nueva España. Ed. by Juan A. Ortega y Medina. México
Jacobsen, N. and Puhle, H.( eds.), (1986): The economies of Mexico and Peru during the late colonial period. 1760 - 1810. Bibliotheca Ibero-Americana 34. Berlin
Mansilla, H.C.F. (1989): Zur politischen Kultur des Autoritarismus in Lateinamerika. In: Iberoamerikanisches Archiv N.F. 15, pp. 371-398
Ouwneel, A. (1988): Onderbroken groei in Anáhuac. De ecologische achtergrond van ontwikkeling en armoede op het platteland van Centraal- Mexico (1730-1810). Latin America Studies 50. Amsterdam
Ouweneel, A. (1989): The agrarian cycle as a catalyst of economic development in eigtheenth century Central Mexico. The arable estate, Indian villages and proto-industrialization in the central highland valleys. In: Iberoamerikanisches Archiv. N.F. 15, pp. 399-417
Salvucci, R. (1987): Textiles and capitalism in Mexico. An economic history of the obrajes. Princeton, New Jersey
Walllerstein, I. (1979): The capitalist world economy. Princeton
Wauchope, R. ( ed.) (1972): Handbook of Middle American Indians. Vol. 12. Austin

# 19

# The Early Modern World-System - Periphery-Core Interactions as Exemplified by Plantation Regions of the Caribbean

David Watts

## 1. Introduction

Immanuel Wallerstein (1974, 1980, 1989) proposed that the modern world-system took the form of a capitalist world economy that had its genesis in Europe in the long sixteenth century, which involved the final stages of change from a feudal system (the 'ancien régime economique' of Braudel, 1978) to a more modern social system. A functioning capitalist world economy was certainly in place in Europe by 1600-1750 (Mauro, 1961), this having been marked during the first 50 years by a general conversion of land to the production of cereals, a general rise in population concomitant with the resulting increase in food supply, the growth of small-scale industry in urban areas, an increase in the money supply, and an increase in the number of marginal entrepreneurs, both rural and urban. By 1650, Holland had achieved its 'golden age' of early modern state development, with a world-wide trade network, and England was not far behind, aided and abetted by its mercantilistic principles, and only these two countries were to assume hegemony within the system prior to 1750, England assuming dominance over Holland by the 1670s. However, following 1650 and until 1700, a slowing-down of the rapid rate of expansion of the preceding years also took place, in a period of economic consolidation, during which the strengthening of state structures, such as the Navigation Laws of England, were greatly enhanced (Schoffer, 1973). Subsequently, the capitalist world economy continued to expand itself geographically eventually to cover the entire globe; it was affected by cyclical patterns of expansion and contraction, along with changing cores, peripheries and semi-peripheries, and at the same time underwent secular transformation, including technological advance, industrialization, and later proletarianisation, and the emergence of structural political resistance to the system itself, a process which is still in train today.

From 1492 to, say 1750, Wallerstein suggests that the Americas were always peripheral to the core of the early modern world-system which was located in Northwest Europe, particularly England, France and the Netherlands, following the sixteenth-century decline in influence of Spain. At first, gold and silver from Central and South America were the most important items of trade from this periphery, in that these commodities were required for the operation of the world economy, and were used as money, providing a solid currency base for the expanding capitalist system. At the same time, any surplus could be used for the growing trade in spices and cloths with Asia. Then, as this trade declined in the late sixteenth century (Kamen, 1983), high demand in the core countries of Europe for tobacco, cotton and sugar led to the establishment of these new, successful plantation items, notably in the Caribbean islands.

It is the role of both groups of these trading elements to the development of the early modern-world system which is to be further examined in this paper. For these purposes, I intend to concentrate on the islands of the Caribbean, rather than adjacent mainland areas, for it was in the islands that adjustments to home country practice and institutions, both in terms of settlement and trade, were first examined and brought into effect, better to facilitate the processes of European expansion, colonization and the generation of capital. This was certainly the case both in respect of the early Spanish, and the later English and French development of the region, though in somewhat different ways. During the seventeenth century, in particular, peripher-core interactions between French and English colonies and their respective homelands undoubtedly helped to determine the shape of early capitalism, the role of individual entrepreneurs being especially important in the years prior to 1660. The Dutch too, then in their hegemonic role, influenced events by the provision of credit and trading facilities. But by 1685, English colonies in the region were substantially helping to replace the Dutch in this role, particularly in respect of their financial contributions from sugar estates to the homeland. But more needs to be known about the periphery-core relations of the time. Certainly, a major primary product (sugar-cane) had by then become well-established in this periphery, involving coerced labour and, later, a low-wage economy, three out of four Wallerstein's periphery activity-structures. But, with the increasing formalization of an effective sugar production model in the eighteenth century, a certain withdrawal of activity on the part of the region from some periphery-core events took place, and the consequence of this then was a diminution of influence, which was to peak during the last quarter of that century.

## 2.    Early Spanish Settlement

One might, to begin with, examine why the beginnings of the full world-system postulated by Wallerstein did not coincide with the re-emergence of Spain as a major European power in 1492. Many of the requirements for this were present, including the re-investment of monies from Venice, as the Venetian empire was increasingly challenged by the Ottomans in the east; and Iberia had the added advantage of an early start in oriental and Atlantic trade by sea compared to the rest of Europe. However, when put to the test, it is clear that early Spanish trade and settlement outside the homeland and especially in the West Indies, was largely unsuccessful in the sense that it was long-term, more destructive than constructive, more exploitatively negative than positive (Watts, 1987). A contributory factor was the decimation of the culturally-rich aboriginal Indian population of c. six million inhabitants of the Caribbean islands within a space of 26 years following the first contact; and very little was left in the place of this other than a handful of settlers, and large numbers of domesticated animals (especially, cattle, horses and pigs), brought in from Iberia, which wrought havoc on the local ecology. 'We should remember', said Bartolomeo de las Casas, the contemporary historian (Historia de las Indias, 1520-1561), writing of Espanola, 'that we found the island full of people, whom we erased from the face of the earth, filling it with dogs and beasts,' an accurate assessment of the de facto position at the end of the third decade of colonization.

What were the reasons for this appalling change in fortune of one of the more inherently fertile areas of the world in such a short period of time? Undoubtedly, one should not underestimate what may best be called an 'ignorance' factor, resulting from the major culture clash between inhabitants of the Old and the New Worlds, which was virtually inevitable considering the conditions of the era - ignorance of the nature of the West Indian environment and how it might best be put into use, ignorance of the culture, agricultural and food-producing systems of the native Indians, ignorance of their knowledge of the sometimes sensitive balance of West Indian ecology, and so on. Arising from this ignorance, misconceptions arose; perhaps the most important of these for the course of Hispanic settlement was set by Colombus himself during his first encounter with the Arawak of Espanola, and this was that their inherent hospitality would make them willing servants of Spain under almost any circumstances. Such a mistaken, basic view was to override all other perceptions and, in so doing, was to misdirect official Hispanic attitudes towards New World peoples for many years thereafter, and contribute indirectly to the decline of Spain as a major imperial power, set in the semi-periphery of Europe.

On top of this, there is very little doubt that certain institutions of settlement, developed to some extent on an ad hoc basis as Spain began to reclaim her own territory from the Moors prior to 1492, played an important role in her lack of success in development of the islands of the Caribbean during the next thirty years. These institutions emerged from a reassessment of older ideas of land division, land use, land settlement and the treatment of captive peoples. Prime among them were the concepts of *repartimiento* and *encomienda*. On the mainland of Spain land recovered from the Moors was put under the initial jurisdiction of the Crown, and thus termed *realenga*. Some of this land subsequently was granted by the Crown to representatives of those who had participated in its recovery, in recompense for services rendered. This meant that a core of trusted settlers could be placed upon it, whose task it was to encourage its further peopling, and at the same time arrange for its government and its defence.

Originally a matter of royal privilege and custom, this style of land partitioning in recaptured territory came to be formalized, during the military operations in Andalucia, within the constraints of *repartimiento*, a legal system which, under the continuing aegis of the Crown, apportioned any 'new' land mainly between the lords of the realm, prelates, particular churches, and the leaders of monastic foundations and military orders, but also left some more available for more junior militiamen: thus, 200 caballeros (horse soldiers) were at various times granted land in Sevilla, 40 in Jerez de la Frontera, and 333 in Murcia (Chamberlain, 1939). *Repartimiento* accordingly was meant to benefit all social classes. Concurrently, the legal authority in recaptured terrain was granted by the Crown to selected members of the older militia. Such authority was held in trust (*encomienda*) for the Crown, which also retained ultimate legal responsibility. But despite the provision of land for social classes, there was not much democracy here: autocracy was the rule.

The aims behind these ideas also included references to the good treatment of captured or subject peoples (though these were sometimes not adhered to); and the major adopted practice for economic activity on these new lands lay in stock-rearing, often on an open-range basis where this was possible. The new methods of land apportionment, coupled with the fact that legal (and also, incidentally, religious) appointments were in Crown hands, effectively ensured that Spain's rulers were able to keep strict control over most aspects of life in her new domains.

Their translation into the circumstances of the New World were not, however, easy. Two factors complicated matters: one, the early presence of so many native Indians; and two, the discovery of placer gold, the exploitative search for which became thereafter the favoured economic activity, at least until the 1520s when the main focus of settlement switched to Mexico. Most settlers took the view of Hernan Cortes: 'I came here to get gold, and not to till the soil like a peasant'. In point of fact, it was not until 1497, during his second return to Spain, that Columbus received official approval to introduce the *repartimiento/encomienda* system into Espanola, the aim being to counteract the lure of gold, and develop agriculture. But, less interested in land in its own right, many other New World settlers took the view that, so as to facilitate gold extraction, *repartimiento* in the Americas should be modified to include the award to recipients of Indian communities and labour rather than land, no territorial limits being specified; and it was this revision which eventually was put into effect in the New World. In time, it was to change totally the patterns of contract between Hispanic peoples and native Americans, not only in the West Indies but throughout the continent as well, becoming notorious in its implications of widespread servitude for Indian groups everywhere, and leading indirectly to the virtualyl total removal of those in the West Indies.

As regards the latter, it is to be noted that this predated the arrival of the first killer disease (smallpox, in 1518) into the region; it can be linked to the decline of native food-producing systems, deaths from forced labour, and increased suicide and inhibition of ovulation from shock arising out of the culture clash (Watts, 1987). In the island Caribbean, the attempt to place within it Iberian land settlement institutions and concepts was unsuccessful, for they proved to be too rigid, too inflexible, and too incapable of counteracting the greed of individuals which the presence of gold had engendered to be effective, or of being adapted without major modification to the totally new cultural and physical environments in which they found themselves. The result in the West Indies was an overwhelmingly negative one in development terms: the Spanish were left with no labour, no investment, no political direction, and no good trading facilities. Such *hacienda* development which later emerged served mainly regional markets. In consequence, the region for many years remained a backwater, once other gold and silver deposits had been discovered elsewhere in the Americas, initially in Mexico. From its lack of success, Spain itself suffered a growing schism in its core-periphery relations which, despite the wealth subsequently brought in from the New World mainland in terms of gold and silver, ultimately was to lead to its political and economic decline, in the ensuing century.

## 3.    The English and French West Indies

Although a certain grand design, attendant upon the search for gold and silver, had prevailed in the colonization of Hispanic America, such could not be said to be the case in the development of the English and French West Indies. Yet paradoxically, core-periphery relationships proved to be more successful and stable in the latter (both French and English), with at the same time the periphery proving to be influential in the economic, and to some extent the political evolution of the core for some years. Central to this success and stability was the distinctiveness of the chosen settlement pattern, and the commodity trade it engendered. The latter contributed directly too to the emergence of capitalism as an emerging world economic driving force, and indirectly, later to the strengthening of the modern world-system as we know it.

It was at the end of the sixteenth century that the broad psychology of northwestern European transatlantic undertakings began to undergo a change, from one raised with Elizabethan notions of daring deeds, exemplified by Fleury, Hawkins and Drake, to an alternative in which the major elements were 'purse-adventurers' (Williamson, 1941), i.e. those willing to underwrite the immense complexities and expense of permanent settlement in the Americas. The change received impetus from 1612/13 onwards, when James Rolfe sent back to England the first crop of tobacco from Virginia, and effectively demonstrated that agricultural colonies in the New World could be both self-supporting and profitable. From that time, the future projected course of new French and English colonies in the New World was set; crops (or other products) exotic to Europe were to be produced by Europeans (albeit at times with slave labour) for European markets, in plantations, which to begin with were on extremely small scale.

It has been emphasized by several writers (e.g. Andrews, 1934) that the whole concept of plantation colonization in the Americas after 1613 was almost entirely without precedent. Certainly, Spain had never been interested in building settlements across the Atlantic in which large numbers of European men, women and children were to be present, organized cultivation begun, regular trade patterns formulated, and a stable social life brought into being, except for very brief, inconsequential periods. Indeed, no clear idea as to how such notions were to be put into effect was immediately forthcoming. Both investors and settlers (very few of whom had an agricultural background, let alone a knowledge of tropical conditions) alike embarked upon a huge gamble that it would eventually succeed, even though a tough phase of hard experimentation initially was unavoidable. It is also to be noted that, for the most part, there was no official government backing (i.e. no official core backing) for these ventures for several years, other than minimal lip service expressing support; capital originated from the emergent merchant centres of London and Amsterdam. Moreover, most of the hard decisions affecting success or otherwise of the new colonies were made by capable individuals (James Drax, e.g., in Barbados), who themselves introduced the new capital, new techniques, and new crops, with new methods of processing, in what formerly had been economically-desolate territories.

To begin with, any land acquired in the region by the English and French was, following Spanish precedent, placed under the ownership of the Crown, and then disbursed subsequently to subsidiary parties. At that point, however, land allocation practices commenced to differ. Under English law and custom, individual islands were usually granted (or 'leased') by the monarch to Court favourites; and under French law, control was handed over to settlement companies. The lessees then appointed on-site Governors to take charge of the day-to-day management of the new colonies, and to these, dues from other settlers were payable. It should be noted that this pattern of allocation differed from that of *repartimiento* in the sense that it was much more selective in terms of the social class of people who were recipients; and indeed the main land grants of whole islands were of a kind no longer valid in the home country and antithetic to the new capitalism, often being quasi-feudal awards to particular noblemen.

Nevertheless, it served both the settlement process and the developing world-system well, encouraging both at all social levels.

Once the grant had been confirmed, the recipient then was free to do as he wished with the land, all legal control passing to him (again in contrast to the *repartimiento* system); and indeed, much was either sold of sublet virtually immediately, in order to repay incurred debts at home, often from gambling. In the case of one island, Barbados, 10,000 acres (4048 ha) were quickly leased by the recipient (the Earl of Carlisle) to a society of London merchants, which had Dutch financial backing; and this society was for a while the effective controller of trade to and from the island. This particular grant gave very favourable terms to the grantees, including the right to choose their own Governor and government, the only colony in which this was the case. Land owners and lessees were then at first required to bring in indentured servants to work the land. Everywhere, dues were collected from secondary lessees and owners, both by the Proprietor and Governor (or by means of a poll tax in the case of the French company land) and, in the absence initially of an effective monetary system, these were payable in tobacco and cotton, the two major plantation crops from the region.

The two first islands to be settled were St.Kitts (1624, by both the French and the English simultaneously), and Barbados, by the English in 1627. From the very beginning, this pattern of land allocation, aided by its labour support structure of servant indenture, and then (after 1645) slavery, worked very well in the new colonies. Credit for development was always available from the Dutch, then the masters in this business. By 1645, both in terms of population (Table 1) and trade, several of the new West Indian colonies served as the most important overseas territories of both the English and French. But it was not until the introduction of sugar-cane in 1645 that the significance of the acquisition of these islands to the early modern world-system really became apparent, and profits on a major scale began to accrue from such ventures.

Table 1: Estimated population of the most developed colonies in the Americas, 1642

| Territory | Population | |
|-----------|-----------|--|
| New England | 40,000 | (English) |
| St. Kitts | 20,000 | (English & French) |
| Nevis | 10,000 | (English) |
| Barbados | 40,000 | (English) |
| Guadeloupe | 9,000 | (French) |

Source: Watts, 1987

The story of the complete transfer of a sugar-producing system from Pernambuco into Barbados is by now well known. Its rapid modification into a more economically-efficient and capital-generating 'Barbados sugar-producing system', which was to be subsequently adopted by other islands, and to remain essentially unchanged for c 250 years, is perhaps less generally appreciated. Most of the initiative for these developments again was generated by individual local entrepreneurs, home governments displaying little interest in such possibilities at the time. In the event, Pernambuco served as the transfer source-area for sugar production, insofar as it was, although economically low-key, still the best-developed production system worldwide during the 1640's. It was also relatively close to hand for Barbados, the main recipient territory. Trade contacts between the two, sponsored by the Dutch, were reasonably good; and entrepreneurs such as Drax had ready capital from the former successful sale at high prices of tobacco and cotton crops to effect the transfer. In Pernambuco, sugar production was entered into by both landowners and sharecroppers, the latter playing a particularly important role. Sugar estates there already used slave labour and, apart from this, were intended to be self-sufficient, not only in terms of their food supply but also in terms of their renewable resources for fuel and building materials, notably wood. A deal of timber and food-producing land was kept on estates, in a multi-cropping system. Under this system, an average of c. 23,000 tonnes of sugar each year were exported back to Europe from Brazil between 1638 and 1645 (Boxer, 1957), a large amount for the time. It was this 'model' of production which Drax sought, and eventually succeeded in bringing into Barbados during the latter year.

The new commodity brought with it, however, many wide-ranging problems. It was never an easy crop to grow optimally, involving a magnum leap in sophistication of the necessary agricultural techniques compared to those previously evolved in Barbados in the production of tobacco and cotton. The subtleties of cane processing also were not always fully appreciated, so that the final product proved at times to be disappointing. On smaller estates, self-sufficiency was not always possible, in view of the shortage of available land; and hairline financing often led to a severe shortage of capital. Political and legal interference, some associated with the Civil War in England, could interrupt development, marketing impediments had to be overcome, and an expanded labour force settled into the framework of society. Further, the whole concept of share-cropping was alien to the ideals of contemporary English and French culture. Fairly quickly, these hurdles gave rise to some evolution away from the original rationale and design of the Pernambuco sugar complex, as it came to be fitted more neatly into its new physical and economic environment. Share-cropping never took in Barbados, the English there always being more capitalistic and less paternalistic in their approach to cane cultivation than were the Portuguese.

The English estate owner also was much more likely to be a 'self-rais'd' man than his Brazilian equivalent, and much less concerned with past precedent; and, as a result, his role in society came to be rather different, and much more multi-structured. He usually presided over a much smaller estate and, at least in the early days, tended to be much more active on it, often supervising all stages of cultivation and milling himself, and making arrangements for the sale of the product as well, a point which Wallerstein (1980) does not concede. Nor does he fully appreciate that slave labour was used from the very beginning in Barbados sugar-cane agriculture, not least to clear the land before the crop could be planted (Watts, 1987). The amount of human energy involved in all of this work was phenomenal, leading one contemporary writer to observe that 'the devil is in the Englishman, that he makes everything work: he makes the negro work, the horse work, the ass work, the wood work, the water work and the wind work' (Anon, 1676). The end product in terms of the Barbados sugar cane estate was one which was more capital-intensive, capital-generative and more 'efficient' than that of Pernambuco, but which nevertheless retained the classic tropical plantation attributes of raising an alien export crop, for sale in temperate lands, through the use of imported labour.

Although it was largely a matter of chance that Barbados became the first overseas European territory to benefit from the new crop, it very quickly, because of it, became the richest colony in the Americas by far, a position which it was to hold for some time. Profits were high, as much as 40-50% on investment at times, in the first years of production (Ward, 1978). Much of the accrued early wealth stayed on the island, being reinvested to assist the further emergence of new estates, or to improve facilities in the first substantial town-port of the West Indies, Bridgetown, which soon attained an entrepôt importance out of all proportion to its very small size. But many profits also went back to the home country, either directly or in the form of tax. Drax himself vowed not to return to England until he could invest £ 10,000 per annum from his profits in Barbados, and there are indications that he did just that. Overall, the monies received by the home country must have been substantial, and helped in no small way to finance aspects of the agricultural, and later industrial revolutions there. Some idea of the scale involved may be gained from Malachy Postlethwaite's judgement in 1774 that England had acquired £ 2 million from Barbados between 1636 and 1656, and another £ 4 million from 1656 to 1676; and from Sheridan's (1974) conclusion that £ 278,000 was realized from the 139,00 cwt of sugar brought from Barbados to England in 1655 alone.

From many points of view, the spectacular financial success of the early sugar-production years in the English West Indies was never fully repeated, despite the fact that, as the Barbados model of sugar production was expanded into other West Indian territories, both English and French, several, for a while, became extremely rich from this one crop. But, after the intial profits surge, several mitigating factors operated against such a repetition at quite the same level. From 1600, both the local and external conditions which had encouraged the emergence and success of sugar-cane agriculture in the Lesser Antilles (and Barbados in particular) were beginning to fade. As strong monarchs, in the shape of Charles II and Louis XIV, assumed the thrones of England and France, their governments were directed to become much more involved than formerly in the active control of West Indian economies, thus effectively reducing the freedom of action of island entrepreneurs (and also, incidentally, legislatures), who had been responsible for the early successes. From 1660, new Navigation Acts annihilated the role of the Dutch in the Caribbean, interfering with trade patterns and the provision of credit, and in the course of time leading England, France and Holland into conflict during the Second Dutch War in 1667, which was to end Dutch hegemony in the developing world-system. Indirectly, too, this set a pattern of military conflict in the region which was to last for a further century and a half. Both sugar and land prices began to wane, as markets were periodically flooded with the continued extension of cane cultivation into other islands. Estate costs began inexorably to rise, not least the replacement costs for slave labour (Ward, 1978), and those for imported food. As soil nutrient losses made themselves felt, yields from cane then were lowered appreciably. One may say that, by 1665, the first golden age of sugar-cane production in Barbados had come to an end, and a new era, in which the Barbados production was formalized by governmental stricture, protected in commodity markets, and transferred to other islands, was about to begin.

During the eighteenth century, difficulties of production were further accentuated (except in Barbados, where the phenomenon was a relative rarity, a point also not conceded by Wallerstein) by the establishment of a strong West Indian absentee estate-owner lobby in home country governments, ostensibly to protect local interests in an expanding, European-controlled world-system. In itself, such absenteeism, while favourable in theory to West Indian/peripheral interests, further prodigiously increased estate costs and reduced profits, especially through the necessary hiring of managers to ensure the continuation of production on their properties. However, despite the efforts of this lobby, West Indies interests in the British parliament were reduced more and more to a side-show as national government interests came to be focused elsewhere, notably in Canada and India, and particularly after the end of the Seven Years War in 1763. By the end of the eighteenth century, industrial activity too, in both Britain and France, was generating capital of its own on an ever accumulating scale; and the influence

of the colonies to the west in the formulation and financing of their part of the
modern world-system was already well on the wane.

## 4.    Conclusions

    Comparisons between the relative success of the colonization patterns of, on
the one hand, the Spanish and, on the other, the English and French, in the West
Indies, particularly in respect of their influence on the evolution of the modern
world-system, are interesting. Both involved a wholehearted exploitation of
resources, mineral on the one hand and agricultural on the other, and the extensive
use of a coerced labour force from outside their home countries (AmerIndian and
African). Both involved the generation of settlement systems, and institutions of
settlement, which were relatively or entirely new, and which evolved further
subsequent to the experience of settlement in the New World. In both, financial
provision from the home country was minimal, and trading patterns unpredictable.
It is true that the scale of settlement in the early French and English colonies was
much greater than that found in Spanish territories, and this may have made the
ultimate difference between the success of the former and the failure of the latter.
But there is more than a hint of suspicion that the institutions of settlement in the
case of the Spanish were deficient as well: moreover, there was an abuse of
available capital, and a general lack of investment.
    Also there is no doubt that the English were lucky in that they introduced a
well-chosen commercial crop (sugar cane) to their territories at exactly the right
time in respect of its potential profitability; and it was well suited to the
environment. They, and the French, perhaps also had a superfluity of successful
entrepreneurs to exploit this crop; and a more than customary number of willing
potential settlers, bearing in mind the high proportion of male children born in the
homeland in the critical years 1620-40 (twice the normal number). Once
established, both the English and French were able to transfer ideas and capital
freely between colonies and homelands and, at opportune times, a good deal of
capital was sent back to reinvest in the latter. More than anything else, this was
the payoff from early colonial development by both home countries; and it was
one which was in large measure unexpected, particularly in terms of the scale of
the monies involved, and the way in which this then was able to finance later
commercial developments, and industrialization within them. There is little doubt
that the early sugar islands of the West Indies aided the evolution of the modern
world-system substantially; the precise modes and methods through which this was
achieved in homelands need, however, further detailed explanation before they
may be fully understood.
    Setting aside the uniquely negative features of early Spanish settlement in the
island Caribbean, how do the trends of development found in the early English
and French West Indies fit in with those of other peripheral areas of the evolving
world-system? The greatest similarities are with Ireland, whose colonial settlement
began in 1590. In both cases, the speed of change developmentally, involving the
establishment of a new capitalist system, was extraordinary; and, in both places, a

'new' estate agriculture and a new labour system were involved in the development. In both cases, development was centred around small ports. In both, a form of passive mercantilism evolved, activated by the Navigation Acts of both countries. Subjugated peoples directly aided agricultural production in both areas; and the latter was impeded from time to time by sequential wars and uprisings. Unlike Ireland, however, the West Indies were a long way from home base, and this distance factor was to lead to greater economic and military interruptions to production there from time to time than in Ireland. Ultimately, the problems of environmental decline in the West Indies also became more severe than in Ireland, as incessant sugar-cane production for European markets took a greater and greater toll.

Clearly, the West Indies, and Barbados in particular, created increased wealth and available capital not only for themselves, but also for the home countries, in a situation which, to begin with, was not noticeably core-driven, but stimulated rather by local entrepreneurs. Most of the early pump-priming loans for development came from the Dutch, not the English or French (Pares, 1960). In this, at least, it does not correspond particularly well to Wallerstein's model. But, later, from the late seventeenth century, the formalization of the sugar production system in the region caused a better adherence to the model, the system then being more core-driven, but with sources of change in production pattern continuing to be stimulated by and located in the periphery. Finally, only the French made determined efforts to export goods (foodstuffs and manufactures) in quantity to the colonies, as required in the Wallerstein model, the English largely being uninterested in a balanced trade and preferring to import most of their required goods from North America into the West Indian colonies.

[Editor's comment: See the article of K. Whelan, in this volume, on the role of southern Ireland in the provisions trade to the Caribbean.]

## References

Andrews, C.M. (1934): The colonial period of American history. 4 vols. New Haven, Yale University Press

Anon (1676): Great newes from the Barbadoes. London

Boxer, C.R. (1957): The Dutch in Brazil. Oxford, O.U.P

Braudel, F. (1978): The expansion of Europe and the 'longue dureé'. In: Expansion and reaction in Asia and Africa (Ed. H.L. Wesselring), pp. 1-27. Leiden, Leiden University Press

Chamberlain, R.S. (1939): Castillian backgrounds of repartimiento-encomienda. Carnegie Institute of Washington, Contributions to American Anthropology and History, 5, pp. 33-52

Kamen, H. (1983): Spain, 1469-1714 : a society of conflict. London and New York, Longman Press

Las Casas, B. de. (1520-61.): Historia de las Indias. Madrid

Mauro, F. (1961): Towards an 'intercontinental model': European overseas expansion between 1500 and 1800. In: Economic History Review, 2nd Ser. 14, pp. 1-17

Pares, R. (1960): Merchants and planters. In: Economic History Review, 2nd Ser. Supplement 4, pp. 1-91

Postlethwaite, M. (1774): The universal dictionary of trade and commerce. London

Schoffer, I. (1973): A short history of the Netherlands. Amsterdam, Albert de Lange lv

Sheridan, R.B. (1974): Sugar and slavery: an economic history of the British West Indies. Bridgetown, Barbados, Caribbean Universities Press

Wallerstein, I. (1974, 1980, 1989): The modern world-system. 3 vols. New York and London, Academic Press

Ward, J.R. (1978): The profitability of sugar planting in the British West Indies. In: Economic History Review, 2nd Ser. 31, pp.197-213

Watts, D. (1987): The West Indies: patterns of development, culture and environmental change since 1492. Cambridge, C.U.P

Williamson, J.A. (1941): The ocean in English history. Oxford, Clarendon Press

# 20

# At the Cutting Edge: Indians and the Expansion of the Land-Based Fur Trade in Northern North America, 1550-1750

Arthur J. Ray

## 1. Introduction

In North America the fur trade was the cutting edge of historical capitalism as defined by Immanuel Wallerstein. Over a three-century period beginning about 1550, the industry expanded from the Gulf of St Lawrence westward and northwestward to the Pacific and western Arctic oceans. The industry linked the capitalist system of Europe to the reciprocal oriented economies of the native world. Wallerstein argues that the capitalist system usually did not penetrate new areas in search of fresh markets for its products because the inhabitants of these external territories most often were 'reluctant purchasers'. He asserts that they did not need European products for their own economic systems and commonly they lacked the means to buy them. According to Wallerstein it was the search for new products and a lowcost labour force that propelled historical capitalism into new realms. (Wallerstein, 1983, p. 39) But, are we able to explain the rapid territorial expansion of the Canadian fur trade in these Eurocentric terms? Did native people play an active part in the process? If so, why? How did this early involvement affect their economic behaviour?

## 2. The Establishment of the Fur Trade

The classic economic history of the industry is Harold Innis', The Fur Trade in Canada (1930). Innis noted that before the late sixteenth century the European market for North American furs was too small and fragmented to support the development of an independent fur trade. Potential traders could not generate enough volume to pay for the very high transportation costs that had to be borne. Consequently, fur trading began as an appendage of the fishing and whaling industries, which developed in the Gulf of St Lawrence during the early sixteenth century.

When the felt hat became an important European fashion beginning about the middle of the century, the economic climate for the fur industry changed markedly. Felt makers made their highest grade product from the underwool of beaver pelts. However, by the sixteenth century beaver had been largely exterminated in northwestern Europe. Only Siberia and North America could supply the growing demand for beaver pelts. As early as the 1580s the market for this commodity had developed to the point that European merchants decided that it was worthwhile to specialize in the trade. After that, the industry ceased to depend on the fishery (Innis, 1930). From the late sixteenth century until the middle of the nineteenth century the felt hat industry remained the driving force behind the beaver-dominant northern fur trade. Other furs were secondary. Of these, martin (a luxury fur known as Canadian sable) and muskrat (a utility fur used for lining material and lower grade garments) were among the most important.

## 3.    Indians in the Fur Trade

One of the most striking aspects of the fur trade is that Native people eagerly participated in the enterprise from the outset; they were the ones who carried the trading frontier across the northern half of the continent. Apparently, a lively trade developed between fishermen and whalers and Indians, mostly Micmac and Montagnais, well before the demand for beaver developed in Europe. Unfortunately, few records of this early exchange survive so we have no way of finding out what its volume was. However, when Jacques Cartier visited the Gulf of St Lawrence and St Lawrence River valley in 1534 and 1535 most of the Indians he met were eager to barter their furs for his exotic goods. It is reasonable to conclude that the Indians would have willingly engaged in much more extensive intercourse had Cartier and other Europeans had wanted to do so.

We do know that Indian entrepreneurs tried to monopolize exchange between their European contacts and the larger Indian population in the hinterland when regular trade commenced. The Montagnais of the Saguenay River valley were the first of the groups to appear. They denied their interior Algonquian neighbours contact with the Gulf of St Lawrence. Instead, the Montagnais carried European goods to Lake St John and other rendezvous points where they exchanged them for the furs of their Algonquian trading partners. When conducting this barter commerce the Montagnais advanced the prices of goods and furs well above their own prime costs. By the early seventeenth century the Montagnais had become so skilled at pitting fur buyers against one another that they had driven prices at Tadoussac to levels that made it difficult for Europeans to make reasonable profits. Also, the Montagnais constrained the expansion of output since they brought only enough furs to meet their own need for goods (Trigger, 1985).

French traders responded to this problem by employing two tactics. Successive merchant groups sought monopoly trading privileges from the Crown and they made repeated efforts to outflank Indian entrepreneurial groups to deal directly with their fur suppliers. Ultimately, these strategies failed. Of particular relevance here, every time the French advanced their trading base farther inland, another group of Indian trading specialists emerged to supplant the displaced one. The process began in 1609 when the French established Quebec City. From that date until the founding of the Hudson's Bay Company in 1670, a succession of Indian entrepreneurial groups (Table 1) in the Subarctic and Great Lakes regions carried the trading frontier forward. The French moved their base of operations westward far behind the Indian trading frontier. Shortly after Montreal was established as a religious settlement in 1642, it emerged as the key trading centre eclipsing Trois Rivieres and Quebec City (see map). By that time Indian trading networks probably reached as far west as Lake of the Woods and Lake Winnipeg and perhaps even beyond.

Table 1: Indian Entrepreneurial Groups, ca. 1580-1670

| Group | Major Period of Control | Initial Travel Distance [km]* | Means of Travel | Travel Time in Months |
|---|---|---|---|---|
| Montagnais | 1680 - 1620 | 0- 800 | Canoe | 0 - 1 |
| Algonkins | 1610 - 1620s | 1,200-1,600 | Canoe | 1 - 2 |
| Nipissing | 1615 - 1640s | 1,600-2,000 | Canoe | 2 - 4 |
| Huron | 1615 - 1649 | 2,000 | Canoe | 3 - 4 |
| Ottawa & Ojibwa | 1650 - 1720s | 2,400-3,600 | Canoe | |
| Assiniboine | 1670s- 1770s | 2,000-2,400 | Canoe | 3 - 6 |
| Western Cree | 1670s- 1780s | 1,200-2,400 | Canoe | 2 - 6 |
| Chipewyan | 1717 - 1800 | 1,800-2,400 | Foot | 12 - 24 |

* As the 'crow flies'

Inland expansion created new economic problems for the French, however. In particular, the lengthening lines of communication added to the overhead costs of conducting the fur trade. The destruction of the Huron by their Iroquois enemies in 1649 brought this problem to the forefront. The Huron collapse sent their Nipissing, Ottawa, and Ojibwa trading partners fleeing to the north and west to escape the dreaded Iroquois. To reestablish regular contact with these people, French traders and explorers travelled to the upper Great Lakes area in the late 1650s and early 1660s. While visiting this region two of these men, Pierre Esprit Radisson and Medard Chouart, Sieur des Groseillers, learned that the prime fur country was located to the north of Lake Superior. They concluded that it would be much less costly to approach the area through Hudson Bay instead of from the St Lawrence. However, French colonial officials were not receptive to the idea of further territorial expansion. Instead, they wanted to consolidate their holdings in New France and reduce the colony's dependence on the fur trade by promoting economic diversification (Ray and Freeman, 1978, p. 24). Not to be denied, Radisson and Groseilliers sought and obtained backing in England. Following a successful expedition to James Bay in 1668, the backers of the scheme obtained a royal charter from Charles II which gave birth to the Hudson's Bay Company. The charter gave the company exclusive trading rights in the territory (called Rupert's Land) drained by waters flowing into Hudson Straits.

This development served to greatly accelerate the spatial expansion of the industry. Previously the western Cree and Assiniboine obtained French trade goods from the Ottawa and Ojibwa. The construction of trading posts in James Bay in the early 1670s and in western Hudson Bay in the 1681 provided them with direct access to this merchandise. The Cree and Assiniboine immediately seized their new strategic advantage and they established their own trading empires. They eventually extended them southwest into the parkland and prairie region and as far northwest as the headwaters of the Churchill River. One decade after the foundation of the English company, the Cree and Assiniboine had cordoned off the bay from their opponents. They dominated the trade of western Hudson Bay for nearly a century afterward. More than three-quarters of the furs they brought they obtained from their inland trading partners (Ray, 1974, 59-61 and Ray and Freeman, pp. 42-51).

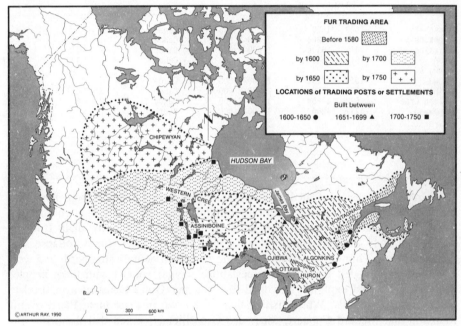

*Expansion of the fur trade, 1580-1750*

Athabascan speaking Chipewyan were one of the groups the western Cree tried to prevent from having access to Hudson Bay. Using their European firearms they were largely successful until 1717. In the latter year the Hudson's Bay Company built Fort Churchill to break the Cree blockade and deal directly with the Chipewyan. This move was successful. Afterward the Chipewyan became important trappers and traders who controlled the trade in the woodlands lying north of the Churchill River as far west as the Mackenzie River valley. Thus, by the middle of the eighteenth century, Indians had carried the fur trade from the Gulf of St Lawrence coast to the Rocky Mountain foothills. The French continued to follow in their wake. From the 1680s to 1712 they had fought against the Hudson's Bay Company to gain control of the maritime entry to the western interior. But, after the Treaty of Utrecht awarded this approach to the English in 1713, the French redoubled their efforts in the interior. Before 1751 they had built a string of posts between Lake Superior and the forks of the Saskatchewan River. Meanwhile, the Hudson's Bay Company 'slept by the frozen sea', as its parliamentary critics charged, and let the natives bring their furs to its bay-side posts. The Indian middlemen welcomed this development because it enabled them to promote competition between the two European networks. These trading specialists brought enough furs to the bay-side to enable the company to conduct a highly lucrative business (Ray, 1985).

The early fur trade attracted native participants for several reasons. Initially Indians believed that many European articles had supernatural powers. Very quickly they learned to appreciate the utilitarian value of certain categories of goods, most notably brass and copper pots, ice chisels (used to hunt beaver), metal knives and hatchets, and firearms. Metal weapons and firearms quickly became prized articles because of their military significance. The differential rates at which groups obtained these commodities upset traditional balances of power. This was a critical issue in the war-torn area of the St Lawrence and Ottawa river valleys throughout the seventeenth century (Trigger, pp. 164-297). Also, Indian entrepreneurs used their European arms to establish their position in the expanding system.

Besides these concerns, the goods obtained at European posts represented new forms of wealth that individuals and kinship groups used to enhance or maintain their status. They did so by redistributing it. This behaviour reflected the fact that aboriginal societies emphasized sharing, or general reciprocity, and treated the hoarding of wealth as antisocial behaviour. The behaviour of Indian trading captains serves as an example. These were the men who led parties of Indians to the European's posts. According to Hudson's Bay Company accounts, trading captains kept few of the goods which they received in trade or as gifts. Instead, they gave most of the articles away to curry the favour of their followers. Besides this consideration, native men sought to be trading captains to gain enhanced status in the eyes of their fellow travellers and in those of their European partners (Ray, 1975, pp.586-602 and Ray and Freeman, pp.63-78).

The land-based fur trade that developed between 1580 and 1700 represented an economic bonanza for the native for additional reasons. Europeans prized coat beaver (*Castor gras*) because it was cheaper to process into felt than parchment beaver (*Castor sec*). This was because Indians turned the hairy side inward and after one or more seasons of use they wore off the long guard hairs leaving behind the wool which the felt makers coveted. In contrast, Indians produced parchment beaver pelts specifically for trading purposes and therefore they did not wear them before selling them. These skins had to be sent to Russia for processing because only the Russians had developed the technological capability necessary to remove the guard hairs cheaply.

The strong demand for coat beaver represented a windfall for the Indians who could barter their used winter clothing for the exotic goods they treasured. The Indians quickly perceived the irony and economic advantage this development presented to them. For example, in 1634 Father le Jeune wrote that a Montagnais told him: "it [the beaver] is the animal well-beloved by the French, English and Basques, - in a word, by the Europeans." (Ray, 1978, p. 9) Le Jeune continued:

"I heard my [Indian] host say one day, jokingly, Missi picoutatau amiscou, 'The Beaver does everything perfectly well, it makes kettles, hatchets, swords, knives, bread; and in short, it makes everything.' He was making sport of us Europeans, who have such a fondness for the skin of this animal and who fight to see who will get it; they carry this to such an extent that my host said to me one day, showing me a beautiful knife, 'the English have no sense; they give us twenty knives like this for one Beaver skin.'"

Unfortunately for the Indians this advantage did not last. By the close of the seventeenth century, important changes were taking place which altered the market for beaver. The west Europeans learned the secret of processing parchment beaver, which was of more uniform quality, and the demand for coat beaver slackened therefore. In the 1690s these furs piled up in Hudson's Bay Company and French warehouses as the demand turned toward parchment beaver. This posed a problem for the native people. If they stopped wearing beaver coats they needed to find a substitute. The Hudson's Bay Company hoped they would adopt imported sheep skin and cloth coats (Ray and Freeman, pp. 159-60). On the other hand, if they continued wear beaver they would have to put greater pressure on this resource to produce coat and parchment beaver. Although they forced the traders continue to accept coat beaver at par with parchment for the time being, the traffic in coat beaver diminished over time. It ceased to be an item of commerce by the middle of the nineteenth century. By then beaver were badly depleted throughout the central subarctic and the Indians were increasingly dependent on European clothing.

Indian attitudes toward travel appear to have been another factor which encouraged groups at the periphery to become involved in the fur trade and carry the frontier forward. Clearly Indians did not shirk from undertaking long voyages. On the contrary, many groups probably relished these ventures. In other words, they did not regard travel as an opportunity cost. As late as the twentieth century, Hudson's Bay Company traders said that Indians were too sensitive to spatial differences in fur prices because they did not calculate travel costs into their trading decisions (Ray, 1990a, pp. 91). This attitude probably partly explains why groups of middlemen often travelled considerable distances over difficult routes to reach their European contacts in the early phases of exchange. Table 1 shows that at the time of their initial involvement with English and French traders most Indian middlemen undertook return voyages of between 1,200 and 3,600 kilometers. Most of them took one to five months to complete these expeditions. The first Chipewyan to arrive at York Factory travelled on foot and took over two years (Yerbury, 1986, pp. 17-50). These great expenditures of effort yielded slim returns. Indians could only carry 200 pounds of furs or goods per canoe, or roughly 100 pounds per person. The Chipewyan backpackers carried less. Hudson's Bay Company men concluded that the early efforts of members of this Athabascan group were not worth the business they brought. Obviously the Chipewyan did not share their viewpoint.

The accommodations that Europeans made to native trading traditions also encouraged Indian participation. One of these involved accepting the Indian custom of exchanging gifts. Gift exchange ceremonies, which often were elaborate affairs, became an integral part of the early fur trading practices; these events preceded barter exchange (Ray and Freeman, pp. 53-62). When competition for the Indians' custom was strong, the English and French gave lavishly in efforts to show that they were more generous than their opponents (Ray and Freeman, pp. 199-201). The Indian leaders received royal treatment. For instance, Hudson's Bay Company officers presented them with special outfits of clothing, tobacco, and exotic food before they dealt with their followers.

Besides exchanging gifts, the traders provided the Indians with goods on credit and credit/barter became an integral part of the fur trade at the outset. The Europeans extended credit for several reasons. It enabled them to stake a claim on the Indians' future returns. If the Indians honoured their debts, which most of them did, it meant that rival traders had no opportunity to obtain a significant portion of their next year's harvest. Fur output was unpredictable. Returns depended on animal population cycles, weather conditions during trapping season, the general health of the native population, and whether intertribal peace or war prevailed. By giving Indians goods on credit, traders made sure their clients always could satisfy their desires and basic needs in spite of the periodic shortfalls in fur returns that occurred. Undoubtedly Indians regarded this credit as a kind of general reciprocity that established a bond of mutual obligation between themselves and the trader who provided it.

## 4.    The Impact of the Early Fur Trade on the Native Economies

Another striking feature of the early fur trade is that native people retained control of the means of production. Although Charles II gave the Hudson's Bay Company title to a large portion of the central Subarctic, this grant meant little before the 1860s. Neither the company nor the French evicted the Indians from their lands. Consequently the Indians remained independent producers who were free to sell their furs to whom ever they pleased. An important reason for this was that a debt peonage system did not emerge partly because strong English and French commercial rivalries developed in the Hudson Bay hinterland after the War of Spanish succession. For these reasons the early post-contact experience of Subarctic Indians was very different from that of most aboriginal people elsewhere in the Americas, Africa and Asia.

Granting that Indians retained considerable control in the early fur trade, did their participation alter their traditional economic behaviour and practices in any fundamental ways? Scholars have debated this question the past two decades. In 1968 British historian E. E. Rich argued that Indian economic behaviour differed sharply from that of Europeans in that they did not seek to maximize the returns from their labour. He contended that Indians traded with Europeans mostly to satisfy their immediate needs, to maintain their political alliances, and to obtain European arms (Rich, 1960, pp. 46-9). Of particular relevance, Rich stated that

Indians did not respond to market pressures as expressed by price fluctuations and in that sense they were not 'economic men'. He cited the remarkable stability of the Hudson's Bay Company's official price schedules for furs and trade goods as evidence that supported his conclusion. Rich interpreted this was an indication that the Indians had forced the Hudson's Bay Company to maintain price stability because they did not understand market forces. Also he pointed out that Indians reduced their per capita production whenever fur prices rose. Rich maintained that this type of behaviour showed that the natives were not interested in maximizing their returns for the purposes of capital accumulation.

The political economist Abraham Rotstein accepted Rich's conclusions and attempted to put them in a theoretical context. Rotstein believed that the early fur trade was little more than an institutional extension of the Indian alliance system. Consequently he argued that it was an example of administered or treaty trade as defined by Karl Polanyi. Therefore, Rotstein said native participation is better understood in terms of 'non-economic' factors such as kinship structures, political/military alliances, and warfare (Rotstein, 1970, pp. 118-20).

Arthur J. Ray and Donald B. Freeman challenged Rich's and Rotstein's interpretation. They agreed that elements of treaty trade certainly were present in the Hudson's Bay Company's trading system before 1763, most notably reciprocal pre-trade gift-exchanges, but market trade also was very important. Their conclusion for this was based on the fact that actual barter rates at the bay-side posts differed markedly from those of the official price lists. The company paid the highest prices for furs when French competition was strong. This shows that Indians did react to price differentials (market pressures). So, Ray and Freeman postulated that the fur trading system of the late seventeenth and early eighteenth century was a transitional one. It incorporated features from both the aboriginal reciprocal economy and the expanding capitalist system.

More recently, anthropologist Bruce Trigger has taken issue with the conclusions of Rotstein, Ray, and Freeman. He believes that Rotstein erred by failing to allow for scale differences between the large ancient civilizations of the middle east and the small band societies of the Subarctic. Little structural differentiation took place in the latter societies nor were they large enough to produce the anonymous crowds of buyers and sellers needed to make a classical market work. Therefore, Trigger states that trade and peace necessarily were synonymous (Trigger, 1985, pp.192-3). He concludes that intertribal trade never was as embedded in the social and political structure as Rotstein envisioned. According to Trigger, this helps explain why Natives haggled with Europeans from the commencement of contact and they advanced prices of furs and goods over their prime costs. Trigger further postulates that profit taking had been an aspect of intertribal exchange for millennia. For this reason he does not think that the Hudson's Bay Company's eighteenth century trading system was transitional contrary to Ray and Freeman's suggestion. Instead, he thinks that the combination of gift-trading and optimizing had been a feature of aboriginal exchange networks for thousands of years.

Also, the fact that Indians reduced their trapping effort in response to fur price increases does not mean that they were not interested in maximizing the returns on their trapping and trading activities. We can account for this behaviour in several terms - a mobile lifestyle, limited transportation capacity, and attitudes toward leisure time. Ray and Freeman noted that the highly mobile life-style of the Subarctic Indians set limits on their ability to consume durable goods in these early years. The anthropologist M. Sahlins predicted this was the case for most "stone age" hunters and gatherers (Ray and Freeman, 1972, pp. 222-3 and Marshal Shalins, pp. 32-33). The great distances Natives travelled to trading posts served as added constraints. Also, fur prices rose when competition was strong and these were the times when Europeans lavishly gave presents to woe Indian bands to their sides. The more goods Indians received as gifts the fewer they had to buy.

Economist Ronald Trosper explored another dimension of the topic (1988, pp. 201-4). He notes that the principal input in fur production was labour and therefore we should regard Indian fur returns as a surrogate measure of their labour supply. Trosper adds that we should regard leisure time as a 'normal good'; consumption levels for these types of goods rises when income increases. By taking this approach we discover that Indians behaved in a way that is very similar to that of twentieth-century North American workers. As their incomes rose, they consumed more leisure time.

Decidedly Indians were economic men in the sense that they sought to satisfy their short-term demands for European goods with the least expenditure of effort. The available evidence does not suggest, however, that they sought to make profits to accumulate more capital. In this important respect they differed fundamentally from their European trading partners.

## 5.   Conclusion

In conclusion, the fur trade represented the frontier of the expanding world capitalist system. Between 1580 and 1750, approximately the time of the so-called 'B-phase of European economic stagnation' (Wallerstein, 1979, pp. 74-5), the industry expanded from the Gulf of St. Lawrence two-thirds of the way across present-day Canada. Indians played a central role in the expansion because they welcomed the opportunity to obtain European merchandise at what they considered bargain rates. Some of these goods, such as metal hatchets, knives, ice chisels and arms, improved the efficiency of hunters; others, chiefly brass and copper pots and cloth, made domestic life less burdensome for women. Equally important, individual Indians and kinship groups redistributed the goods they obtained through trade in traditional ways to enhance or sustain their social position.

The French made a concerted effort to outflank Indian trading groups to gain direct access to their suppliers, but they did not attempt to displace the native trappers. This "unproletarianized" labour force provided furs at rates of exchange which enabled French and English traders to turn substantial profits. Indeed, instead of displacing these Indian producers, European traders provided them with outfits of staple merchandise for encouragement.

While the native economies retained most of their pre-contact characteristics during this period, particularly their reciprocal orientation, they did become intermeshed with the advancing world capitalist system as Indians began to replace their traditional tools of production for those of European manufacture. The process accelerated in later years when the expansion of the trading post frontier and the proliferation of posts eliminated the constraint on Indian consumption caused by limited transportation capacities. Increased consumption of staple and luxury goods led to resource depletion. These developments made Indians increasingly dependent on European traders and more vulnerable to manipulation and exploitation by the latter group in subsequent years (Ray, 1984, pp. 1-20). By the early late nineteenth century, many Indians in the central Subarctic had to supplement their trapping activities with part-time wage labour and they needed increasing amounts of economic relief provided as aid for the sick and destitute and as credit which they could not repay (Ray, 1990a, pp. 207-21 and Ray, 1990b). The Hudson's Bay Company exploited this situation to keep natives in the bush and it was in this manner that EuroCanadians eventually gained control of the native labour force.

## References

Eccles, W. J. (1974): The Canadian Frontier, 1534-1760. Albuquerque

Innis, H. (1970 [1930]): The Fur Trade In Canada. Toronto

Ray, A. J. (1974): Indians and the Fur Trade. Toronto

Ray, A. J. (1975): The Factor and the Trading Captain in the Hudson's Bay Company fur trade before 1763. National Museum of Man Mercury Series, Canadian Ethnology Service Paper 28, Ottawa, pp.586-602

Ray, A. J. (1984): Periodic Shortages, Native Welfare, and the Hudson's Bay Company. In: S. Krech III (ed.): The Subarctic Fur Trade: Native Social and Economic Adaptations. Vancouver, pp.1-20

Ray, A. J. (1985): Buying and Selling Hudson's Bay Company Furs in the Eighteenth Century. In: D. Cameron (ed.): Explorations in Canadian Economic History: Essays in Honour of Irene M. Spry. Ottawa, pp.95-115

Ray, A.J. (1990) The Fur Trade in the Industrial Age. Toronto

Ray, A. J. and Donald B. Freeman (1978): Give Us Good Measure: An Economic Analysis of Relations Between the Indians and the Hudson's Bay Company Before 1763. Toronto

Rich, E. E. (1960): Trade Habits and Economic Motivation Among the Indians of North America. Canadian Journal of Economics and Political Science, 26, pp.35-53

Rotstein, A. (1970): Karl Polanyi's Concept of Non-market Trade. Journal of Economic History, 30, pp.118-20

Sahlins, M. (1972): Stone Age Economics. Chicago

Trigger, B. (1985): Natives and Newcomers. Montreal

Trosper, R. (1988): That Other Discipline: Economics and American Indian History. In: C. Galloway (ed.): New Directions in American Indian History. Norman, pp.201-4

Wallerstein, I. (1979): Underdevelopment and Phase-B: Effect of the Seventeenth-Century Stagnation on Core and Periphery of the European World Economy. In: W. Goldfrank (ed.): The World-System of Capitalism: Past and Present. Vol 2, Beverly Hills, pp.74-5.

Wallerstein, I. (1983): Historical Capitalism. London

Yerbury, J. C. (1986): The Subarctic Indians and the Fur Trade, 1680-1860. Vancouver

# Part VI   External Arenas

# 21

# The Role of the Eastern Mediterranean (Levant) for the Early Modern European World-Economy 1500-1800

## J.Malcolm Wagstaff

## 1.   Introduction

The eastern Mediterranean basin may be defined as lying east of a line between the Strait of Otranto and the island of Malta, but with an important extension into the Aegean Sea. The coastlands framing the sea, together with the islands of Crete and Cyprus, are the principal concern of this paper.

I begin with a sketch of the region's physical diversity and then outline its political history between 1500 and 1800. The case for the region's incorporation into the European world-economy in this period is presented briefly before attention is drawn to some neglected aspects of the Levant's historical geography and to different interpretations of the available economic information. Finally, the paper suggests an alternative view of the Levant to that of incorporation into the European world-economy.

Although mountains are the predominant landform in the Levant, strips of coastal plain are found, as well as more extensive areas of lowland, some of it desert but parts of it swampy in the period under review. The Aegean Sea and its southern approaches are sprinkled with islands, each with its own characteristics. On this scale, Crete and Cyprus are virtually mini-continents. Whilst the climate of the region is 'Mediterranean', terrain and cyclone tracks produce considerable local variety. In the period under review the mountains provided timber, gathered products (like valonia, for example) and summer grazing; silk came from some of the hilly areas, which also provided cereals, grapes, olives and cotton, though some of the plains were also important sources. The sea provided ready access to all parts of the region, but navigation in the eastern Mediterranean and the Aegean seas was never as easy as is sometimes supposed. Rocks and sudden squalls are hazards at all seasons, and especially so in our period when charts were defective, whilst the gales and mountainous seas of winter effectively closed the region in the days of sailing ships and galleys.

At the beginning of the period covered by this paper much of the Levant was not under Ottoman control. With the exception of the Ionian Islands, which remained in Venetian hands until almost the end of the eighteenth century, the Levant was finally incorporated into the Ottoman Empire only between 1500 and

1669 (Fig.1).[1] In the west, most of the Morea (Peloponnese) had passed under
Ottoman control by the end of 1500, though the Venetian fortresses of Napoli di
Romania (Nafplion) and Napoli di Malvasia (Monemvasia) were not ceded until
1540[2]. Selim I conquered Syria in 1516 and added Egypt to the Empire in the
following year. Cyrenaica passed into Turkish hands in 1521 and Tripolitania in
1551. Kanun-i Süleyman (Suleiman the Magnificent) obtained Rhodes in 1522.
Cyprus followed in 1571. The Cyclades came under direct Ottoman rule in 1579,
after being granted to the financier, Joseph Nasi, at their initial conquest in 1566.
The great War of Candia (1645-69) ended with the incorporation of Crete, though
the offshore fortresses of Grabusa and Spinalonga, as well as the Cycladic island
of Tinos, did not pass to the Ottomans until 1718. Even then the Serene Republic
retained Kithira (Cerigo) and Antikithira (Cerigotto), islands positionally within
the domain of Ottoman seapower, whilst the other frontier outpost of
Christendom, Malta, was held until 1798 by the Knights of St. John following the
unsuccessful Turkish siege of 1565. With these exceptions, the whole of the
Levant had become a single trade area by the end of the seventeenth century.

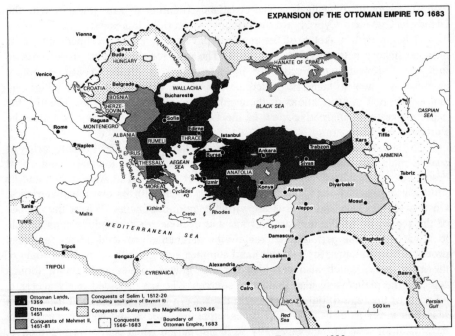

Fig.1: Expansion of the Ottoman Empire to 1683

Ottoman control of the Levant was disputed on two occasions before the French invasion of Egypt in 1798. Taking advantage of Ottoman involvement on the Danube, Venice was able to reconquer the Morea (1684-87) and retained the peninsula until 1715, when the Turks retook it. The Russo-Turkish War of 1768-74 saw Russia instigate a number of diversionary activities in the Levant, as well as inflict a resounding defeat on the Turks at the battle of Çeçme (6/7 July, 1770). The diversions included an uprising in Mt. Lebanon, a brief occupation of some of the Cycladic Islands, and a revolt in the Morea which plunged the peninsula into almost a decade of anarchy[3]. Russia obtained significant concessions in the Levant as a result of the Treaty of Küçük Kayarnca (1774) which ended the war, including the right of access to Ottoman ports for ships flying the Russian flag; this allowed the Greeks to take over much of the carrying trade of the region. The eighteenth century closed with the French occupation of Egypt (1798) and the alliance of Britain and the Porte for its recovery[4].

## 2.    The Case for Incorporation

Whilst the Ottoman Empire was expanding across the Levant and enlarging its internal market, it was also either being incorporated into the European world-economy or on the eve of incorporation. At least, this is the theory advanced by Immanuel Wallerstein and his associates. The theory need not be outlined here, but it is necessary to summaries some points which bear on the Ottoman Empire.

Wallerstein concedes the existence of world-empires other than that associated with capitalist Europe[5]. The Ottoman Empire, in fact, is a classic case for Wallerstein[6]. As a world-empire it was part of the "external arena" of the European world-economy. According to Wallerstein's theory commerce between the two should have been in luxury goods[7]. At some point, though, the Ottoman Empire lost its economic independence and was incorporated into the periphery of the European world-economy. The chronology of these events is uncertain. Wallerstein originally suggested two possible dates for incorporation.[8] The first was early in the seventeenth century but the alternative was the late eighteenth-early nineteenth century. In 1982 Hopkins and Wallerstein suggested that incorporation took place in the period 1750-1815[9]. By 1983 Wallerstein was definite that it had happened between 1750 and 1839[10]. In 1987, however, a note of uncertainty had crept back. Wallerstein and his colleagues then admitted that the date of incorporation still remained to be resolved; it would vary, they thought, for different parts of the Empire[11].

Assuming for the moment that incorporation did take place somewhere between 1600 and 1839, the precise chronology will be decided ultimately - as Wallerstein himself pointed out - on the basis of "the real economics of the situation", by looking at actual production processes and particularly at the character of trade[12]. Thus, resolution of the problem depends upon the statistical series for trade between the Ottoman Empire and Europe which the economic historian can put together. So far they have depended largely upon western sources for the period. These show that there was a rise in the volume of commercial crops being exported to Europe. Wallerstein and his associates lay great stress on the export of cereals and cotton, but also mention maize and tobacco. During the seventeenth and into the first half of the eighteenth century, though, the most valuable item exported from the Levant was silk[13]. At first Iranian silk dominated. It was brought by caravan to Aleppo and Smyrna (Izmir) before being shipped on to Europe. After 1725, however, locally produced, poorer quality silk was exported to Europe. Returning to Wallerstein, he and his associates also note how the goods imported by Europeans to the Ottoman Empire changed from specie to woollen cloth and other manufactured items, as well as colonial produce such as coffee and sugar; they appear to ignore such raw material as base metals which were also important[14].

Wallerstein and his colleagues have noted how the changes in trade coincided in time with various socio-political developments in the Ottoman Empire itself. These included the granting of the first Capitulations (with France in 1569, England 1580, Netherlands 1612), and the establishment of the first French, Dutch and English consulates in the Ottoman Empire (French 1523, England 1581, Netherlands 1613). It is principally the chronology of these events, and their coincidence with the first phase of expansion for the European world-economy (put by Wallerstein at 1450-1650)[15], which suggested that the incorporation of the Ottoman Empire into the European world-economy could have taken place in the seventeenth century. The case can be reinforced by noting that this early period saw the foundation of the various joint-venture-capital companies trading with the Levant (the English Levant Company in 1581; the Chambre de Commerce de Marseille in 1652) and thus a change in the style of trading carried on in the region by Europeans[16].

Other events to which great significance has been attached by the Wallerstein school for the period 1600-1800 are the decay of the "classical" Ottoman system of administration and taxation based upon "livings" (dirlik), the emergence of a system of tax farming, the development of the çiftlik and the appearance of virtually independent rulers in different parts of the Ottoman Empire. These developments have usually been interpreted by western scholars as the symptoms of decay[17]. For Wallerstein they also demonstrate a shift in productive relations occasioned by a change-over from something like subsistence agriculture to commercialized production. An important catalyst, to put it no higher, was western demand for cereals and raw materials. These views can be contested.

## 3.   Points for Consideration

The basic notion of incorporation seems to depend upon detecting a change in the volume of Ottoman trade with Europe and in the types of commodities being traded[18]. So far this has been done largely from a European stand-point and using European sources[19]. Whilst it is high time the Ottoman sources were brought into the debate, it is clear that some of the European evidence is capable of being interpreted differently. In particular, it points to the relative unimportance of trade with Europe for the Ottoman Empire. Several examples may be noted.

The commercial duties paid by Europeans in the Ottoman ports c.1750 are estimated to have been worth £100,000 per annum when the total income of the Porte probably exceeded £1,000,000.[20] At the same date, Egypt exported an estimated £100,000 worth of goods to France, the major European trading power in the Levant at the time, but sent £500-800,000 worth of goods to Syria. In 1776 Egypt's trade with Europe was estimated at 13,000,000 francs, compared with 67,000,000 francs for trade with the rest of the Empire (equivalent to 19 per cent)[21]. Seven years later (1783) Trécourt estimated the total external trade of Egypt at 1,609,260,664 paras, of which only 14 per cent (235,598,053 paras) was carried on with Europe, though imports were slightly greater than exports (15 per cent of the total compared with 14 per cent). Trade with the Red Sea and Indian Ocean region was much more important with a total value of 573,750,000 paras or 25 percent of all Egypt's external trade[22]. The implication of these figures is that Egypt at least should be regarded as still part of a world-economy different to that centred in north-western Europe.

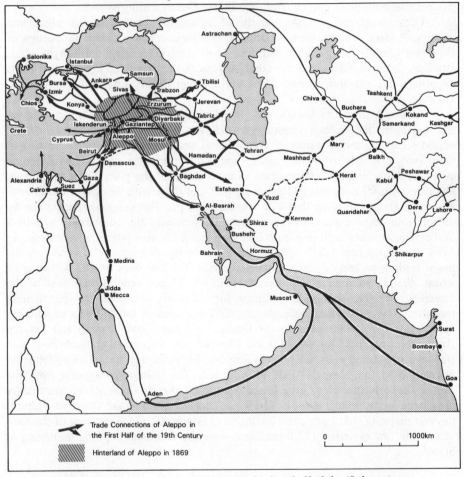

*Fig.2:Trade connections of Aleppo in the first half of the 19th century*

This conclusion is substantiated to some extent by the experience of Aleppo, Crete and the Morea. Raw silk was the major European export from Aleppo down to 1725-50, most of it coming from Iran and 'a good deal' of it actually used by Syrian manufacturers rather than sent on to Europe[23]. From 1750 onwards locally produced cotton became an important export to Europe and the silk sent there was also locally produced. Whilst these changes can be interpreted as indicative of incorporation into the periphery of the European world-economy, they must be set against other evidence. Silk simply ceased to arrive in Aleppo because of internal upheavals in Iran and alternative export commodities had to be sought locally. This strengthened, rather than weakened Aleppo's position as the economic focus for a wide region in Syria and southern Anatolia (Fig.2)[24]. Whilst the caravans bringing goods from the Indian Ocean through Iraq appear to have declined in frequency and importance after about 1750, nonetheless they continued to arrive and if they did not reach Aleppo it was often because they were diverted to Damascus[25]. Moreover, colonial goods imported by Europeans were clearly in competition with those coming from the Indian Ocean trading system.

Crete produced a wide variety of items for export in the eighteenth century.[26] Most of them went to Ottoman territories. Thus, a large proportion of the Great Island's raisins, wine and honey was shipped to Egypt and Libya, whilst Istanbul, Smyrna, Chios and the other Aegean Islands received a good deal of its exported olive oil and cheese. Most of the soap exported from Crete went to Ottoman ports[27].

Evidence from the Morea, the 'Jewel of the Levant'[28], is also instructive. For example, the French consul reported that only eleven ships reached the region's capital, the city-port of Napoli di Romana (Nafplion), in 1756 and of these only one originated in France, whilst the other ten came from Ottoman ports. In 1763 20 ships are reported to have called at the port, 18 of them from Ottoman ports[29]. Both sets of figures suggest that trade with the rest of the Empire was more significant than that with western Europe. The same conclusion can be drawn from some figures on wheat exports, precisely the commodity to which Wallerstein attaches such importance in supporting his incorporation hypothesis. In 1758, the Morea sent 1,540,000 okes of wheat to Istanbul but in the same year the major trading partner, France, imported either 233,200 okes or 462,000 okes of wheat, that is an amount equivalent to 15 or 30 per cent of that sent to the capital[30]. Finally, the trade statistics for 1794-95 produced by the Sicilian traveller, Xavier Scrofani, indicate that 62.2 per cent of the Morea's imports by value came from other parts of the Ottoman Empire, with Istanbul and Smyrna (Izmir) the principal sources (22.0 and 19.6 per cent respectively). Some re-export was involved (London camelots from Smyrna, for example), but again an Ottoman orientation to the Morea's trade is indicated. The pattern of exports from the Morea supports this view to a lesser extent. Of the 92 destinations specifically mentioned by Scrofani, only 10 (10.9 per cent) lay within the Ottoman Empire; the vast majority (65.2 per cent) lay in Italy. Foodstuffs (56.2 per cent of the total value) and raw materials (42.8 per cent) were the dominant exports, according to Scrofani[31].

Another argument in favour of the incorporation hypothesis and its comparatively early date is the coincidence in time of the alleged shifts in trade with evidence which Wallerstein has construed as indicating changes in productive relations in the Ottoman Empire. Here I will simply draw attention to the changes in the taxation system of the Empire and the emergence of the *çiftlik*[32].

The 'classical' Ottoman administrative system fused responsibility for law and order with military rank and responsibility. It was underpinned by a financial system which provided 'livings' for the soldier-administrators, theoretically on a set scale and without attached rights of inheritance. The 'livings' might be derived from the working of a corn mill or the custom duties of a port, but in most cases came directly from the produce of landed properties (*ziamets* and *timars*) and were often paid in kind. It is misleading, but commonplace to call these arrangements 'feudal'. The holders of the 'livings' were regarded as the *kuls* (slaves) of the Sultan, not the liegemen of a particular lord in a hierarchy of hommage and fealty. These arrangements were widespread in the Levant, though not in Egypt[33]. They began to be replaced by tax-farming during the seventeenth century.

The reasons for the moves away from the 'classic' arrangements are complex. They involve changes in the nature of warfare which required more full-time professional troops rather than the part-time cavalry supported from the 'livings' system, as well as the financial strain of fighting wars on several fronts almost simultaneously. The end of expansion to the Empire removed plunder and booty on any scale from its finances and made warfare less self-financing, whilst loss of territory reduced the number of 'livings' available. Another factor was inflation, arguably imported from the West with American silver[34].

The system of tax farming which emerged during the seventeenth century allowed wealthy individuals to acquire control not only of the taxes of particular districts but also of the land itself and the farmers. The *ayan* were also able to acquire coercive power through building up their own private armies and they fought each other for power and territory. Power in the Empire became more decentralized and powerful individuals in the provinces were less easily controlled from Istanbul, though they could always be removed by the imperial government when strictly necessary. It is usually claimed that the increasing burden of taxation and its arbitrary nature forced the actual tillers of the soil into becoming sharecroppers on *çiftliks*[35]. In other words, changes in taxation forced a change in the system of production.

I have tried to present the modifications to the Ottoman land system in as neutral a way as possible and at a high level of generalization. Their spatio-temporal unfolding still has to be established in detail. However, in linking them to the commercialization of agriculture, Wallerstein and his associates make a number of assumptions which, because they appear to be derived from a Eurocentric historiographical tradition, distort the interpretation of the limited evidence currently available. Western historical writing about the Ottoman Empire has tended to accept any indication of change as also evidence of decline or decay[36]. The decline may be from some high point of 'pure' Islamic civilization under the United Caliphate or, following Ottoman scholar-bureaucrats themselves[37], from the period of the classical 'Turkish-Islamic' synthesis of the fifteenth-sixteenth century. It is a typically orientalist stance which has been sharply criticized in recent years, notably by Roger Owen[38]. Acceptance of this view, linking change with decline, seems to lead naturally to the assumption that changes in the system of production must also be changes for the worse.

Within a context of such ideas, three further assumptions can be detected. The first is that share-cropping necessarily implies coercion. The second is that share-cropping was a new feature of the land system in the Levant. The third assumption is that commercial production, production for a market, whilst not perhaps new in the region, was aimed at inter-regional trade for the first time. Each assumption may be questioned, though there is space here only to sketch the grounds for doing so.

Share-cropping has existed in the region for as far back as records go and it may have been an important element in land-holding systems from at least late Roman times. In a situation of land surplus and labour shortage, share-cropping is essentially a mechanism for keeping fields under cultivation where debt might otherwise force the farmer to abandon his holding. Whilst an element of coercion must be conceded (through foreclosure, for example), the system contains some benefits for the share-cropper. He is provided with land and often with the means of cultivating it, including working capital in the form of loans. In return his 'partner' takes only a share of the harvest rather than a fixed amount. The share-cropper has at least something left. It would have been foolish to leave him with less of the crop than would be needed to maintain his family. Moreover, share-cropping works best where the main crops produced are cereals for these are easily assessed and shared on the threshing floor. Whilst the principle of shares may have been applied to cotton production, it is less easy to envisage it being applied to other Levantine produce such as silk, olive oil and dried fruit. In any case, the commercial crops which featured in the European trade of the Levant in the seventeenth and eighteenth centuries (cotton and silk) had long been produce on a commercial scale in the region[39]. As Wallerstein is aware[40], they were even sent to Europe during the Middle Ages [41], though then as later it is difficult to establish the proportions of the total output exported to Europe, to regions bordering the Levant, to regions further east or, indeed, to different parts of the Levant itself. Tobacco is a little different. It was new to the region and, though widely grown in the seventeenth and eighteenth centuries, was destined for local consumption or trade within the Levant. West European demand was met from the Americas.

Ever since Braudel adopted Busch-Zanter's notion that çiftlikisation in the Ottoman Empire was analogous to the re-enserfment which apparently affected parts of north-eastern Europe in the sixteenth century[42] there has been a tendency for scholars to stress the coercive side of the çiftlik. It was, thus, an obvious step for Wallerstein to accept the emergence of çiftliks as prima facie evidence for the change in productive relations associated with incorporation into the European world-economy. The argument was strengthened by Stoianovich's paper on the Balkan merchant which stated that during the second half of the eighteenth century Macedonia and Thessaly exported 40 per cent of their cereal production and 50 per cent of their cotton production, and that almost all of the exports came from *çiftliks*[43]. The whole subject, however, is one of considerable debate by economic historians of the Ottoman Empire[44].

*Çift* is simply the Turkish work for farm, but *çiftlik* has been adopted to describe a plantation-like estate, largely worked by share-croppers, and in the Balkans at least it has been associated with a particular form of settlement (Fig. 3)[45]. In Macedonia a typical *çiftlik* ranged between 25 and 50 ha[46]. *Çiftliks* appear at different times in areas bordering the Aegean Sea. Thus, the first ones multiplied near Istanbul during the early seventeenth century[47]. In the Balkans, McGowan has distinguished two waves of *çiftlik* formation, one in the late seventeenth century (which may have missed the Morea because it was under Venetian control) and the other towards the end of the eighteenth century[48]. *Çiftliks* may not have appeared in western Anatolia until the early nineteenth century and may have been confined to the hinterland of Smyrna[49]. They seem to have been absent from other parts of the eastern Mediterranean region[50], at least until after the introduction of the Ottoman Land Code of 1858. Whilst an orientation towards commercial crop production may have been important in the formation of *çiftliks*, their principal markets may have lain within the Ottoman Empire rather than outside it. In any case, other factors have also been recognized as involved in *çiftlik*isation. These include the long-established tradition of the state granting unused land (*mevat*) to anyone who would cultivate it, the establishment of tax farming and the forced dispossession of small farmers (peasants) to seize their assets, perhaps chiefly because of debt. Ready access to the coast also seems to have been significant[51].

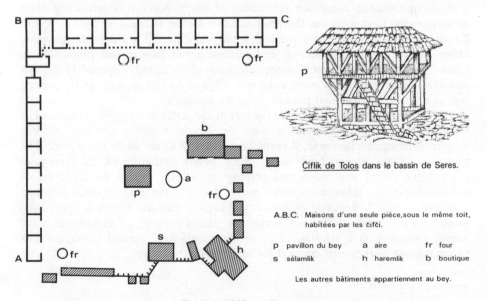

Čiflik de Tolos dans le bassin de Seres.

A.B.C.  Maisons d'une seule pièce,sous le même toit, habitées par les čifči.

p  pavillon du bey    a  aire     fr  four
s  sélamlik         h  haremlik    b  boutique

Les autres bâtiments appartiennent au bey.

*Fig.3:A çiftlik settlement*
*Source:Cvijić 1918,p.223*

The importance of *çiftlik*-produced crops in regional exports has been questioned on the grounds that exported items may have been the yield of tax farming rather than the produce of agricultural systems designed to produce crops principally for export[52]. In any case, the system of 'livings' was introduced into Crete after the fall of Candia (Irakleion) when 17 *ziamets* and 2,550 *timars* were created[53], whilst various attempts were made to reform the whole system during the eighteenth century before it was officially abolished in 1834[54].

Finally, share-croppers are not serfs, though they might be treated as such on occasion, and the Ottoman government never formally accepted or required the enserfment of its cultivators in the territories directly under its authority; the Turkish-Islamic tradition believed that care of the *reaya* was an important duty[55].

To sum up this discussion: the *çiftlik* is not a feudal estate; its origins are differentiated in both space and time; it was not found everywhere in the Levant in the seventeenth and eighteenth centuries; and its emergence is not prime facie evidence for the development of new social relationships in the Levant or for the peripheralisation of the Ottoman Empire.

So far little has been said of those parts of the Levant which were under Venetian rule for all or part of the period 1500-1800. It is not clear, to begin with, where Wallerstein would place the Republic's overseas territories in his spatio-economic scheme for the period. Secondly, as undoubted colonies, the production and trade of the overseas territories were tightly controlled in the interests of the Serene Republic[56]. The results may be illustrated from Crete.

The Regno di Candia was under Venetian control from 1209 to 1645. For much of the sixteenth century its economy was dominated by viticulture and wine was exported as far as England (to Southampton, in fact) and Flanders[57]. This was a result of Ottoman expansion into some of the major cereal producing areas adjacent to the Levant, in the Danube lowlands and the Balkans, as well as Egypt. Cereals became so cheap as a result that the Republic did not have to provision either Venice or its ships sailing in Levantine water from cereal production in Crete. This had been necessary earlier. Instead, more lucrative specialization was possible and foreign ships were evidently allowed to sail directly to the island, instead of collecting colonial produce from the mother city.

The situation changed towards the end of the sixteenth century. The prices of Ottoman cereals rose, relations between the Serene Republic and the Ottoman Empire deteriorated into war, Cyprus was lost and Crete itself threatened. The island could no longer feed itself. State policy shifted back in favour of encouraging cereal production and curbing viticulture, though the problems of provisioning Crete remained a major worry down to the long-expected Ottoman invasion of 1645[58]. Removal of the colonial regime soon put an end to specialized production and seems to have encouraged a greater diversity of exports, as noted above, though with a tendency for olive cultivation to expand under the Pax Ottomanica to meet market demand for soap within the Levant.

## 4.    Conclusion

The economy of the Levant in the period 1500-1800 was more differentiated in areal terms than originally envisaged by Wallerstein. This much emerges from adopting a Levantine standpoint for the consideration of the available evidence but it also results from taking into account recent investigations, some of which have used Ottoman sources. Ottoman sources are obviously fundamental to deciding anything about the relationships of the Empire to the European space-economy. Amongst these the most important, especially for geographers, are the registers of taxable resources compiled at intervals for the Ottoman government. A virtually complete series for the Empire exists from the reigns of Selim II (1566-74) and Murad III (1574-95), whilst new surveys were carried out following the conquest of Crete and the reconquest of the Morea[59]. The considerable detail in the registers allows the reconstruction of geographies not only of land holding but also land use and production levels[60]. A diachronic series of comparisons should reveal precisely what changes took place over space, as well as through time, and should reveal, in particular, how they relate to productive relations.

Even without this painstaking work, however, the evidence reviewed above does suggest that, if incorporation took place at all, it may not have been until after 1800. In fact, the volume of inter-regional trade within the Levant itself during the eighteenth century points to the integrity of a trading system within the Ottoman Empire at that time. Possibly only those items surplus to the Levant's own needs were exported to Europe. It is also clear that the Levant was part of a wider trading system which was orientated towards the East, as well as to the West. The export of bullion eastwards suggests a trade deficit in that direction[61], similar to the one which existed between Europe and the Ottoman Empire for much of the period reviewed here. These points perhaps indicate that the Levant was part of a world-economy still separate from that centred around London, Amsterdam and Paris, though linked to it.

The degree of urbanization may be indicative of a region's economic status and Issawi has estimated that in 1800 towns of 10,000 or more inhabitants contained about 10 per cent of the population of Egypt and 20 percent of that of Greater Syria[62]. The proportion for Anatolia was probably not less than 20 per cent[63]. As early as 1582, the urban population of Crete was at least 17.8 per cent of the total of 208,117[64], while that of the Morea in 1700 was at least 14.8 per cent out of a total of 176,844[65], though in both cases individual towns had populations numbered in hundreds rather than thousands. By comparison, 24 per cent of the population of England and Wales (arguably the dynamic centre of the European world-economy at the time) lived in towns of 10,000 or more inhabitants at the census of 1801[66]. Thus, on this criterion, the Levant was not as 'developed' as England and Wales but nor was it economically 'backward' either. London had a population of about 960,000 at the time, that is more than twice the estimate for greater Istanbul (Istanbul itself, together with Eyub, Galata and Üsküdar); c.1700, however, Istanbul had an estimated population of 700-800,000 when London contained something over 600,000 inhabitants (Table 1).

|                          | Population Estimates | | | |
|--------------------------|----------------|----------------|----------------|------------------|
| Cities in<br>The Levant  | 1500           | 1600           | 1700           | 1800             |
| Aleppo                   | 67,000[1]      | 200,000[2]     | 115,000[2]     | 80-120,000[3]    |
| Cairo                    | 150,000[4]     | -              | -              | 263,000[4]       |
| Istanbul                 | 97,956[5]      | -              | 700-800,000[6] | 426,000[7]       |
| Izmir                    | -              | -              | c.100,000[8]   | 100,000[8]       |
| Cities in<br>Western Europe | | | | |
| Amsterdam                | 11,000[9]      | 105,000        | 200,000        | 215,000          |
| London                   | 60,000[10]     | 250,000[11]    | >600,000[12]   | 960,000[13]      |
| Paris                    | 300,000[14]    | -              | 500,000[15]    | 547,736[15]      |
| Venice                   | 100,000[16]    | 100,000[16]    | 100,000[16]    | 100,000[16]      |

JMW/MBK/W4/CITIES
April 1990

*Tab.1: Population of selected cities, 1500-1800*

Istanbul was the heart of the Ottoman world-economy. At the hinge of Europe and Asia, as well as on the link between the Black Sea and the Aegean Sea, it was not only the largest city in the Empire but, throughout the period 1500-1800, it was also one of the largest cities in the Muslim world and, as we have seen, in Europe as well. Foodstuffs and raw materials were drawn to the capital from a wide region with a travel-time radius of more than the critical 60 days noted by Wallerstein, not all of which was under Ottoman political control. As well as Istanbul, the core of the Ottoman world-economy included the major entrepôts for the eastern trade, Aleppo and Cairo, together with Smyrna (Izmir) which came to prominence in the eighteenth century as the major centre for the Levant's trade with Europe. These were large cities in contemporary terms (Table 1), with complex economic bases involving trade and manufacturing, as well as administration and services. They could certainly be described as 'growth poles' in the sense of centralizing supply systems as Hopkins, Wallerstein et alii envisaged[67], but the dendritic mercantile systems which focused on them channelled flows not only to Europe but to other parts of the Levant, whilst the cities themselves were major consumers of foodstuffs, raw materials and luxury goods. The periphery of their commercial activity lay in the Balkans, north Africa, the Arabian peninsula, Iraq and the borderlands of eastern Anatolia. In the last three regions, the Ottoman world-economy impinged upon the economic systems of Persia and India.

Geographers attracted to Ottoman economic history are particularly well placed to accept some of the major challenges set by Hopkins, Wallerstein et alii[68]. These are, first, to delineate the fluctuating boundaries of the hinterlands to the cities located in the core of the Ottoman world-economy; second, to establish the spatial patterns of trade which radiated from them; and, third, to investigate the relationships between "world-economic boundaries and those under state control". A related task is the reconstruction of land use patterns to determine the nature of the productive systems in use and thus to decide the status as core, periphery or semi-periphery of the different city-regions which made up the Ottoman Empire. With its diversity and varied source material, the Levant offers particularly rich opportunities for geographical investigations.

## Notes to Table 1

1.  Barkan, Ö.L. (1970): 'Research on the Ottoman fiscal surveys' in Studies in the Economic History of the Middle East edited by M. Cook, London, pp. 162-71, Table 1.
2.  Masters, B. (1988): The Origins of Western Economic Dominance in the Middle East: Mercantilism and the Islamic Economy in Aleppo, 1600-1750, New York and London, p.41.
3.  Raymond, A. (1984): The Great Arab Cities in the 16th and 17th Centuries: An Introduction, New York and London, p.7.
4.  Ibid.
5.  Barkan (1970), op. cit., note 1, The area described as Istanbul usually included the city, together with Eyub, Galata and Üsküdar, though it may not have done so in 1500 or 1700.
6.  Mantran, R. (1962): Istanbul dans la Seconde Moitié au XVIIe Siècle, Paris, p. 47.
7.  Karpat, K.H. (1985): Ottoman Population, 1830-1914, Madison, p. 103, Table 5.3.
8.  Frangakis, E. (1985): 'The raya communities of Smyrna in the 18th century (1690-1820): demography and economic activities' in Praktika tou Diethnous Symposiou Istorias Neoellenike Pole tis Eteireias Meletis Neou Ellenismoubd, Athens, pp. 27-42
9.  Data for Amsterdam kindly provided by A.C. Duke.
10. Butlin, R.A. (1978): 'The late Middle Ages, c.1350-1500'. In: An Historical Geography of England and Wales edited by R.A. Dodgshon and R.A. Butlin, London and New York, pp. 119-150 (142).
11. Emery, F.V. (1973): 'England circa 1600'. In: A New Historical Geography of England, edited by H.C. Darby, Cambridge, pp. 248-301 (298).
12. Darby, H.C. (1973): 'The age of the improver'. In: A New Historical Geography of England, edited by H.C. Darby, Cambridge, pp. 302-88 (386).
13. Prince, H.C. (1973): 'England circa 1800'. In: A New Historical Geography of England, edited by H.C. Darby, Cambridge, pp. 389-464 (463).
14. Pounds, N.J.G. (1979): An Historical Geography of Europe, 1500-1840, Cambridge, p. 118.
15. Clout, H.D. (1977): 'Urban growth, 1500-1900'. In: Themes in the Historical Geography of France, edited by H.D. Clout, London, pp. 483-540.
16. Lane, F.C. (1973): Venice: A Maritime Republic, Baltimore and London, pp. 19-21.

## NOTES

1.  Pitcher, D.E. (1968): An Historical Geography of the Ottoman Empire, Leiden.
2.  Miller, W. (1908): The Latins in the Levant, Cambridge.
3.  Anderson, R.C. (1952): Naval Wars in the Levant, 1559-1853, Liverpool, 278-304; Finlay, G. (1877): A History of Greece (ed. H.F. Tozer), Vol. 5, Oxford; Gritsópoulos, S.A.. (1967): Ta Orofika, Athens; Salibi, K.S. (1965): The Modern History of Lebanon, London,

14-16; Shaw, S.J. (1976): History of the Ottoman Empire and Modern Turkey, Vol. 1, Cambridge and New York.

4.    Mackesy, P. (1974): Statesmen at War. The Strategy of Overthrow, 1798-1799, London; Rodger, A.B. (1968): The War of the Second Coalition 1798 to 1801, Oxford.

5.    Wallerstein, I. (1974): The Modern World-System, New York and London, 17.

6.    Wallerstein, I. (1979): 'The Ottoman Empire and the capitalist world-economy: some questions for research', Review 2, 281- 98 [also published in The Social and Economic History of Turkey (1701-1920) edited by O. Okyar and H. Inalcik, Ankara 1980, 117-22].

7.    Wallerstein (1974), op. cit.,note 5, pp. 301-02.

8.    Wallerstein (1979), op. cit, note 6.

9.    Hopkins, T.K. and Wallerstein, I. (1982): 'Structural transformations of the world-economy', in their World-Systems Analysis: Theory and Methodology, New York, 121-42.

10.   Wallerstein, I. and Kasaba, R. (1983): 'Incorporation into the world-economy: change in the structure of the Ottoman Empire, 1750-1839', in Economie et Sociétés dans l'Empire Ottoman (fin du xviiie-debut du xxe siècle) edited by J.L. Bacque-Grammont and P. Dumont, Paris, 335-43.

11.   Wallerstein, I., Decdeli, H. and Kasaba, R. (1987): 'The incorporation of the Ottoman Empire into the world-economy'. In: The Ottoman Empire and the World-Economy, edited H. Islamoglu-Inan, Cambridge and Paris, 88-97.

12.   Wallerstein, I. (1979), op. cit, note 6.

13.   Davis, R. (1967): Aleppo and Devonshire Square: English Traders in the Levant in the Eighteenth Century, London, 134-41; Frangakis-Syrett (1988): 'Trade between the Ottoman Empire and Western Europe: The case of Izmir in the eighteenth century', New Perspectives on Turkey 2: 1-18.

14.   Parry, V.J. (1970): 'Materials of war in the Ottoman Empire', in Studies in the Economic and Social History of the Middle East, edited by M. Cook, London, 219-39; Shaw, S.J. (1964): Ottoman Egypt in the Age of the French Revolution by Huseyn Effendi, Harvard Middle Eastern Monographs II, Cambridge, Mass.

15.   Wallerstein, I. (1979) op. cit, note 6.

16.   Steensgaard, N. (1974): The Asian Trade Revolution in the Seventeenth Century, Chicago and London.

17.   For example, Gibb, H.A.R. and Bowen, H. (1950): Islamic Society and the West, London.

18.   Hopkins and Wallerstein (1982), op. cit, note 9.

19.   For example, Kremmydas, B. (1972): Tò Emborio tis Peloponnisou sto 18 Aiona (1715-1792), Athens; Svoronos, N. (1956): Le Commerce de Salonique au xviiie siècle, Paris. This is largely true also of the nineteenth century, eg. Kasaba, R. (1988): The Ottoman Empire and the World-economy: The Nineteenth Century, New York; Pamuk, S. (1987): The Ottoman Empire and English Capitalism, 1820-1913, Cambridge.

20.   Davis (1967), op. cit, note 13.

21.   Owen, R. (1981): The Middle East in the World-economy, 1800-1914, London, 52.

22.   Quoted by Raymond, A. (1973): Artisans et Commercants au Caire au XVIIIe Siècle, T. 1, Damascus, 193. See also pp. 193-98.

23.   Davis (1967), op. cit, note 13, p. 159.

24.   Gaube, H. and Wirth, E. (1984): Aleppo, Beihefte zum Tübinger Atlas des Vorderen Orients, Reihe B, Nr. 58, Wiesbaden, Fig. 66, p. 259.

25.   Masters, B. (1988): The Origins of Western Economic Dominance in the Middle East: Mercantilism and the Islamic Economy in Aleppo, 1600-1750, New York and London, 31-32.

26.   Triantafyllidou-Baladié, Y. (1981): 'Dominations étrangères et transformations de l'agriculture crétoise entre le xive et le xixe siècle', Greek Review of Social Research, Special Number: Aspects du Changement social dans la Campagne grecque, 180-90.

27. Triantafyllidou, Y. (1975): 'L'industrie du Savon en Crète au xviiie siècle: aspects economiques et sociaux', Etudes Balkaniques 4: 75-87.
28. An expression of the French ambassador to the Porte, Des Alleurs, quoted by Kremmydas (1972), op. cit., note 19, p. 142, n.1.
29. Kremmydas (1972), op. cit., Table 3, p.23.
30. Kremmydas (1972), op. cit., Table 20, p.176.
31. Scrofani, X. (1801): Voyage en Grèce de Xavier Scrofani, Sicilien fait en 1794-95, Paris et Strasbourg, Tables 10A and 10B.
32. Wallerstein (1979), op. cit., note 6; Wallerstein and Kasaba (1983), op. cit., note 10; Wallerstein, Decdeli and Kasaba (1987), op. cit., note 11.
33. Gibbs and Bowen (1950) op. cit., note 17, pp. 46-52, 236-44, 258-59; Inalcik, H. (1955): 'Land problems in Turkish history', Muslim World 45: 221-8; Lewis, B. (1979): 'Ottoman land tenure and taxation in Syria', Studia Islamica 50: 27-36.
34. Cvetkova, B. (1977): 'Problèmes du regime Ottoman dans les Balkans du seizième au dix-huitième siècle', in Studies in Eighteenth Century Islamic History edited by T. Naff and R. Owen, Carbondale and London, pp. 165-83; Barkan, Ö.L. (1975): 'The price revolution of the sixteenth century: a turning point in the economic history of the Near East', International Journal of Middle East Studies 6: 3-28; Inalcik, H. (1977): 'Centralisation and decentralisation in Ottoman administration' in Studies in Eighteenth Century Islamic History edited by T. Naff and R. Owen, Carbondale and London, pp. 27-52; Rafeq, A.K. (1977): 'Changes in the relationship between the Ottoman central administration and the Syrian provinces from the sixteenth to the eighteenth centuries' in Studies in Eighteenth Century Islam History edited by T. Naff and R. Owen, Carbondale and London, pp. 53-73.
35. Hourani, A.H. (1957): 'The changing face of the Fertile Crescent in the eighteenth century', Studia Islamica 8: 89-122; Inalcik (1977), op. cit., note 34.
36. Gibbs and Bowen (1950), op. cit., note 17; Lewis, B. (1958): 'Some reflections on the decline of the Ottoman Empire', Studia Islamica 9: 111-27.
37. Lewis, B. (1962): 'Ottoman observers of Ottoman decline', Islamic Studies 1: 71-87.
38. Owen, R. (1976): 'The Middle East in the eighteenth century - An "Islamic" society in decline? a critique of Gibb and Bowen's Islamic Society and the West, Bulletin of the British Society of Middle Eastern Studies 3: 110-17; Owen, R. (1977): 'Introduction to Part II' in Studies in Eighteenth Century Islamic History, edited by T. Naff and R. Owen, Carbondale and London, 133-51.
39. For example, Ashtor, E. (1976): A Social and Economic History of the Near East in the Middle Ages, London.
40. Wallerstein (1974, 1979), op. cit., note 5 & 6.
41. Ashtor, E. (1976): 'The Venetian cotton trade in Syria in the later Middle Ages', Studi Medievali 17: 675-71. Reprinted in Ashtor, E. (1978): The Medieval Near East: Social and Economic History, London.
42. Braudel, F. (1972): The Mediterranean and the the Mediterranean World in the Age of Philip II, London, Vol. 2, 724-25; Busch-Zanter, R. (1938): Agrarverfassung, Gessellschaft und Siedlung in Südosteuropa unter besonderer Berücksichtigung der Türkenzeit, Leipzig; Wallerstein (1974), op. cit., note 5, p. 302.
43. Stoianovich, T. (1960): 'The conquering Balkan merchant', Journal of Economic History 20: 274-313.
44. Faroqhi, S. (1987): 'Agriculture and rural life in the Ottoman Empire (ca. 1500-1878)', New Perspectives on Turkey 1: 3-34 (especially 19-24).
45. Cvijié, J. (1918): La Péninsule balkanique: Géographie humaine, Paris pp. 222-23; McGowan, B. (1981): Economic Life in the Ottoman Empire: Taxation, Trade and the Struggle for Land, 1600-1800, Cambridge; Wagstaff, J.M. (1969): 'The study of Greek rural settlement: a review of the literature', Erdkunde 23: 306-17.

46.    McGowan, ibid., p. 72.
47.    Inalcik, H. (1983): 'The emergence of big farms. Çifliks: state landlords and tenants', in Contributions à l'Histoire Economique et Sociale de l'Empire Ottoman edited by J-L. Bacque-Grammont and P. Dumont, Louven, pp. 105-26.48.McGowan, (1981), op cit., note 45, pp. 75, 148.
49.    Faroqhi, S. (1981): Der Bektaschi-Orden in Anatolien (vom späten fünfzehnten Jahrhundert bis 1826), Vienna, p. 50f; Nagata, Y. (1976): Some Documents on Big Farms (Çiftliks) of the Notables in Western Anatolia, Studia Culturae Islamicae 4, Tokyo; Veinstein, G. (1976): 'Ayân de la region d'Izmir et commerce du Levant (deuxième moitié du xviiie siècle)', Revue de l'Occident Musulman et de la Méditerranée 20: 131- 46.
50.    For example, Cohen, A. (1973): Palestine in the Eighteenth Century, Jerusalem; Slot, B.J. (1982): Archipelagus Turbatus. Les Cyclades entre Colonisation Latine et Occupation Ottomane, c.1500-1718, Nederlands Historisch-Archaeologisch Instituut te Istanbul, Istanbul.
51.    Faroqhi (1987), op. cit.: McGowan (1981) op. cit., note45, pp. 60-71, 78-79.
52.    Veinstein (1976), op. cit, note 49.
53.    Miller, W. (1904): 'Greece under the Turks', Westminster Review 162: 195-210.
54.    McGowan (1981), op. cit., note 45, p. 147.
55.    Darke, H. (1978): The Book of Government or Rules for Kings, revised ed., London; Faroqhi (1987), op. cit.
56.    Brown, H.F. (1907): 'The economic and fiscal policy of the Venetian Republic' in his Studies in the History of Venice, London Vol. 1. pp. 335-61.
57.    Margaritis, S. (1978): Crete and the Ionian Islands under the Venetians, Athens; Ruddock, A.A. (1951): Italian Merchants and Shipping in Southampton, 1260-1600, Southampton Record Series 1, Southampton, pp. 72-73, 104, 138, 221-23, 247-48, 256; Triantafyllidou-Baladié (1981), op. cit., note 26.
58.    Clutton, E. (1978): 'Political conflict and military strategy: the case of Crete as exemplified by Basilicata's Relatione of 1630', Transactions of the Institute of British Geographers n.s. 3: 274-84.
59.    Barkan, Ö.L. (1970): 'Research on the Ottoman fiscal surveys' in Studies in the Economic History of the Middle East edited by M. Cook, London, pp. 162-71.
60.    Hütteroth, W-D. Abdulfattah, K. (1977): Historical Geography of Palestine, Transjordan and Southern Syria in the late 16th Century, Erlanger Geographische Arbeiten 5, Erlangen.
61.    Kremmydas (1972), op. cit., note 19, pp. 110-20; Masters (1988), op. cit., note 25, p. 148; Raymond (1973), op. cit., note 22, pp. 136-37.
62.    Issawi, C. (1969): 'Economic change and urbanisation in the Middle East' in Middle Eastern Cities, edited by I.M. Lapidus, Berkeley and Los Angeles, pp. 102-08.
63.    Wagstaff, J.M. (1985): The Evolution of Middle Eastern Landscapes: An Outline to A.D. 1840, London, p. 200.
64.    Calculated from figures in Margaritis (1978), op. cit., note 57, Table 5, pp. 52-55.
65.    Urban population of the Morea calculated on the basis of places identified as città, borgo or terra in Libro Ristretti della Famiglie e Animi Effettive in Cadaun Territori[o] del Regno [di Morea] Archivio di Stato, Venice, Archivio Grimani dai Servi, B.54, N. 158. It is probably an underestimate.
66.    Law, C.M. (1976): 'The growth of the urban population in England and Wales, 1801-1911', Transactions of the Institute of British Geographers 41: 125-44.
67.    Hopkins, T.K., Wallerstein, I. et alii (1982): 'Patterns of Development of the Modern World-System' in World-Systems Analysis: Theory and Methodology edited by T.K. Hopkins and I. Wallerstein, New York, pp. 41-82.
68.    Ibid.

# 22

# The Role of the Ottoman Empire in the Early Modern World-System

Wolf Hütteroth, Erlangen

Geographers working in the Islamic Orient have had some difficulties in applying theories of peripherisation or dependencia. The political and economic development since the 19th century, fairly well investigated for the lands between the Atlas and the Hindukush, the Altai and Yemen, may perhaps be stressed in one-dimensional theories like these. But as a consequence, important aspects of social and economic development, which are not connected to the world economy, will be lost as well as all the individual character of regions and countries. The more "general" a system is ("Asian mode of production", "Hydraulic civilization", "Dependencia", "Capitalist world-system") the more trivial it becomes when applied to individual countries at a specific point in time.

First of all it is pertinent to stress that the historical relations of the Islamic Orient with Europe are different from those of all other cultural worlds (Black Africa, South Asia, both Americas) in four important ways:

Firstly, ever since the 7th and 8th century, to the south and east, Europe has been opposed by a powerful alien world with high self-consciousness, separated not by oceans, deserts or unpopulated forest-wilderness, but immediately "in front of the door".

Secondly, contacts with this "anti-world" have never disappeared since the common culture of the old Mediterranean world and classical antiquity. The Muslims were known in Europe, there were diplomatic connections as well as military conflicts, and there was always a permanent trade with them. The traditions of economic exchange between Venice, Genoa, Pisa, Ragusa etc. with Byzantium were immediately transferred to the Ottomans, when they became the leading power.

Thirdly, the popular classifications of each other as "pagan" led both sides to a strong feeling of superiority. "We" are the better ones, in religion, in battle and generally all aspects of civilization. With the Turkish conquest of Constantinople in 1453 it seemed absolutely clear to whom the future belonged. European, "pagan" states could, at the upmost, be tolerated for a limited period.

And finally, such consciousness of superiority was usually kept alive among the population masses even after political, military and economic conditions began to change. Vezirs and pashas might have sorrowfully recognized European superiority in the 18th and 19th century, but for the simple man Europeans remained miserable creatures, with ragged rural Greek, Bulgarian or Serbian subjects being considered as typical for Christian Europeans. It was quite natural that one could not expect anything of value from these despised people. Even in

the middle of the 19th century the simple Ottoman soldier refused to salute
Prussian officers in his own army (IV Moltke 1893).

At least through the 15th and 16th century the Ottoman Empire was, even in
using Wallerstein's concept, a world-empire. Kreiser (1979) first considered the
centre and periphery inside this world-empire, or what could be called a "core"
area of the state. The most "Ottoman" of provinces were the original *eyalets* of
Rumeli and Andalu, e.g. approximately west and northwest Anatolia, Thrakia,
Makedonia and Bulgaria. Bursa was the metropole on the Asiatic side and Edirne
on the European side, even after Constantinople became the dominating centre of
political power and economic gravity.

This core becomes apparent as the area where relatively much money was
allocated when compared to that extracted. Here are the rich Ottoman Islamic
foundations (*wakf*) of the great Sultan-mosques of Constantinople, the foundations
for the charitable institutions and the *medresses*. Beyond that core, pre-Ottoman
*wakf* foundations of mere local and regional importance appear, besides those with
dedication to the holy places in the Hejaz. In the same core area, the most
important monuments (mosques, schools, bridges, caravanserais, fountains etc.) of
Sinan and his disciples are found. Beyond this area, the more representative
buildings were taken over from pre-Ottoman periods (Cairo, Damascus, Aleppo,
Diyarbakir, Erzurum, Sivas etc.). East Anatolia, North Africa and the Danube
countries are comparatively "provincial" in taste.

Furthermore, the core area is the "secure belt", where the high officials of
the Sublime Porte tried to get their "living" (*hass, zecamet, timar*) in the 15th and
16th century and where, in later centuries, the acquisition of quasi-property in land
became most desirable.

In the field of economics, and especially trade, Constantinople was surely
something approaching a primate city from the Ottoman conquest onwards. The
bazar of this town was the destination of the most valuable goods from all over the
empire and beyond. In addition, a high proportion of all taxes came together here
and was needed to nourish the court, bureaucracy and army.

Nevertheless, the medieval transport conditions (which remained "medieval"
on overland routes until the late 19th century) must have caused the relative
economic autonomy of peripheral regions, focussed each on their own centres.
This was surely the case with Egypt and its capital Cairo, but also with southern
Syria centred on Damascus. Aleppo dominated northern Syria, south-east Anatolia
and upper Mesopotamia; Baghdad, Basra, Diyarbakir, Van, Erzurum, Tabriz,
Malatya and several other places were regional metropoles with very specific
character each. Evliya Celebi presents a vivid picture of the relatively rich and
individual economic life of all these towns in the 17th century. The towns of
Rumeli must have been comparable, whereas in the north-western frontier-
provinces the major towns were more like big garrisons than centres of economic
activity (Belgrade, Buda, Temeswar etc.). Luxury goods were exchanged all over
the Empire and beyond. Since the Middle Ages Aleppo had been the great staple
and distribution point for goods from Persia, India and beyond. Anatolian carpets
were sold on the Hungarian markets, while money flowed mainly in the opposite

direction. The Danube countries had little to offer to the East, except perhaps slaves and booty from the "Üç", the military frontier zone (cf. E. Wirth 1984, p. 223 pp).

Thus agreement can be made with Wallerstein's concept of the Ottoman Empire being a world-empire with an economy of its own. Ideas can even be added about its interior differentiation into core, secondary cores and interior periphery. All this undoubted applies to the 15th and 16th centuries, but for the period under discussion, the "early modern" time, the beginning of European expansion in the 17th and 18th centuries some special conditions unique to the Ottoman Empire are to be considered.

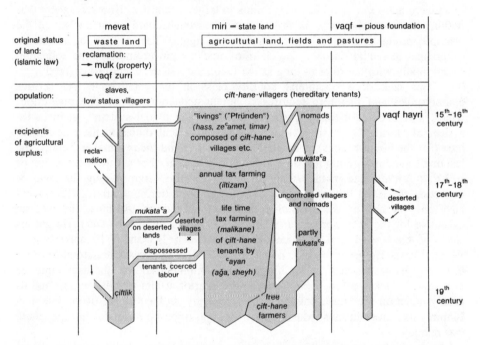

*Graphic demonstration of the probable development of different forms of 'landownership' and their share (width of arrows) in the agricultural land. This diagram may stand for western Anatolia, for other regions the proportions may be different.*

Opposite to the more or less feudalised states of Europe, the Ottoman Empire was built on quite different principles of state and ideology. Centrally, it was the duty of the Sultan respectively the Khalif to care for the welfare of all his subjects in accordance with Islamic law. This was the equivalence and justification for his claim to absolute power. To achieve this "welfare", the villagers - i.e. the majority of the population - were (theoretically) to be preserved in their productivity and protected from unjust burden. The consumer prices in the towns had to be kept moderate, even by the introduction of fixed prices (narh). Sufficient supply for the towns, especially Constantinople, had to be secured; craftsmen had to be supplied with sufficient raw materials to be distributed by the guilds. Law and order had to be kept all over the Empire according to centralized control and legislation. This type of organization is sometimes called a "redistributive-tributary mode of production" (e.g. Wallerstein 1980, p. 121) to be placed systematically somewhere between subsistence and capitalistic production.

The basis of the whole economic life was, as Inalcik has pointed out several times (most recently 1983, p. 105), the so-called *çift hane-system*, which may be compared with German *Hufenverfassung* in the early rural society. The average rural household (*hane*) was considered to have a size of land workable with a pair (*çift* = lat. *iugum*) of oxen. This was the basic unit of fiscal control. Of course there were half-*çifts*, and even lesser units in larger quantities. However, there was nothing like a rural gentry in the countryside, which could claim a share of the peasants production according to their own rights. This is one of the main differences to feudal Europe: The claim of absolute power and revenue collection rested totally with the ruler as long as the Ottoman system functioned correctly.

This lack of a traditional nobility class in the countryside had serious consequences and therefore requires comment and comparison. The institution, which is sometimes incorrectly called the "Ottoman feudal system", is in fact a system of "livings", classified according to merits and duties in the size-classes *hass* (for the highest ranks), *zecamet* (bigger living) and *timar* (smaller living), the last one sometimes giving its name to the whole system. The entitled *beg*, *zacim* or *sipahi*, however, had in the village or villages of his "living" only the right of collecting the fixed tithe and some dues for himself during lifetime. He did not have any judicial or administrative rights and functions. At most, he was an executive for the governor (*mir liva* or *sancak beg*) or the judge (*kadi*). He got his share of the agricultural produce, usually between 1/3 and 1/10 according to regional laws. He got it in kind and had to sell it on an accessible market in order to obtain the cash necessary for his military equipment and for the campaigns he was bound to take part in. His interest in the interior affairs of "his" village and its rural population was thus minimized. Consequently, in the villages of the Ottoman Empire, like those in most Muslim countries and opposite to feudal Europe, there was nobody
- who could try to raise the productivity of "his" village with legal or illegal methods
- who could lay claim on unused land, pastures and forests, to increase his private landholding and, therefore, income;
- who would make investments to raise efficiency and production, e.g. by irrigation or melioration techniques, forest clearance or road improvements;
- who was obliged to protect his villagers in time of need, to care for the conservation of his *çift hane*'s productivity, and to take care for his people's religious and educational sustenance;
- who, because of his class-consciousness, tried to build a representative residence and brought specialized craftsmen to the village solely for this purpose;
- who, because of this same class-consciousness, kept contacts with people of his class all over the Empire, thus achieving a refined life-style, wide contacts and

chances for innovativeness; and finally
- who, at the end, could do all this with the security of legitimate hereditary property, which had to be the prerequisite for achievement of all of these.

The lack of a rural gentry perhaps saved Ottoman rural society from some class conflicts. It had, on the other hand, the disastrous effect of preventing any innovation or autochthonous progress in the countryside. In the later centuries it did not even prevent over-exploitation, riots, starvation and village desertion.

The rural economy was more or less "static": The entitled recipients of rural income had to sell their share on local markets, and this was the main supply of these markets. The share that remained in the hands of the villagers was necessary for their subsistence, and only a very small proportion could have reached the market. For the southern Syrian livas, the present author once calculated (Hütteroth & Abdulfattah 1977, p. 107) that there was a total of marketable surplus of production of roughly 22 % in 1596, and this, approximately, was the amount which the urban population of the area would have needed (ca.18 %), plus that being sold to the bedouins.

For a noteworthy volume of exports there would not have been much left in most parts of the state. Additionally, the state took measures to stop or to forbid exports, being anxious to supply the capital sufficiently. Correspondingly, the size and importance of most Ottoman port towns was small up to the early 19th century. The ports of the southern Levantine coast yielded annual custom dues of the magnitude of one moderate village's taxes. Of course there must have been smuggling, of which we know from fermans against it. But only grain or cotton producing regions near the coast could have been involved in illegal exports: More than four to six days of caravan transport would have caused the price to become prohibitive, bearable only rarely on occasions of extreme grain shortage in the Mediterranean. Long distance overland trade was, up to the late 19th century, a trade in luxury commodities (including textiles), or raw material used to manufacture them (raw silk, mohair, alum). Egypt, of course, must have had an exceptional position in this respect because of its river transport facilities.

Such was the system of the "redistributive-tributary" economy: taxes in kind to local markets, their exchange in money and services (military) to the centre, salaries (partly) and some gains from trade back to the periphery. This caused a certain disadvantage for provinces far from the centre (or from ports) and a certain advantage for the "old core" provinces, which could make profit from direct supply of Constantinople with agricultural products. If, at all, the periphery could take part in this supply, it was through high-priced commodities like rice, dried fruits etc.

This initial early modern situation must be included in discussions on the development that followed. The "high Ottoman" state organization is fairly well documented. Production and redistribution is known from laws and detailed registers, but what can only be estimated is local and regional trade. Something is known about the directions of trade, for the later Ottoman centuries we even have data for foreign trade, but we do not have any quantitative information about interior trade within the Empire.

    This lack of quantitative information on interior trade is the main hinderance
for comparative calculations. Some export has existed at all times; how much
export, however, is necessary (in relation to interior consumption) to classify a
state as export-oriented and thereby dependent on the world market? When
Wallerstein commented for the first time on the specific role of the Ottoman
Empire (1977, at a congress on social and economic history of Turkey, publ.
1980), he put his opinion in the form of a simple question: During the 16th
century, in its heyday, the Ottoman state surely was an economically independent
world-empire. In the 19th century, however, it had obviously become a peripheral
area. "In the end", he argues, "it is perhaps a quantitative question. How much
quantitative change in the production systems ... adds up to a definite qualitative
shift?"

    In different contributions, Wallerstein offers several periods, between the
early 18th and the middle of the 19th centuries, which could be indicative for a
"shift" in interior political and social conditions. But he is unable, like others, to
give an answer to his own quantitative question. It is at this point, that the *çiftlik*s
come into play, as they did earlier in Braudel's work and even in the first
systematic description of this institution by Busch-Zantner (1938). The appearance
of the estate-like *çiftlik*s seems to be a "qualitative shift", which Wallerstein is
looking for in the system of production. *Çiftlik*s were thought of as producers of
cash crops oriented to the world-market. They used coerced labour and were
owned by powerful people with certain politically centrifugal tendencies.
Wallerstein speaks about increased exploitation of the peasantry, widespread
peasant flight and the institutionalization of private property in land (1987, p. 93)
caused by the commercialization of agriculture.

    Since the *çiftlik*s are the only really visible innovation in agricultural
production between ca. 1600 and 1860 (except Egypt's cotton boom), they deserve
some attention. Did their appearance indeed mark a qualitative shift, as
Wallerstein would have it and Busch-Zantner tried to demonstrate as early as
1938? This question has to be posed again, since some important new
investigations, especially by Inalcik (1983) and Nagata (1976), have appeared in
recent years. Inalcik stressed the fact that special conditions for reclaiming land
from the "waste" existed in the Ottoman Empire since its very beginning.
Whosoever invested energy and capital to cultivate unused (*mevat*) lands outside
the *çift-hane*-system, and unclaimed by others, could enjoy special rights: This
land became *mulk* (property), in accordance with Islamic law. Nevertheless,
private property was always precarious in the east, and usually owners were wise
enough to convert it to the status of *wakf*, a pious foundation for the benefit of
their own successors (*wakf zürri*, familiy *wakf*, as opposed to the real beneficial
*wakf hayri* for the use of the public).

    "Unclaimed land" in the aforementioned sense was to be found mainly in
open, often swampy areas. For the necessary drainage work slaves were   used
usually, not normal villagers. That meant that only those people who could afford
the relatively expensive manpower of slaves, could try to establish *çiftlik*s.
Normally, they were members of the *askeri* (officers, governors) and *ulema*

(judges, clergy) class. Inalcik supposes (1983, p. 109) that this procedure was the origin even of the majority of the *çiftlik*s. This is guessing, of course, but it can be substantiated by identifying all places mentioned as *çiftlik*s on the first topographical maps of the 19th century. This would be a genuine geographical contribution, but except for Wilhelmy's study of High Bulgaria (1935) and Soysal's investigation of the Çukurova (1976) no further detailed studies are known to this author to exist. For Anatolia at least, the first detailed map (Kiepert 1906, 1:400.000) confirms this assumption. Places mentioned as *çiftlik*s are usually found in the coastal plains of the Aegean, less frequently near the southern coast and very rarely in the basins of the interior.

Besides the development of *çiftlik*s on coastal plains since the 17th century, there was still another chance for upper class people to acquire wealth out of agriculture in these later centuries of the Empire. Crucial for its understanding is the military situation around 1600 and thereafter: The cavalry troops, the *sipahi*, lost their tactical importance. More and more the rulers relied on standing infantry armies, not only the Ottomans but the European monarchs as well. But these standing armies had to be paid in cash. The state had an increasing demand for money on the one side, but was lacking a new financial system and trained officials for direct taxation on the other. The convincing solution seemed to be a system of renting out the taxes district-by-district (*iltizam*) against payment in advance of the highest bidder (*multezim*). Cash was immediately flowing to the treasury. The tax-farmers then had to regain it from their respective districts - with some surplus of course.

Soon it became apparent that the annual change of tax-farmers at highest bidding had serious consequences: Overtaxation led to riots and even desertion of settlements. A solution was again found in the *malikane*-system: Tax farming rights lasted lifelong and sometimes even became hereditary. These tax-farmers were usually absentee notables from the *askeri* and *culema* class living in Constantinople. Islamoglu-Inan (1987) points to the fact that accumulation of capital in their hands during the preceding period of "livings" was the precondition. The old "redistributive-tributary" system had enriched these classes.

For practical reasons, these big tax-farmers employed local agents with the task of tax collecting. These were the *ayan*, the rural notables, and these people were to become the most important social group in the countryside from the late 17th to the early 19th century. Thus there were two strata of profit motivated people between the tiller of the soil and the state treasury, but the "lower" ones, the local *ayan*, were nearer to the sources of income and hence dominated in the long run.

Up to this point, ca. 1760 to 1810 according to Inalcik, the whole development is a purely interior one: Even without any influence from Europe this could have happened. It may even be interpreted as a "historically normal" shift from central to peripheral influence, political power and economics. The local *ayan* did not change any of the conditions of agricultural production. The traditional village farm, the *çift hane*, remained the basic unit of taxation and its production techniques were the same as ever. Wallerstein (1987) is not quite right when he identifies the growing class of the *ayan* with *çiftlik*-owners. The *ayan* may have owned *çiftlik*s too, but their main source of income and power basis were the taxes and dues taken (or more realistically: exploited) from the villages. There was no need for any change in the production system. The towns of the Empire had expanded in both size and number since the 16th century, and they offered markets and demand for agricultural products. All this was due to interior dynamics, not foreign influence.

Nevertheless, with the *ayan* arose a decentralized political power group. Whoever had the right of tax collecting for lifetime, or even hereditary, would have political power too. The majority of the *ayan* may have been loyal to the Sultan in general terms, but they surely had an interest in raising their income. Since the taxation percentage could not be raised indefinitely in the *malikane*-system, trade in agricultural products offered options which were officially forbidden: selling to foreigners. Thus, the establishment of export-oriented agriculture was necessarily connected with the rise of semi-independent authorities in peripheral regions. Naturally enough the Sultanic government saw this, and saw it with mistrust, because the legitimate central authority was challenged. The *ayan* were recognized nolens volens as local governors, but nevertheless they were expropriated whenever possible.

Anyway, the so-called "*çiftlik*isation", viewed as the only "visible" development (which by some authors is taken as a symptom of economic peripherization like plantations in the Indies) has become increasingly doubted by historians in recent years. Their arguments coincide partly with our geographical viewpoints and may be summarized as follows:

1. The rural *ayan* class, new local authorities since the 17th century, received the greater part of their income from revenues which were at least formally legitimate taxes. They might have had estates (*çiftlik*s) too, and they might have increased them at the expense of the villagers, especially by occupying land given up by villagers for one reason or the other. But their main function was tax farming. Any quantification, however, is not yet possible.

2. The privately owned *çiftlik*s, of the *ayan* class as well as of other owners, did not arise primarily as a consequence of world market demands (grain, animals, cotton), instead they arose primarily from the need to invest money in secure property and to take part in the interior market supply of the region's growing towns.

3. A fair percentage of the 17th and 18th century *çiftlik*s was de facto only devoted to animal raising, because of shortage of manpower and because of technical difficulties in swampy areas that had been so easily acquired. Judicial documents, European travel reports and the location of the *çiftlik*s according to early maps point to this fact. Thus many *çiftlik*s contributed more to agricultural extensivation than to intensivation. This seems an argument for i n t e r i o r peripherization rather than world market integration.

4. *Çiftlik*-production for world market demands could be possible only in areas where transport costs to ports were not prohibitive. The Dobruca, the Makedonian plains, the Bulgarian basins and the lands at the lower Danube, also

the west Anatolian ovas were the areas of çiftlik formation. In interior Anatolia they were very rare and in the Fertile Crescent they did not exist. Large agricultural units do not appear there before the time of railway construction. Even Serbia, Slavonia and the Banat were hardly affected. The middle-Danube lands were a cul-de-sac down to the Iron Gate (McGowan 1987).

5. A fair number of the çiftliks of the 17th to 18th century must have been - according to the *kadi*-protocols evaluated by Nagata (1976) - of rather moderate size. Despite large areas owned, the land actually cultivated may be estimated from the number of pairs of oxen, and these vary around an average of between 10 and 30 only. There are two exceptions among 20 investigated *çiftliks* with 71 and 117 pairs, respectively. The *çiftliks* can have made up just a small proportion of the agricultural land.

Now, since the *çiftlik* seems to loose its character as a favoured index of world market integration, some further questions should be posed: 1) Which changes at all may be connected with the undoubtedly growing influence from outside? 2) Can regional differentiation among the parts of the Empire be detected, and in what sequence did the different provinces or landscapes show reactions to foreign influence? 3) Is it possible to find arguments for the late incorporation of the Ottoman Empire into the world economy, despite its proximity to Europe?

The first and second questions are a challenge for historical geography. Contributions which regionally differentiate the very broad "thinking in state economies" are to be expected from this side in the future. The third question seems necessary as a consequence of the Wallerstein-concept: If a large region with a very traditional economy is n o t incorporated into the capitalist world economy in the early modern period, despite its position close to the "centre", then this requires special explanation.

Referring to the first question, the export of specific commodities for the world market (silk, raw silk, coffee, carpets, dry fruits, wool, cotton, grain) requires further investigation. Any argumentation however, which takes increased export of these commodities as a given proof of world market integration or even "dependency", suffers from two deficiencies: First, we do not have any comparable data for interior consumption of these items, and secondly, economic historians have not yet supplied the criteria for a reasonable percentage at which, in Wallerstein's words, "quantitative change adds up to a qualitative shift" (1980, p. 121).

The settlement expansion since the second half of the 19th century in Thrakia, Anatolia and in the Levant countries is for some regions well documented. In most cases it is clearly not a reaction to world market connections, but an improvement of villagers' subsistence conditions, nomad settlement processes or re-settlement of Muslim refugees. Any external influence can only be detected very indirectly: The modern, "western" equipment of the gendarmerie helped to restore law and order as an essential precondition to the extension of agricultural activities, and "modern" legislation of the *tanzimat* ("reform") period of the 19th century was partly based on western models. The increased foundation and growth of local market towns and new administrative centres point in the same direction (Höhfeld 1977).

The only "visible" change which may by attributed to increasing external orientation of the Ottoman lands is the rise and rapid increase of coastal towns in the 19th century. Izmir was a frequented port still in the 18th century, but its importance for the Empire was considered so small that this town had the rank only of a *kaza*, not a *sancak*. Alexandretta was, at the beginning of the 19th century, a landing place for Aleppo, not a town; Beirut was an unimportant small country town; Haifa was a mere village, Jaffa an embarking place for pilgrims; Trabzon at the Black Sea just started its modern career as port for the Persian trade and Samsun was of only local importance. The rise of all these places did not begin before the 19th century, but mainly in its second half. The decisive impetus did not come until railway construction into the hinterland improved the transport facilities.

Addressing the second question, it is important to realize that to ask for a special point in time of the integration of the Ottoman lands into the world economy means to put the wrong question. Specialized trade in Istanbul, Cairo, and Aleppo was at a very early stage connected with the world market, probably without major interruption since the Middle Ages. Wallerstein is surely right when he expects great regional differences in the time of the beginning of foreign influence. Quasi-independent rulers in the Levant, Albania etc. with the advantage of easy access to the sea, as well as to mountain refuges immediately behind the coast, could even, in the 18th century, stabilize their position by cotton export to the West. Egypt must be conceded a special situation: In the early 19th century (and even before) the accessibility of almost the whole country for river traffic allowed an orientation to western markets besides that to Constantinople. In other parts of the Empire those regions proceeded which already had, besides access to the sea, conditions of relative interior security. The Aegean coast of Anatolia had this long before its southern coast, where uncontrolled nomads roamed up untill to the First World War. The plains and basins of Central Anatolia did not join the market for staple goods before the advent of the Anatolian railway (whereas mohair as a mere luxury item had been exported long before). The interior steppes of Syria were opened up in the last quarter of the 19th century, and those of Upper Mesopotamia not before the First World War. People in many parts of Kurdistan and mountainous western Anatolia, probably did not even take notice of any changes up to the interwar-period. Subsistence production and revenue collection (of those rightly entitled or not) simply continued.

It is a temptation for a geographer, to construct a map which shows the areas of earlier dominance of market production. Physical aspects of landscapes would have to be combined with those of accessibility, security and distribution of local central places. Such a map would show at the same time the backward, inaccessible, poor, insecure landscapes of the "interior periphery". This will be possible one day, but it still requires much research. The concept of peripherization however, imagined (regionally) mainly in broad belts around a

European centre, will be dissolved into a puzzling mosaique of relatively more or less "advanced" small areas, as soon as someone really dares to put it on a map.

The third question, the "why" of the late incorporation of the Ottoman lands into the world economy, leads back to the introductory remarks: Many physical, social and psychological aspects have to be considered, which can be combined to the phenomenon of "cultural resistance", as Braudel has put it (engl. ed. 1973, Vol. II, p. 763 ff). The core lands of the Orient, the Ottoman and the Persian Empire, have never been conquered and totally occupied. The ideology of Islamic superiority was never doubted by the average Turk, Arab or Irani, even up to this century. For traditional Muslims, the concept or even the imagination of something like "progress" had a pejorative connotation: As the Holy Koran and the *Hadith* had regulated human life once and for ever, any "progress" could only mean deviation form the right way. This must have resulted in an unfavourable intellectual climate for any kind of innovation. Religious laws and local social rules are among the most resistant objections to cultural change. Trade, on the other hand, is not affected, it was acknowledged at any time. Consequently, increasing export trade had to rely on traditional ways of production and supply. Production, in the form of modern plantations such as in the Indies, could not be expected in the Orient. Change of methods of agricultural production would have to wait until the middle of the 20th century. Manufactured goods, on the other hand, were adapted with great skill and flexibility to the demands of the market.

The chances of identifying an index for the integration of the Ottoman Empire into the "Early Modern World-System" are not yet very good, or at least not in the sphere of economic production. Still it is easier to find some reasonable "point in time" in the fields of diplomacy, legislation or military organization. Wallerstein's question remains open. It will continue to be a challenge for scholars of many disciplines, and the geographers among them will have to contribute methods of regionalization of the problem.

## References

Braudel, F. (1973): The Mediterranean and the Mediterranean World in the Age of Philipp II. - Vol. I, II., engl. ed.: London/New York

Busch-Zantner, R. (1938): Agrarverfassung, Gesellschaft und Siedlung in Südosteuropa. - Leipziger Vierteljahrsschrift für Südosteuropa, Beiheft 3. Leipzig

Faroqhi, S. (1977/1979): Rural Society in Anatolia and the Balkans during the Sixteenth Century. - Turcica, Revue d'Etudes Turques. I: vol. IX, pp. 161-195; II. vol. XI, pp. 103-153

Gaube, H. und Wirth, E. (1984): Aleppo - Historische und geographische Beiträge zur baulichen Gestaltung, zur sozialen Organisation und zur wirtschaftlichen Dynamik einer vorderasiatischen Fernhandelsmetropole. - Tübinger Atlas des Vorderen Orients, Beiheft B 58. Wiesbaden

Inalcik, H. (1983): The Emergence of Big Farms, Çiftliks: State, Landlord and Tenants. - In: Collection Turcica III; Contributions à l'histoire economique et sociale de l'Empire Ottoman. - Inst. Français d'Etudes Anatoliennes d'Istanbul. Louvain

Islamoglu-Inan, H. (1987): State and Peasants in the Ottoman Empire: A Study of Peasant Economy in North-Central Anatolia during the 16th Century. - In: H. Islamoglu-Inan (Ed.): The Ottoman Empire and the World Economy. Cambridge

Islamoglu-Inan, H. (1987): Introduction - Oriental "Despotism" in World System Perspective. - In:
    H. Islamoglu-Inan (Ed.): The Ottoman Empire and the World Economy. Cambridge
Islamoglu-Inan, H., and Faroqhi, S. (1979): Crop Patterns and Agricultural Production Trends in
    Sixteenth Century Anatolia - Review. - Revue d'etudes interdisciplinaires II, 3, pp. 401-
    436.
Islamoglu- Inan, H., and Keyder, C. (1987): Agenda for Ottoman History. - In: H. Islamoglu-Inan
    (Ed.): The Ottoman Empire and the World Economy. Cambridge
Höhfeld, V. (1977): Anatolische Kleinstädte - Anlage, Verlegung und Wachstumsrichtung seit dem
    19. Jahrhundert. - Erlanger Geographische Arbeiten, Sonderband 6, Erlangen
Hütteroth, W. and Abdulfattah, K. (1977): Historical Geography of Palestine, Transjordan and
    Southern Syria in the Late 16th Century. - Erlanger Geographische Arbeiten, Sonderband 5,
    Erlangen
Kreiser, K. (1979): Über den "Kernraum" des Osmanischen Reiches. - In: K.-D. Grothusen (Ed.):
    Die Türkei in Europa - Beiträge des Südosteuropa-Arbeitskreises der Deutschen
    Forschungsgemeinschaft zum IV. Internationalen Südosteuropa-Kongreß. Göttingen
McGowan, B. (1981): Economic Life in Ottoman Europe. - Cambridge
McGowan, B. (1987): The Middle Danube cul-de-sac. - In: H. Islamoglu-Inan (ed.): The Ottoman
    Empire and the World Economy. Cambridge
Moltke, H. von (1893): Briefe über Zustände und Begebenheiten in der Türkei aus den Jahren
    1835-1839. - VI, Berlin
Nagata, Y. (1976): Some Documents on the Big Farms (Çiftliks) of the Notables in Western
    Anatolia. - Institute for the Study of Languages and Cultures in Asia and Africa. Tokyo
Pamuk, S. (1987): Commodity Production for World Markets and Relations of Production in
    Ottoman Agriculture. - In: H. Islamoglu-Inan (ed.): The Ottoman Empire and the World
    Economy. Cambridge
Soysal, M. (1976): Die Siedlungs- und Landschaftsentwicklung der Çukurova, mit besonderer
    Berücksichtigung der Yüregir-Ebene. - Erlanger Geogr. Arbeiten, Sonderband 4, Erlangen
Sunar, I. (1987): State and Economy in the Ottoman Empire. - In: H. Islamoglu-Inan (ed.): The
    Ottoman Empire and the World Economy. Cambridge
Veinstein, G. (1976): Ayan de la region d'Izmir et commerce du Levant (Deuxieme moitie du
    XVIIIe siecle). - Etudes Balkaniques 12/3, pp. 71-83
Wallerstein, I. (1980): The Ottoman Empire and the Capitalist World Economy: Some Questions
    for Research. - In: Okyar/Inalcik (eds.): Social and Economic History of Turkey 1071-1920.
    Ankara
Wallerstein, I., Decdeli, H., and Kasaba, R. (1987): The Incorporation of the Ottoman Empire into
    the World Economy. - In: H. Islamoglu-Inan (ed.): The Ottoman Empire and the Worl
    Ceonomy. Cambridge, pp. 88-97
Wilhelmy, H.(1935): Hochbulgarien, vol. 1: Die ländlichen Siedlungen und die bäuerliche
    Wirtschaft. - Kiel
Wirth, E. (1986): Aleppo im 19. Jahrhundert. Ein Beispiel für Stabilität und Dynamik
    spätosmanischer Wirtschaft. - Erlangen (Revised reprint from H. G. Majer (ed.):
    Osmanistische Studien zur Wirtschafts- und Sozialgeschichte. Wiesbaden)

# 23

# Plan and Reality in the Cape Colony

A.J. Christopher

## 1.    Introduction

The rounding of the Cape of Good Hope by Bartholemeu Diaz in 1488 was one of the crucial exploits of the Age of Discoveries. Thereafter southern Africa was of little consequence in the era of conquest, colonization and exploitation which followed. Indeed no permanent settlement by Europeans was undertaken until the middle of the seventeenth century and this remained a neglected backwater for a further 200 years. Immanuel Wallerstein (1974, 1980, 1989) in the first three volumes of his magisterial survey of the modern world-system makes only passing reference to the region. This obscurity is deceptive as it reflects more the limited scope of governmental plans than the reality of significant European adaptation to a strange environment. If the European powers viewed the Cape of Good Hope only as a waystation to the East, the European colonists in 200 years created a remarkable settlement extending 2000 kilometres into the interior of Africa, with much of the economy only loosely linked to the world-system. Official plan and reality in the Cape Colony often bore little relationship to one another.

Only necessity forced the Dutch East India Company to intervene on the African coast between Angola and Mozambique in the mid-seventeenth century as the replenishment station on the island of St Helena became inadequate. In 1652 the Company established the first permanent European settlement on the subcontinent at Cape Town with limited objectives and limited resources. From the first, plan and reality differed. They were to do so for the following 200 years, before the era of nineteenth century imperialism transformed the perceived value of the colony within the global concerns of the British Empire.

## 2.    The Cape Settlement

The image of the southern African coastline as inhospitable and useful only for occasionally obtaining water supplies was common to all the European powers plying the Cape route to India and the Indies. The indigenous inhabitants were regarded by the early explorers as hostile and lacking in products which were thought valuable in world trade (Raven-Hart, 1967). However, in the mid-seventeenth century a reassessment of the Cape peninsula indicated its suitability not only for obtaining water but for growing fruit, vegetables, and cereals to supply passing ships.

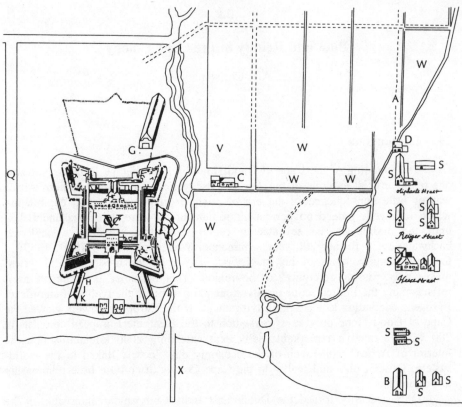

| | | | |
|---|---|---|---|
| A | Demarcation line of 50 roods round the fortress (line of future Heerengracht) | K | Blacksmith shop |
| | | L | Kitchen & bakery for servants & slaves |
| B | Company's cable house or store | Q | Canal |
| C | Company's gardener's house | S | Freeburghers' houses |
| D | Company's water corn mill | V | Company's vegetable garden |
| G | New shed for cows & sheep | W | Gardens |
| H | Hospital | X | Jetty |
| J | Plough & cartwright shop | *VOC* | Trademark of United East - India Company (Vereenigde Oost - Indische Compagnie) |

*Fig.1:First Dutch settlement at Cape Town 1660*
*Source:Christopher 1976*

The Dutch East India Company established the refreshment station in 1652 with a fort for protection and a Company garden in which the fruits, vegetables, and crops would be cultivated (Fig. 1). It is significant that the Company's instructions to the first governor, Jan van Riebeeck, stressed the importance of maintaining cordial relations with the indigenous inhabitants and obtaining livestock through trade, not through Company ranching.

Despite this limited objective, within ten years the settlement had departed substantially from the initial plan. The Company servants proved to be incapable of growing the volume of supplies required, while the indigenous population showed reluctance to trade away their animals. In consequence several significant decisions were taken. First, the Company recognized the importance of individual effort and released a number of its servants as free citizens to work the land for their own account and sell the produce to the Company, if at controlled prices. European agricultural colonization was thus encouraged. Secondly, as the settlement was planned and organized by men who had worked in the Dutch possessions in the East Indies, customs and practises from there were introduced, notably the importation of slaves to undertake the heavy work of development. Thirdly, the Company and later individuals began the enterprise of raising livestock.

## 3.    European Colonists

The introduction of free European agricultural settlers was of the utmost importance to the development of southern Africa. The first free settlers in 1657 were located on the Liesbeeck River to the east of Cape Town on plots of 10-20 hectares, with the requirement to grow crops, notably wheat, for the Company (Christopher, 1976) (Fig. 2). The numbers were small with only 40 farmers on the land after twenty years of endeavour. The first governor when he initiated the scheme had envisaged 1,000 families closely settled on small arable farms in the image of the Netherlands or Java (Leibrandt, 1900). Reality was such that the conditions for intensive agriculture were not reproduced at the Cape. Extensive cultivation practises, allied to mixed crop-animal husbandry replaced the intensive systems proposed (Guelke, 1982). Thus crop rotation, heavy manuring and careful preparation of the ground were abandoned. High yields were not required where land was plentiful and labour expensive. A 'New Lands' economy came into being. It is noteworthy that nearly 200 years later, the British administration at the recommencement of European immigration had the same expectations which were similarly unrealistic and unfulfilled.

In 1679 an expansion of the area of colonization took place beyond the Cape peninsula with the opening of the Stellenbosch region. This was followed by Franschhoek in 1688 for Huguenot refugees and areas yet further from Cape Town in the following decade until the limits set by transport costs were reached (Fig. 3). Farms of 50 hectares were offered in order to attract settlers. Thus by 1699 some 1200 settlers had 'occupied' an area of 4,000 square kilometres, that is over 300 hectares per man, woman and child. Only 5,000 hectares had been granted as

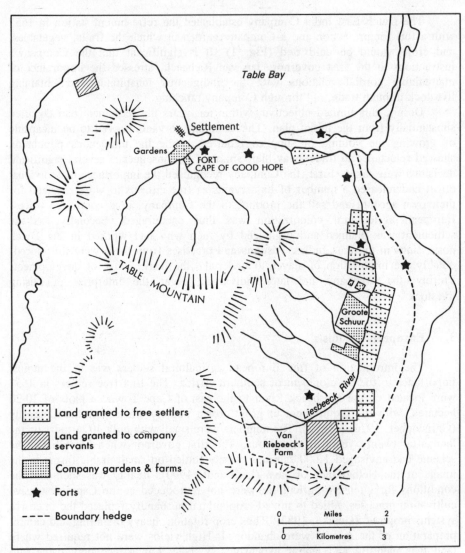

*Fig.2:The first farms granted on the Liesbeeck River*
*Source:Christopher 1976*

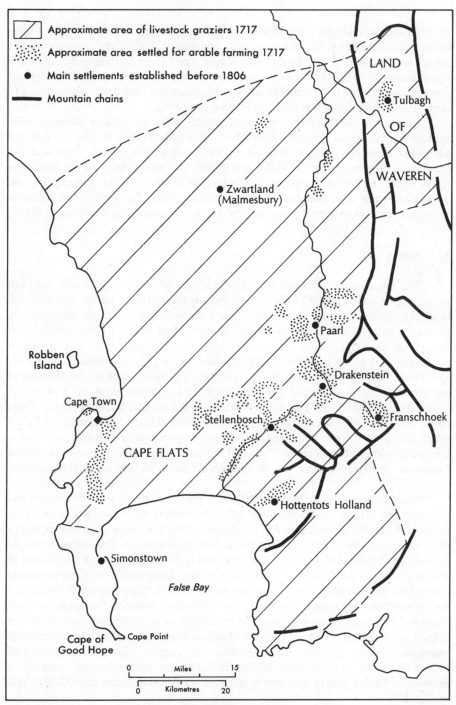

Fig.3: *The settlement of the South-western Cape*
Source: Christopher 1976

freehold farms, with the remainder of the land remaining in Company possession but effectively used by the settlers as free commonage. One of the consequences of the sparsity of settlers was the lack of villages, apart from Stellenbosch. Again the reproduction of a European settlement pattern was defeated by the low level of demand for services from a highly scattered population. Each farm was reduced to supplying its own needs mainly from its own resources.

Production on the farms increased rapidly with the Colony attaining self-sufficiency in wheat by the mid-1680s. Experimentation in the search for additional commercial crops resulted in the development of vineyards and the production of wine in the late 1680s and 1690s. However, the wine produced was not of export quality and production was largely limited to local consumption and the passing trade until the late eighteenth century, when the quality of Constantia wines was such as to permit entry into the European trade (Van Rensburg, 1954).

## 4. Slaves

The introduction of slaves into the Cape was of major significance for both the economic development of the colony and for future race relations (Elphick and Giliomee, 1979). In this manner the Cape Colony is placed within Wallerstein's periphery with its emphasis upon slave labour. Most works have suggested that slavery at the Cape was of limited commercial importance, but a recent reassessment has indicated the development of highly capitalized plantation agriculture, akin to other contemporary plantation colonies, with all the subsequent problems involved (Worden, 1985). Slaves as a component of the production system in the Netherlands East Indies were accepted as such in South Africa. Once slaves had been introduced, the European population refused to undertake tasks for which the slaves were employed, effectively preventing the development of a European colony of settlement. Thus in 1717 the Company Commissioners halted European immigration and the decision was taken to pursue a slave colony economy.

Slave numbers by the early 1710s were greater than those of Europeans and this state of affairs continued until 1820. However, no organized slave trade developed and the competition for the limited slave and indigenous labour forces was such that there was frequently a labour shortage. The dependence of agriculture in the south-western Cape on slaves is evident in the high proportion of enterprises owning slaves. By the late eighteenth century nearly all Cape peninsula enterprises and two-thirds of those in the south-western Cape employed slaves, but the proportion fell rapidly towards the colonial frontier. Furthermore, few employed large numbers of slaves. The mean size of slave holding only increased from 7 to 16 in the course of the eighteenth century and no dominant large-scale enterprises were apparent. The system of partible inheritance prevented accumulation of considerable wealth but by the mid-eighteenth century a prosperous landed gentry had emerged based on the slave economy (Guelke and Shell, 1983).

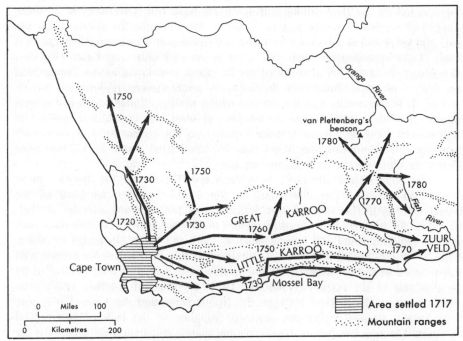

*Fig.4: The spread of the graziers*
*Source: Christopher 1976*

## 5.    The Frontier

The problem of meat supply was to present the colonial authorities with major difficulties. The indigenous population was unwilling to sell sheep or cattle, which constituted their wealth, while conflict over grazing grounds between settler and indigene was a problem which was to bedevil South African society everafterwards. Seizure of animals, seizure of grazing areas and the effective enslavement of the indigenous population on appropriated lands on the one side, and the quest for new grazing grounds for a growing population on the other, resulted in a major expansion of the limits of the colony in the eighteenth and nineteenth centuries.

Initial grazing activities as an adjunct to arable farming gave way to the practise of transhumance over the mountains of the south-western Cape in the early eighteenth century, to avoid the summer drought. Soon the movement in some cases was permanent with settlers taking their livestock to new farms beyond the officially settled areas. The small freehold farms offered by the Company were clearly inappropriate in a remote and semi-arid land. The colonists therefore devised a scheme, subsequently officially endorsed, whereby the farmer appropriated all the land within half-an-hour's walk (on horseback) of a centre point. The resulting circular area of 2,500 hectares became the standard size for over 200 years and was laid out by European frontiersmen to beyond the Zambesi River. Land appropriation thus proceeded at an ever more rapid rate (Fig. 4). Accordingly by 1806 the majority of the European population of the Colony was not settled in Cape Town and the adjacent south-western region, but in the interior. It is noteworthy that the interior with a healthy climate and few diseases, together with the effects of the settler diet, of meat and vegetables, resulted in high natural increase among frontier families in the eighteenth and nineteenth centuries. The European population thus increased from 2,000 in 1721 to over 25,000 by 1806, despite limited immigration.

The considerable distances between Cape Town and the frontier areas resulted in only a loose attachment to the world-economy, as most of the production of the isolated interior farmers was not destined for the market. However, the itinerant pedlar and the herder maintained contact for essentials such as guns and gunpowder and luxuries such as cloth and coffee, in return for hides, skins, ivory etc. Such trade was not strictly profitable for the frontier farmer who was reduced to a state of semi-self-sufficiency, with activities little controlled by the demands of the market (Neumark, 1957; Norton and Guelke, 1977). The arguments over the linkage between the frontiersmen and the Cape Town core have usually stressed either the need for supplies or the lack of commercial products. The apparent lack of regulation and market orientation were vital for the creation of a new society no longer Dutch, but Afrikaner. Questions concerning the influence of the frontier thus arise which again have been subject to varying interpretations (Legassick, 1980; Smith, 1988).

## 6.    The Indigenous Population

The transformation which befell the indigenous population depended upon the relative position with regard to the settlers. European diseases, notably smallpox, were demographically disastrous in many societies in the western and central Cape. Population numbers were in some cases decimated, with the total destruction of the structure of indigenous society. The consequence was the incorporation of the remnants of the population into the colonial state. As their lands were expropriated so they were reduced to a labour force on the farms of the incoming settlers. Thus even in the Stellenbosch and adjacent areas as much as a quarter of the labour force in 1806 consisted of the indigenous Khoikhoi. In the interior and frontier regions the proportion was much higher where the European population was too poor to purchase slaves.

Incorporation, if not legal enslavement, was one option forced upon the indigenous population. Others chose to flee and maintain a semblance of independence. Hence various fugitive societies came into being, particularly beyond the northern colonial boundary. Groups, including the Griquas, established independent governments borrowing upon the model of the Cape frontier (Ross, 1976).

In contrast on the eastern frontier in the 1770s the colonists came into conflict with a semi-sedentary farming and herding population capable of resisting incorporation. Conflict over grazing grounds and resources began a century of frontier wars only resolved in the 1870s. Thus the expansion of the colonial frontier was temporarily halted in this direction.

## 7. Change in the Early Nineteenth Century

The occupation of the Cape of Good Hope by Great Britain between 1795-1803 and permanently after 1806 resulted in few immediate changes. The strategic importance which had led to the first settlement was enhanced as the Royal Navy assumed a dominant role on the coasts of Africa. Cape Town as the way-station prospered and grew and the south-western Cape's commercial agriculture also expanded. However, with time the British administration sought to exercise greater control over its new colony and regulate what appeared to be an ill-organized enterprise. In the course of the 1830s and 1840s a series of events effectively ended the first phase of the Cape Colony's involvement with the world-economy.

An incipient urban hierarchy had begun to form in the course of the early nineteenth century with the active encouragement of the British administration, which viewed the development of an urban culture as essential to the development of trade, industry and good administration. Thus the skeletal framework of administrative and other settlements, numbering only 10 in 1806 was expanded to over 45 by 1845.

European immigration was recommenced in 1820 and attempts were made to raise agricultural production through the creation of a mid-latitude colony of settlement in the eastern part of the Colony. Although crop production was initially unsuccessful, the development of the wool industry in the 1830s provided the staple export product suited to the southern African physical environment. Profits from the raising of woolled sheep further provided the much needed finance to boost the colonial economy and break down the self-sufficiency of the interior districts. The reform of the land grant system in 1813 was significant in that it was intended to provide security of occupation and thus encourage pastoral commercialization and a greater incentive for individual effort (Christopher, 1984). Security of tenure also allowed farmers to gain access to loans through mortgage finance in order to institute these improvements. The reforms of 1829 further recognized that commercial pastoralism required tracts of land far in excess of 2500 hectares ending the apparent equality of frontier farming entitlements.

The abolition of slavery in 1834 and the final freeing of the slaves in 1838 was another measure of modernization. The economy was not disrupted markedly as the European farmers of the south-western Cape continued to produce wheat and wine with now freed labour.

The Great Trek in the later 1830s by approximately 15,000 dissatisfied colonists was of major significance as the trekkers penetrated deep into the interior occupying parts of the now Orange Free State, Transvaal, and Natal. These areas were yet further removed from world markets and their links with the world-economy, particularly in the Transvaal, were tenuous. It should be noted that much of the motivation for this colonization was dissatisfaction with the British administration and its modernization programme. An alternative society was thus established beyond the colonial boundary based on eighteenth century concepts.

That this was possible was the result of a series of convulsions, the Mfecane, in indigenous societies attendant upon the rise of the Zulu military monarchy in the early years of the century. This had disrupted established patterns and social formations and so provided the opportunity for European penetration into lands apparently little settled. The Great Trek involved the nominal incorporation of large numbers of settled Africans into a series of European controlled polities. The Cape Colony followed in the annexation of settled African areas, resulting in the emergence of a new form of colony, the protectorate. The age of imperialism had arrived.

## 8.    The Cape Colony in the 1840s

The Cape Colony thus was divided into three parts by the 1840s, conforming closely to the basic premises of von Thünen's land use model. First, Cape Town and vicinity was the key to the whole enterprise. Until the 1840s Cape Town was still the only settlement worthy of the description, 'urban', with a population of approximately 20,000. The commercial and administrative functions of the Colony were focussed upon the city. Half the Colony's total manufacturing and commercial employment was concentrated in Cape Town and its environs (Cape of Good Hope, 1838). Secondly, the south-western Cape with its distinctive commercial crop production based on wheat and wine was linked firmly to the world-economy. Thirdly, the frontier districts by contrast were economically dependent upon extensive pastoralism and were only beginning to be firmly integrated into the world-economy after over a century of virtual self-sufficiency. Fourthly, beyond the colonial boundary lay indigenous polities little affected by the world economy.

Some 70,000 Europeans and a somewhat larger number of indigenous people and freed slaves occupied a colony of approximately 400,000 square kilometres. The export trade amounted to only £ 300,000 - £ 400,000 per annum, of which wine accounted for a third. The results of 200 years of European colonization were thus meagre when comparison is made with other parts of the world. Plans formulated in Europe had largely been abandoned in the light of the little understood African reality. The term 'periphery' is maybe too positive a concept to use when placing much of the Cape Colony prior to 1840 into the Wallerstein framework. The majority of the European colonists had retreated into the external arena.

# References

Cape of Good Hope (1838): Blue Book of the Colony. Cape Town: Government Printer

Christopher, A.J. (1976): Southern Africa: studies in historical geography. Folkestone: Dawson

Christopher, A.J. (1984): The Crown Lands of British South Africa 1853-1914. Kingston: Limestone Press

Elphick, R. and Giliomee, H. (1979): The Shaping of South African Society 1652-1820. Cape Town: Longman

Guelke, L. (1982): Historical understanding in geography: an idealist approach. Cambridge: Cambridge University Press

Guelke, L. and Shell, R. (1983): An early colonial landed gentry: land and wealth in the Cape Colony 1682-1931. In: Journal of Historical Geography 9, pp. 265-286

Legassick, M. (1980): The frontier tradition in South African historiography. In: Marks, S. and Atmore, A. (eds.): Economy and Society in pre-industrial South Africa. London: Longman, pp. 44-79

Leibrandt, H.C.V. (1900): Precis of the Archives of the Cape of Good Hope: Letters Despatched from the Cape 1652-1662. Cape Town: Government Printer

Neumark, S.D. (1957): Economic influences on the South African Frontier. Stanford: Stanford University Press

Norton, W. and Guelke, L. (1977): Frontier agriculture: subsistence or commercial? Commentary and reply. In Annals of the Association of American Geographers 67, pp. 463-467

Raven-Hart, R. (1967): Before van Riebeeck: callers at South Africa from 1488 to 1652. Cape Town: Struik

Ross, R. (1976): Adam Kok's Griquas: a study in the development of stratification in South Africa. Cambridge: Cambridge University Press

Smith, K. (1988): The Changing Past. Johannesburg: Southern

Van Rensburg, J.I.J. (1954): Die geskiedenis van wingerdkultuur in Suid-Afrika tydens die eerste eeu. In: Archives Yearbook for South African History 17(2), pp. 1-96

Van Zyl, D.J. (1968): Die geskiedenis van graanbou 1795-1826. In: Archives Yearbook for South African History 31(1), pp. 167-290

Wallerstein, I. (1974): The Modern World-System I: Capitalist Agriculture and the Origins of the European World-Economy in the Sixteenth Century. New York: Academic Press

Wallerstein, I. (1980): The Modern World-System II: Mercantilism and the Consolidation of the European World-Economy. New York: Academic Press

Wallerstein, I. (1989): The Modern World-System III: The Second Era of Great Expansion of the Capitalist World-Economy 1730-1840s. New York: Academic Press

Worden, N. (1985): Slavery in Dutch South Africa. Cambridge: Cambridge University Press

# 24

# Commercial Manufacturing in Southeastern India in the Early Modern Period

Brian Murton

## 1. Introduction

It is now widely accepted that regional studies of industry in late medieval and early modern India are essential not only for the development of a pre-colonial economic history, but also for placing India in the comparative perspective of the emergence of commercial capitalism on an international scale in the course of the seventeenth and eighteenth centuries (Perlin, 1983, p.33). In this paper I argue that India, like Europe, was affected by profound and rapid change in the nature of its societies and economies, and state forms, from at least the sixteenth century, and that a fundamental aspect of that development was a local merchant capitalism which emerged within a common international framework of societal and commercial changes. I shall try to show in the case of India that the rise of commercial manufacturing can be best understood within the context of a complex nexus of interdependencies taking form in early modern times.

Until the relatively recent studies of Brennig (1975, 1986, pp.333-355), Ramaswamy (1985), and Subrahmanyam (1986; 1990) regional studies of commercial manufacturing had concentrated on trade rather than on the industry itself (see Arasaratnam, 1978, pp.42-53; 1980, pp.257-281; 1986; Raychaudhuri, 1962; Chaudhurri, 1974, pp.127-182; 1978; 1985). This may have been due to the large size of the geographic units chosen for study, such as the Coromandel Coast, which is too areally differentiated for a detailed study, and also because of the nature of the evidence, usually Company records, which detail the perspective of the European trading companies. However, both Arasaratnam (1986) and Brennig (1986) use these records to reveal all they can about the nature of the textile industry along and in the immediate hinterland of the Coromandel Coast. Ramaswamy (1985) uses epigraphical materials, and literary sources, as well as weaver folk traditions (1982, pp.47-62), to build up a comprehensive account of the textile industry between the tenth and seventeenth centuries. She concludes that the heyday of the industry was in the Vijayanagara period in the fifteenth and sixteenth centuries, and that during the seventeenth and eighteenth weavers began to suffer as they increasingly focused on production for the European companies. Her work exemplifies the integration of what Perlin (1980, pp.267-306) has called the "two antipathetic universes of history" which deal with very different types of documentation. On the one hand there are records emerging from the indigenous societies themselves, informed by an indigenous framework of institutional, conceptual, and linguistic reference and representation. On the other hand there are the records produced by outsiders from A.D. 1500 onwards, most of them European records of one type or another. In the past this tended to produce two distinctive types of historical synthesis, one on land-based states, and the other dealing with long distance trade.

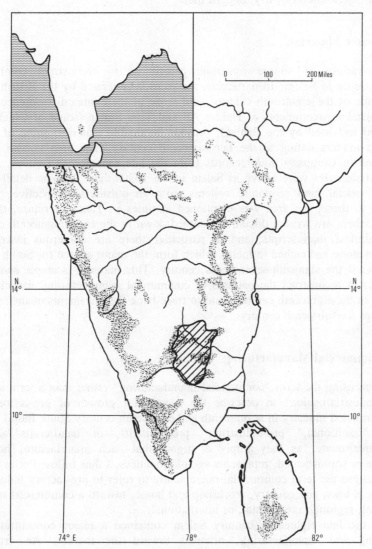

*Fig. 1: Location of Salem*

The remainder of the paper initially takes one small part of interior Tamilnadu, known in the late eighteenth century and today, as Salem (Figure 1), and examines the characteristics of commercial manufacturing in the area in the eighteenth century and earlier. The final section deals with the relationships between the internal social and economic networks of South India and the "external" dimension of early modern India.

## 2.    Source Materials

A variety of source materials permit us to reconstruct commercial manufacturing in Salem. In particular, the records generated by the British in the last decade of the eighteenth century, and those of their immediate predecessors, the Karnataka governments of Haidar Ali and Tipu Sultan (some of which were translated and used by the British), provide invaluable material, some of which refers to matters dating to the late sixteenth century. The area is also briefly mentioned in Company trade records from the late seventeenth century, and as Jesuit missionaries were active in Salem in the 1620s there is some detail of the political, social, and economic systems from an outsiders perspective. But in addition to these, the area was mentioned at times in Tamil literature, there is folklore, there are written documents copied down in the early nineteenth century from palmleaf manuscripts, and in particular there are numerous inscriptions carved in stone and etched in metal, which form the major source for South Indian history until the sixteenth-seventeenth century. This material is ample enough to enable us to reconstruct the patterns of commercial manufacturing, especially of textiles, in the eighteenth century, and to trace back the lineaments of the industry to perhaps the thirteenth century.

## 3.    Commercial Manufacturing

I am using the term "commercial manufacturing" rather than a term such as "proto-industrialization" to describe the spectacular growth of pre-factory and market-oriented industry in seventeenth and eighteenth century South India. Labels such as "traditional," "pre-industrial," "pre-capitalist," or "handicrafts" also are totally inadequate, as they imply a stigma that such manufacture, however extensive or sophisticated, represents age-old practices. I thus follow Perlin (1983, p.43) and use the term commercial manufacture to refer to pre-factory industry of all kinds in town and country, workshop and home, toward a commercial market, be it local, regional, continental, or international.

In the late eighteenth century Salem contained a major concentration of processing and manufacturing, oriented toward the market. An array of agricultural products, including wild sesamey, castor, and cultivated sesame were processed to make cooking oil. Arrack and toddy, intoxicating beverages were widely manufactured from the juices of the palmyra palm. The manufacture of coarse sugar (*jaggery*) was located in the southern and eastern parts of the area.

*Ghi* was processed in sufficient quantities to enter into the tax and trade statistics, especially in the west, where there was a breed of cow noted for its milk production. Twine and tape making was important in parts of the south and indigo was processed in the east in small quantities. Mineral processing included the manufacture of saltpetre, and most importantly, iron and steel, located in the eastern and southern parts of the region. A considerable amount of both iron and steel was exported from Salem, both to the coast and the interior.

But there is little question that textile manufacturing was the pre-eminent commercial activity in Salem in the early modern period, just as it was throughout India. Chaudhuri (1974, p.126) has stated that before the discovery of machine spinning and weaving in the second half of the eighteenth century, the Indian subcontinent was probably the world's greatest producer of cotton textiles. Although textile, especially cotton goods, were manufactured everywhere, this widespread dispersion should not be allowed to mask the vital qualitative difference between production for purely local markets and production for export and inter-regional trade. By the seventeenth century four great industrial regions, specializing in export goods had developed in India: Punjab, Gujarat, Bengal, and the Coromandel. A number of location factors had operated to the advantage of these areas, including the availability of raw materials, access to markets, transport costs, the scale of output, social values, and political conditions. Especially in regard to textile exports, the problems of distance directly entered into the question of profitability, and Chaudhuri (1974, pp.132-133) strongly suggests that the key factors were: 1. transportation costs, and 2. market orientation. The major producing areas, therefore, were located where costs were minimized in relationship to their markets, with each area having a primary market of its own with its products oriented to consumer preference. The weaving regions of Coromandel relied on trade with Southeast Asia for their overseas outlet, and the types of goods produced reflect this.

Within Coromandel there were a large number of textile regions of which Salem was one of the more significant in the interior. To understand the textile industry in Salem it is necessary to analyze both its spatial distribution and interdependence of tasks and functions, and how the industry was organized, both socially and industrially.

## 4.   The Textile Industry in Salem

The cotton textile industry involved a range of activities, all of which need detailed consideration: cotton production, trade in raw cotton, spinning, weaving, bleaching, dyeing, painting.

## 4.1   Cotton Production

Cotton was grown in the western part of Salem (Figure 2) in limited quantities ("Products", The Baramahal Records, 1912, p.69). Two varieties,

*Nadan paruti* and *Uppan paruti* (cotton) were cultivated. The former was a bushy perennial which preferred red soils, and the latter was a smaller annual, best suited to black soils which were limited in Salem. *Nadan* was used to manufacture coarse cloth and *Uppam*, finer products. Partial data from eastern Salem indicate that 45,700 *tulam* (just over 300 tons) of thread were made from locally grown cotton (a *tulam* is 15.5 lbs.) in 1795, while 16,201 *tulams* (112 tons) were imported from the west, and 20,500 *tulams* (142 tons) from the south and east. In other words, perhaps 55 percent of the cotton required by the weaving industry was locally grown.

## 4.2  Cotton Spinning

Thread production, therefore, was an important local activity. In the villages of the west there was a flourishing thread making industry, carried out by cultivators and their wives ("Imposts", The Baramahal Records, 1920, p.28-30), mostly of the Kongu Vellalar (Gavundar) caste, the dominant landed group in this part of Salem. According to Ramaswamy (1980, pp.227-241) the technology used to prepare the yarn included the bow for carding the raw cotton (probably introduced into South India in the early first millennium A.D.), the spinning wheel first introduced in the fourteenth century from West Asia, and the hand spindle, of earlier origin, and used in the area until the nineteenth century along with the wheel. Gold thread was manufactured in Salem town, and red thread produced in small quantities in the east.

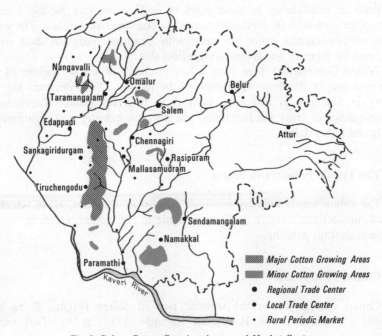

*Fig. 2: Salem: Cotton Growing Areas and Market Centers*

## 4.3 Trade in Raw Cotton and Thread

A distinguishing feature of the export sector of Salem's textile industry was its dependence on supplies of fine cotton brought in from Coimbatore to the west and Tiruchchirappalli to the southeast, where there were areas of black soil. Bullocks carried the relatively bulky and relatively low-value raw cotton and the more valuable thread to bazaar towns scattered throughout Salem. Thread produced in Salem was marketed by cultivators at the periodic markets found throughout the region (Figure 2). Not all of the markets depicted were involved with thread marketing, as a number dealt with other agricultural products (grain was an important trade item), cattle, and hill products, but many of those in the west and south were important thread markets. Weavers could purchase thread directly at these markets, but more commonly merchants did the purchasing by traversing a market circuit, and carrying the thread to the weavers. Unfortunately, we have no information concerning the days of the week on which these markets met so it is impossible to decipher the spatio-temporal synchronization of integration of periodic market place meetings. But it is clear that there were clearly defined circuits centering on Salem, Taramangalam, Nangavalli, Edappadi, Sankagiridurgam, Tiruchengodu, Paramathi, Namakkal, Rasipuram, and Sendamangalam. These places all contained major concentrations of weavers. The markets were oriented both to traders and consumer demands: agricultural produce was "retailed by farmers at the santas directly to the consumers" (Imposts", The Baramahal Records, 1920, p.110), and the markets were also held "at the most eligible places for riots to dispose of their products, and the merchants of the bazaargrams to barter with them and with one another. ("Imposts", The Baramahal Records, 1920, p.130). In Salem, in contrast to further north in Karnataka and the Deccan where the Banjara community dominated trade in cotton and grain (Brennig, 1975, pp.335-338), trade in raw cotton and thread was handled by either weavers who had become merchants, or by other chettiyar (merchant) groups, such as the Komati and Nagarattar.

## 4.4 Industrial Capacity, Products, and Location

According to the data from the 1790's there were just over 4,000 looms in Salem. This compares favorably with Brennig's estimate of 7,500 looms in the East Godavari delta region (1986, p.343), Arasaratnam's figures for Madurantakam of 700 looms (1986, p.570), and Buchanan's (1807, 3 volumes) figures for places in Coimbatore, (for example, Satyamangalam 800 looms (Buchanan, 1807, II, pp.239-242), Coimbatore 459 looms (Buchanan, 1807, II, pp.261-262), Perindur 800 looms (Buchanan, 1807, II, 287), and Erodu 2000 looms (Buchanan, 1807, II, pp.287-288). How much cloth could the looms of Salem produce if fully employed? The technology of the throw shuttle loom, the type probably employed, according to Brennig (1986, p.343), would have permitted the average weaver household to produce as much as 1,600 yards of cloth per year. But as weaving was an outdoor activity carried out under trees usually in front of temples patronized by weavers, work could not proceed during the rainy season. Thus a single loom produced perhaps 1,200 yards of cloth a year, with the maximum output possible from the looms in Salem being about five million yards. Much of the output, in fact between 80-90 percent ("Products", The Baramahal Records, 1912, pp.108-110), was coarse cloth, a staple for domestic consumption as well as for export to Southeast Asia. However, in addition to coarse white cloth (*khadi*) coarse muslins (*selas pegris*), and tent material (*moto khadi*), finer muslins (*salempores*), and a variety of plain and colored *sari* pieces were produced. Many of these, plus male cloths (*veshtis*, *dhotis*), had silk borders, the silk being imported from coastal Madras ("Imposts", The Baramahal Records, 1920, pp.64-65). Probably about 10 percent of the output was destined for export markets, a figure in keeping with estimates for other areas (Arasaratnam, 1986, p.96). A number of the types produced in Salem were those popular in various Southeast Asian markets, (Arasaratnam, 1986, pp.96-105), especially *salempores*, (later important in the European market as well), and *selas*, which were blue striped or checkered.

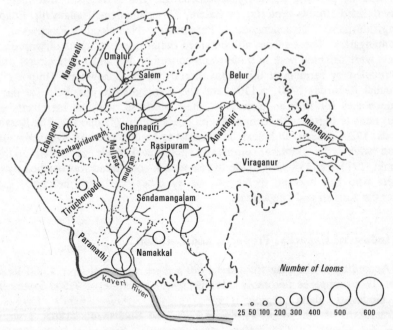

*Fig. 3: Looms in the late 18th Century*

Within Salem weaver households were significantly concentrated in the small market towns of the region, and in villages nearby to these towns (Fig.3). Places such as Edappadi, which was the "greatest manufacturing village" in the west ("Police", The Baramahal Records, 1923, p.154), Nangavalli, Tiruchengodu, Mallasamudram, Amarakundi, Vennandur, Rasipuram, Sendamangalam, and Namakkal are mentioned as weaving centers in the records. In the Salem town area, the Shevpet, Gugai, and Ammapettai sections were weaving centers. This type of distribution pattern accords with other parts of Coromandel where weaving was found scattered in both towns and villages, in contrast to Gujarat and Punjab where it was an urban activity (Chaudhuri, 1974, pp.140-141). It should be noted that the major centers all had important local temples, or had been the capitals of warrior chieftains in the sixteenth and seventeenth centuries.

## 4.5  Bleaching, Dyeing, Painting, and Printing

The only mention of any of these activities in the records is to the use of locally grown indigo to dye thread blue and the use of imported vegetable materials to dye thread red. The painting and printing of cloth appears to have been carried out in areas closer to the coastal ports (Gittinger, 1982). However, it appears that every weaving village had a cadre of washermen who washed and beat finished cloth for the weavers ("Imposts", The Baramahal Records, 1920, p.28, p.51).

## 5.  Organization of the Industry

While one would like to ask a number of questions about the organization of the industry in Salem before the late eighteenth century, the lack of detailed and comprehensive data makes this difficult. However, the fragments of information that do exist, when used in conjunction with the findings of studies of other areas in South India (for example, Arasaratnam, 1980, pp.257-282, 1986; Ramaswamy, 1980, pp.119-140, 1985a, 1985b, pp.294-325; Brennig, 1986, pp.333-355) permit us to gain some insight into social differentiation among weavers, how the industry was organized, and how the system operated to get finished textiles to the ports of the east coast.

## 5.1  Social Organization

Weavers held a central position in the complex inter-relationships between different groups involved in the textile industry. Like all other occupational groups in India weavers were organized into a number of castes. Weaving castes in Salem, according to the "Inhabitants" volume of The Baramahal Records, included Kaikolar (18520), Saliwar, (7500), Janapar (5976), Devanga (2594), Patnulkar (2200), a number of small cotton weaving castes. A shepherd caste, the Kurubar, wove woolen *kamblis*, and the Janapar wove the gunny sacks used to carry goods on the backs of bullocks. Kaikolars operated 1833 looms and Devangas 1034, or nearly 70 percent of those in the region. Kaikolars dominated weaving in Tiruchegodu, Edappadi, Mallasamudram, Attur, Sendamangalam, Namakkai, and Paramathi. Devangas and Kaikolars were found in almost equal numbers in Salem and Devangas dominated Nangavalli, Sankagiridurgam, and Rasipuram

("Imposts", The Baramahal Records, 1920, p.18, p.27, p.63). The Saliwar and
Kaikolar are Tamil weaving castes, the Janapar are Telugu in origin, and the
Patulkar came originally from Gujarat. Devangar tradition states that they
migrated from Vijayanagara, settling in Salem originally at Amarakundi, probably
in the sixteenth century (Richards, 1918, I, 1, p.181).

Saliyar weavers were the major weaving community until the fourteenth
century in South India (Ramaswami, 1985, p.14), but thereafter the Kaikolar, who
seem to have functioned as soldiers earlier, became fulltime weavers. The earliest
mention of loom taxes on Kaikolars anywhere comes from the thirteenth century.
The thirteenth century was a major period of agrarian development in Salem, and
according to an oral tradition, now written, the Cholun Purva Pattayam, various
groups including, Vellalar and Kaikolar, settled in the region at this time
(Ramaswamy, 1982, pp.47-62). A few inscriptions also document the history of
the Kaikolar in Salem in the late medieval and early modern periods. They are
especially abundant in the early sixteenth century, during the reigns of the
Vijayanagara rulers, Krishna Deva Raya (1509-1529) and Achyuta Raya (1529-
1542). One at Aragalur (Annual Report on South Indian Epigraphy, Inscription
409 of 1913) and another at Tiruchengodu (Annual Report of South Indian
Epigraphy, Inscription 14 of 1915-1916) tell of the remission of taxes on looms.
The inscription at Aragalur relates that to encourage the rehabilitation of deserted
villages all taxes were to be reduced and that the new rate on looms was to be
three *panam*. The inscription at Tiruchengodu, located in Kaikolar Street, also
reduces the loom tax to three *panam*. Another at Tiruchengodu (Annual Report on
South Indian Epigraphy, Inscription 622 of 1922) refers to the founding of a new
village called Samasamudram, named after Sama Nayanar, the agent of
Triyambaka Udaiyar, who was governing at Mulavay. Kaikolars, Chettis, and
other artisans settling there were to be exempt from taxes in the first two years and
were to pay three *panam* thereafter. During Achyuta Raya's reign (Annual Report
on South Indian Epigraphy, Inscription 452 of 1913-1914) at Aragalur, a weaver
was given land, a loom, and certain ritual privileges for instituting the car festival
at the Karivaradarajan Perumal temple. Also, during Achyuta Raya's reign the
major trading guild of the time, the Ayyavole, made a contribution of raw cotton,
thread, *saria*, *chintz*, and *sela* cloth to the temple in Taramangalam (SII, South
Indian Inscriptions, Vol. VII, Inscription No. 21). From this evidence I conclude
that by the early sixteenth century weaving was well established in Salem, and that
the Kaikolar were pre-eminent, with Devangar and Saliyar playing lesser roles.

## 5.2  Social Differentiation

Most weavers in Salem in the late eighteenth century were relatively poor.
This was demonstrated in the 1790's when the English East India Company forced
weavers to work directly for company agents and were unwilling to pay prices that
exceeded the costs of production ("Imposts", The Baramahal Records, pp.16-17).
This created great hardship among most of the weavers, who were forced deeper
into debt. It is also possible to conclude that most weavers only had one loom. As

weaving was a full-time occupation restricted to male members of families, and as the numbers of male weavers and the numbers of looms were basically in balance everywhere in Salem we can conclude that most production came from single loom family operations.

However, some weavers had up to ten looms ("Imposts", The Baramahal Records, 1920, p.26), and there was a small group of wealthier weavers. Other evidence tells us that most weavers worked at their trade, "but there is a numerous class of head weavers, chettis, who are less weavers, than (sic. tax) farmers and merchants, and who deal in oil and grain and every article that the country produces" ("Imposts", The Baramahal Records 1920, p.25). By the eighteenth century therefore, there was economic differentiation among weavers in Salem very similar to that described by Ramaswamy (1985b, pp.294-325) for the fifteenth and sixteenth centuries throughout South India and that described by Arasaratnam (1980, pp.257-282; 1986, pp.265-273) for coastal Coromandel in the eighteenth century. Ramaswamy tells us that during the fifteenth and sixteenth centuries master-weavers emerged everywhere in the south. Such people either supplied primary weavers with money to purchase thread and food, or had control of considerable numbers of looms, or did both. Some of these people had emerged from the ranks of ordinary weavers, but they could also belong to merchant groups involved in the cloth trade. Arasaratnam (1980) describes a variety of different categories of people involved in the industry. There were primary weavers ("cooly" weavers in the European records), and above them, and in some position of nebulous authority over them, were head-weavers, a group who had differentiated itself on the basis of wealth and economic position, and who in many places had become hereditary. However, while they exercised paternal control they had no economic control over primary weaver output. All of the evidence, therefore, points to an early and significant differentiation among weavers. Most were poor, but a few became wealthy and played crucial roles in both production and marketing.

## 5.3  Industrial Organization

The previous discussion has hinted at how textile production was organized. Most of the looms were operated by single loom families. There also were larger scale operators, with many looms and involved in trade, including the cloth trade. Thus the cloth trade was not only in the hands of merchant castes, such as the Komati and Nagarattar, but of wealthy weavers, especially Kaikolar, who took the title *chetti*, and who can be distinguished through their use of the cast honorific, *mudali*.

The fundamental fact of textile production for export in the late medieval period and early modern period in South India was that there was a vertical linkage between marketing and industrial production (Chaudhuri, 1974, p.147). Between the large export merchant located at a port and the primary weaver there was a line of intermediaries (ibid, p.151). From the fourteenth to the early sixteenth century the textile trade in South India was in the hands of merchant guilds (Ramaswamy, 1980, pp.125-126), whose membership probably excluded weavers. However, it was during this period that the wealthier weavers (the master weavers) began to market their own products and assume the title *chetti*. They appear to have operated at the local level as small-scale middlemen, who dealt with larger merchants (either from weaving or merchant castes) operating from major inland market centers or from the coastal ports. By the seventeenth century some of the latter large operators were involved in procuring cloth for the European Companies. In some places there were "brokers", *copudaru* in Telugu ("copdar" in the English records) or *careedar* in Urdu, who acted as local intermediaries. Brokers were the agents involved in negotiating between outside merchants and primary weavers. It appears that they could be hereditary "head weavers" or simply small master weavers.

Regardless of the precise local details, the market oriented system from late medieval times operated on a monetary basis. In Salem there is abundant evidence to demonstrate that this was the case ("Imposts", The Baramahal Records, 1920, pp.15-35). Merchants placed orders with weavers, through some kind of "broker", and an advance of cash was given to the weaver. This advance was made on the basis of an agreed price for the finished goods, and provided the weaver with money to buy thread and other essentials for the manufacturing process, as well as food. It is important to note that though the weaver took advances and was in debt to the merchant, the merchant had no right to expropriate the weavers product. The weaver could sell to anyone, provided he could settle his debt.

This was not, therefore, a putting-out system, in the sense that the merchant provided the weaver with thread. Weavers purchased their own thread at the periodic markets, as we have described earlier. Neither was it a truly cottage industry, with the weaver making all of his decisions independently. Merchants did control the ordering process, and although they endeavoured to keep weavers in debt, weavers could and did abscond. Head weavers stood between merchant and weaver. Merchants found it difficult to translate control over capital into control over production (Brennig, 1986, 352). But there was a strong element of social control wielded by both hereditary weaver caste leaders, master weavers, and by wealthy middlemen, usually from the same castes as the primary weavers.

## 5.4  Spatial Organization

Until recently it was generally believed by scholars that only landowning castes defined territorial units, or *nadu*, as they are called in the Tamil country. Research has now demonstrated that artisan and merchant castes, from at least the fourteenth century, began to organize their own territorial systems (Ramaswamy, 1980, 130-131; Mines, 1984). Mines (1984,76), who studied Kaikolar social organization in modern Salem, states that these socially defined territories were integrated into supra-local pyramidal hierarchies, and were originally organized to administer trade. This type of system replaced the itinerant artisan-merchant associations of early medieval times, when such organizations were incorporated into local agrarian territorial structures (Hall, 1980, 202). By Vijayanagara times the territorial system of the Kaikolar had developed to replace the earlier system for the marketing of textiles.

The Kaikolar system of the early modern period had four major territorial divisions (the *desai* or *thisai nadu*) and 72 *nadu* in total (Ramaswamy, 1980, p.130). Kanchipuram was at the apex of the system, and was known as the Mahanadu. The Salem area, today as in the past among Kaikolars, was one of the four major divisions, known as the "Seven City Territory" (Elu-ur-nadu) or the "Seven Division Nadu" (Elu-karai-nadu) (Figure 4).

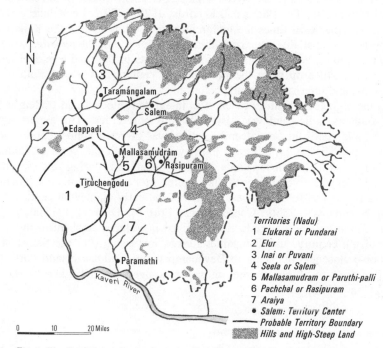

Territories (Nadu)
1 Elukarai or Pundarai
2 Elur
3 Inai or Puvani
4 Seela or Salem
5 Mallasamudram or Paruthi-palli
6 Pachchal or Rasipuram
7 Araiya
• Salem: Territory Center
——— Probable Territory Boundary
▓▓▓ Hills and High-Steep Land

0    10    20 Miles

*Fig. 4: The Traditional Seven Area (or City) Territory of the Kaikolar Weavers*

Here, as elsewhere, Kaikolar social organization provided the structure for spatial organization. In villages it centered on common ground by a Murugan temple used for stretching the warps (*pavu*, thus called *pavathi*). In the small market towns these areas were in front of the caste's Murugan temple. These temples served as the meeting places for Kaikolar local councils. A number of such councils comprised a *nadu*, each centered on a place with a major Siva temple (for example, Taramangalam, Tiruchengodu, Rasipuram). Each *nadu* had two villages (one known as the West (*mel*) and the other as the East (*kil*) whose village councils were more important than those in other villages. Today, Tiruchengodu is the place where all the nadu councils meet, and it has supreme jurisdiction, but Taramangalam may have played such a role in the past (Richards, 1918, 1, 1, p.181).

That this system is of considerable antiquity in the region is attested to by an inscription of the early sixteenth century from Taramangalam which relates how the road taxes of Elu-Karai-nadu were to pay for the upkeep of the *matt* in Chidambaram (Annual Report on South Indian Epigraphy, Inscription 21 of 1900), and a copper plate grant, dated to 1717 A.D. states that "the district dependent of Sankagiri is called the Elu-karai-nadu" (Richards, 1918, I, 2, p.282). Strikingly, there is also evidence from the late eighteenth century of a territorial unit called *paykat* or *paykit* in the Company records in Chingleput, North and South Arcot. Vennandur, the West Village of Salem Nadu is referred to as a *payakari* ("Imposts", The Baramahal Records, 1920, p.29). I conclude tentatively that from late medieval times the hierarchical council system of the Kaikolar was also a system for spatially organizing trade. In this system, the *pavathis* were local production centers, and higher levels formed points in the textile market network that brought textiles up from the producers to centers that traded cloth outward into regional and international markets.

In the eighteenth century Salem was integrated into a broad trading network within southern India, and was also linked to port cities. The export trade of southern India, until the seventeenth century was directed almost entirely toward Southeast Asia. Salem lay 20 days journey inland from the ports of Cuddalore, Porto Novo, and Devanampatnam, which developed during the seventeenth century. Merchants from the coast had agents in Periya Salem, and Salem long cloth was reputed to be of the best quality in the early eighteenth century ("Price Lists of Merchants", Public Consultations, Fort St. George, 13 March, 1737). Further, as the volume of production declined in villages adjacent to the coast in the eighteenth century, the Companies looked increasingly to the Salem market. The relative closeness of Salem to Devanampatnam, Cuddalore, and Porto Novo, gave advantage to shippers from these ports over those further to the north (Arasaratnam, 1986, p.58).

## 6.    The Indian Ocean and Beyond

In recent years the research carried out on manufacturing and trade in South India has led to a conceptual deconstruction of earlier interpretations of both the European entry into the area and the kind of society and economy which the Europeans encountered. Late medieval and early modern India was subject to a fundamental process of change, of which the growth of commercial manufacturing of all types, and especially the textile industry, is but part. To summarize the basic argument, from at least the fifteenth century there appears to have been a significant acceleration of population growth and of new agrarian settlements. Extensive new tracts of country came to be populated by settled village people. In South India, this was particularly true of the drier tracts, the areas Baker (1984) calls the "Plains" and "Kongu." Salem formed part of this later tract. In such areas, as Ludden (1985) has documented for Tinnevelly, crops such as cotton became important. It was also a period of accelerated urban growth at all levels from the small market town, to major urban places, such as Vijayanagar,

Seringapatnam, Madurai, Tanjore, and Jinji. Centers for worship, trade, defence, and manufacturing, towns grew up along major routes everywhere, and especially in the drier tracts.

Politically, South India between A.D. 1336 and A.D. 1642 was under the overlordship of Vijayanagar. However, this rule was tenuous at times, and especially after the disastrous defeat of Vijayanagar in A.D. 1565, much of the south was ruled by great and lesser chieftains. This was true in Salem, which through the sixteenth and seventeenth centuries was divided between three chiefly families. In the late seventeenth century these families were destroyed by the expansion of the Udaiyar family of Mysore, who established a period of stable rule which lasted until the middle of the eighteenth century, at which time Haidar Ali, an opportune adventurer, effectively assumed rule in Mysore. Haidar's son, Tipu Sultan succeeded him. Stein (1989) has characterized all of these regimes as patrimonial, defined as ones in which access to core authoritative roles and offices depended on personal relations to rulers. There were at least two levels of authority extent in these patrimonial regimes, state level sultans and local lordships. The first were formed around control of large scale military organizations and the second around control of communal institutions.

From about A.D. 1500, with the introduction of new military technologies, money was required to pay fulltime and trained soldiers. This was achieved throughout South Asia by demanding cash payments instead of military contingents form subordinate lordships, by the creation of revenue bureaucracies, by serious attention to foreign trade, and to currency and money management, including minting operations and the supply of metal for coins (gold, silver, copper). These processes have been labelled "military fiscalism", but more comprehensively they are manifestations of "mercantilism" (Stein, 1989, p.13). All of the evidence now suggests that the growth of military fiscalism, in conjunction with increasing commerce, created a level of monetization that penetrated deep into the countryside. At the same time we see the appearance of relatively sophisticated monetary and financial institutions, which permitted numerous monetary transactions to occur without the exchange of actual coin and which provided wide access to credit for commercial operations and money lending.

In summary, then, what we see happening in Salem is a local manifestation of a subcontinental wide phenomenon. The organization of manufacturing had passed under the logic of mercantile capital by about A.D. 1500, just as the first Europeans were arriving, and Perlin (1983, pp.30-95) has argued convincingly that mercantile capitalism emerged in India independently of that of Europe, but within a common international framework of social and commercial changes.

Furthermore, by the early modern period local, regional, sub-continental, inter-Asian, and inter-oceanic developments in trade and commerce can hardly be separated from one another. Internal, or trans-regional trade, within South Asia is now known to have been extremely large, not only that in luxury items, but that in foodstuffs, raw materials, and capital goods as well (Varady, 1979, 1-18; Washbrook, 1988, p.66). Additionally, growing interests in "maritime" history has led to a redrawing of the "external" dimension of early modern South Asia. The research by K. N. Chaudhuri (1985) modelled on Braudel's (1972) work on the Mediterranean World has demonstrated the importance of a unitary view of the Asian-European World from at least the eighth century A.D. This work and that on the social history of Islam has demonstrated that influences from Europe, filtered through and modified by West Asian Islamic regimes, affected the nature of South Asian society and economy (patrimonial state, military fiscalism, mercantile capitalism), from at least A.D. 1500. (Washbrook, 1988, pp.57-96; Stein, 1989, pp.1-26).

Certainly, one of the most significant trends in the writing of Indian history during the 1980s has been the attempts to link processes of socio-historical transformation within the subcontinent to processes of large-scale change operating on global, or supra-regional levels (Palat and Wallerstein, 1990). From strong manufacturing and mercantile bases in western India and southeastern India stretched tentacles of capitalist activities from the Persian Gulf to Southeast Asia. Indeed, there is much to suggest that Southeast India should be understood as part of an economic system embracing Southeast Asia. Similarly, Western India was drawn strongly into the orbit of West Asia. All of these connections, including those with Europe, brought stimulation and development as much to South Asia as to Europe or elsewhere. South Asia, therefore, was a central actor in the early modern "world economy". It had a pivotal position in international trade, and was the hub of several trading circuits. The area was responsible for a much larger share of world trade than any other comparable zone and between the sixteenth and eighteenth centuries may have possessed upwards of one-quarter of the world's total manufacturing capacity.

Prior to the eighteenth century, there would seem to have been a number of different "mini-world" or regional systems, bearing many of the characteristics (such as core-periphery relations) which Wallerstein (1974; 1980; 1984; 1986; 1988) attributes to the European-centered system, but either having little to do with Europe or treating Europe as an essentially peripheral zone. For example, in looking at the effects of the cloth trade between South Asia and Europe in the seventeenth century, it would be difficult to decide what is core and what is periphery. South Asian exports threatened to "de-industrialize" European textile industries to the point at which state action, in the form of tariff barriers, had to be taken against them. Basically, before the eighteenth century Europe represented no more than one of several trading systems feeding into various South Asian regional economies. Growth in these regional economies actually facilitated the establishment of the preconditions for the development of a system of international exchanges and dependencies in which Europe established an increasing hegemony in the late eighteenth century.

However, as Palat and Wallerstein (1990) point out, despite the temporal contemporaneity of post-1500 expansion of networks of exchange and intensification of relational dependencies in Europe and in the world of the Indian Ocean, the processes of large-scale socio-historical transformation in the two historical systems were fundamentally dissimilar. In one zone, it led to the emergence of the capitalist world-economy. In the other, to an expanded petty commodity production that did not lead to a real subsumption of labor. Why the extended circuits of commodity production and exchange linked closely to the fiscal structures of patrimonial regimes, and the vital participation of merchants and moneymen in both extended exchange and state finance failed to create the conditions for extensive, self-producing capitalist relations in South Asia has now become a focus for research during the 1990s, (Stein, 1989).

## REFERENCES

Arasaratnam, S. (1978): Indian Commercial Groups and European Traders, 1600-1800: Changing Relationships in South-Eastern India. In: South Asia 1, pp. 42-53

Arasaratnam, S. (1980): Weavers, Merchants, and the Company. In: The Indian Economic and Social History Review 17, pp. 257-281

Arasaratnam, S. (1986): Merchants, Companies and Commerce on the Coromandel Coast 1650-1740. Delhi: Oxford University Press

Baker, C.J. (1984): An Indian Rural Economy 1880-1955. The Tamilnad Countryside. Delhi: Oxford University Press

Braudel, F. (1972): The Mediterranean and the Mediterranean World in the Age of Philip II., New York: Harper and Row, 2 vols

Brennig, J.J. (1975): The Textile Trade of Seventeenth-Century Northern Coromandel: A Study of a Pre-Modern Asian Export Industry. Ph.D Dissertation, University of Wisconsin at Madison

Brennig, J.J. (1986): Textile Producers and Production in Late Seventeenth Century Coromandel. In: The Indian Economic and Social History Review 23, pp. 333-355

Buchanan, F. (1807): A Journey from Madras Through the Countries of Mysore, Canara, and Malabar. London: Bulmer, 3 volumes

Chaudhuri, K.N. (1974): The Structure of the Indian Textile Industry in the Seventeenth and Eighteenth Centuries. In: The Indian Economic and Social History Review 11, pp. 127-182

Chaudhuri, K.N. (1978): The Trading World of Asia and the English East India Company, 1660-1760. Cambridge: Cambridge University Press

Chaudhuri, K.N. (1985): Trade and Civilization in the Indian Ocean: An Economic History of the Rise of Islam to 1750. Cambridge: Cambridge University Press

Gittinger, M. (1982) Master Dyers to the World. Washington D.C.: The Textile Museum

Hall, K.R. (1980) Trade and Statecraft in the Age of the Colas. New Delhi: Abhinav

Ludden, D. (1985): Peasant History in South India. Princeton: Princeton University Press

Madras. Government of Madras (from 1887): Annual Report of South Indian Epigraphy. Madras: Government Printer

Madras. Government of Madras. Public Consultations. Fort. St. George, 13 March, 1737. Madras: Government Press

Madras. Government of Madras (1897): South-Indian Inscriptions, vol. VII, Inscription No. 21. Madras, Government Press

Madras. Government of Madras. The Baramahal Records. "Products", 1912, "Inhabitants", 1918, "Imposts", 1920, "Police", 1923. Madras: Government Press

Mines, M. (1984): The Warrior Merchants: Textiles, Trade, and Territory in South India. Cambridge: Cambridge University Press

Palat, R.A. and Wallerstein, I. (1990): Of what World System was Pre-1500 'India' a Part? Paper presented at the International Colloquium on "Merchants, Companies and Trade," Paris

Perlin, F. (1980): Pre-Colonial South Asia and Western Penetration in the Seventeenth to
    Nineteenth Centuries: A Problem of Epistemological Status. In: Review 4, pp.267-306
Perlin, F. (1983): Proto-Industrialization and Pre-Colonial South Asia. In: Past and Present 98,
    pp.30-95
Ramaswamy, V. (1980) Some Enquiries into the Condition of Weavers in Medieval South India.
    In: The Indian Historical Review 6, pp.119-140
Ramaswamy, V. (1980): Weaver Folk Traditions as a Source of History. In: The Indian Economic
    and Social History Review 19, pp.47-62
Ramaswamy, V. (1985a): Textiles and Weavers in Medieval South India. Delhi: Oxford University
    Press
Ramaswamy, V. (1985b): The Genesis and Historical Role of the Masterweavers in South Indian
    Textile Production. In: Journal of the Economic and Social History of the Orient 28,
    pp.294-325
Raychaudhuri, T. (1962): Jan Company in Coromandel, 1605-1690. A Study in the Interrelation of
    European Commerce and Traditional Economies. S'Gravenhage: Martinus Nijhoff
Richards, F.J. (1918): Salem, Madras District Gazetteers. Vol. 1, Part 1, Vol. 1, Part 2. Madras:
    Government Press
Stein, B. (1989): Eighteenth Century India: Another View. In: Studies in History. N.S. 5, pp.1-26
Subrahmanyam, S. (1986): Trade and the Regional Economy of South India, c. 1550-1650. Ph D,
    Dissertation, University of Delhi
Subrahmanyam, S. (1990): The Political Economy of Commerce: Southern India, 1500-1650.
    Cambridge: Cambridge University Press.
Varady, R.G. (1979): North Indian Banjaras: Their Evolution as Transporters. In: South Asia,
    N.S. 2, pp. 1-18
Wallerstein, I. (1974): The Modern World-System, I, Capitalist Agriculture and the Origins of the
    European World-Economy in the Sixteenth Century. New York: Academic Press
Wallerstein, I. (1980): The Modern World-System, II, Mercantilism and the Consolidation of the
    European World-Economy, 1600-1750. New York: Academic Press
Wallerstein, I. (1984): The Politics of the World-Economy: The States, the Movements and the
    Civilization. Cambridge: Cambridge University Press
Wallerstein, I. (1986): Incorporation of Indian Subcontinent into Capitalist World-Economy. In:
    Economic and Political Weekly, 21, "Review of Political Economy," January 25
Wallerstein, I. (1988): The Modern World System, III, The Second Era of the Great Expansion of
    the Capitalist World Economy, 1730s-1840s. New York: Academic Press
Washbrook, D.A. (1988): Progress and Problems: South Asian Economic and Social History.
    c.1720-1860. In: Modern Asian Studies 22, pp.57-96

# 25

# The Changing Role of Port Cities as Part of the European Expansion in the Southeast Asian Archipelago: Outlining an Approach

Wolfram Jäckel

## 1 Introduction

In a recent article in *Asian Business* the following can be read: "Infrastructure: Key to the '90s. Asia prepares for a massive - and expensive - refit" (June, 1989:16). There it is argued, that *infrastructure* is not something which just influences a region's economic development, but which rather plays a key-function in this very process. In an economic environment of advanced capitalism, revealing a globalization of capital accumulation and more or less efficient mechanisms of national and supra-national regulation, nobody will be particularly surprised when this "refit" is successfully realized in the end.

On the other hand, looking back on 17th century Southeast Asia a "massive - and expensive - refit" of the infrastructural system can be observed to have been undertaken. The region then saw new lines of communication, new port settlements, fortifications and garrisons being placed into its centuries-old international trading world. It is the Dutch *United East India Company*, the VOC (*Verenigde Oost-Indische Compagnie*), which seems to be of particular importance in this respect, because investment in infrastructure only then became part of a certain project called *monopolizing trade*. What was the function of infrastructure within this project? Was the strategy of monopolizing trade in effect underpined by a concomitant "strategy of infrastructural development"? Alternatively, was the investment in infrastructure, at least at some time, even counter-productive? Which criteria could be employed to assess the impact of infrastructural investments on the VOC-history? Was infrastructure if not *the*, then maybe *one* "key to the future" at that time? Being aware of not dealing with developed nation-states, what were the institutional pre-conditions for the infrastructural development Southeast Asia experienced in those days? Such questions should guide any research which attempts to understand the dimension and underlying causes of the region's infrastructural readjustment and its impact on the European expansion in this part of the world.

The term *infrastructure* has a wide range of connotations of which the material side will be stressed here. *Infrastructure* thus means buildings, communication networks of every kind etc. which have the property of durability or, in economic terms, a long period of invested capital remuneration. Therefore, it is nowadays usually the state that invests in infrastructure, hoping that these investments will be payed back in the long run in terms of economic growth. Moreover, at least in its modern meaning, infrastructure usually represents a kind of a collective good.

Returning to the present context of 17th century Southeast Asia, *infrastructure* takes on special meanings, because nation-states which could provide infrastructural investments did not exist yet, and economic growth was not a collectively defined end to which action could be adhered to. In order to keep things simple, and to find a starting point, this paper will focus on one element of the material infrastructure which was of some significance in the context of European expansion in the maritime world of Southeast Asia, the port city. It was these nodes of trading networks, the *emporia*, which gradually became the units from which huge economic systems were managed. Thus, something which is essentially intertwined with the opening phase of the *modern world economy*.

The literature on trade in South and Southeast Asia is quite expansive. But it is only very recently that the function of port settlements has been directly addressed by a number of scholars. Adding to the classical studies of Boxer (1965), Glamann (1958), van Leur (1955) or Meilink-Roelofsz (1962), authors like Arasaratnam (1984), Blussé (1988), Broeze (1989), Kathirithamby-Wells (1986b, 1990), Villiers (1990), Manguin (1988), Pearson (1988), Subrahmanyam (1988), Sutherland (1989) and many others are now particularly concerned with the port city. Its function, and its internal ethnic, social, political and physical structure are now scrutinized in great detail.

Instead of making an empirical contribution to these important and enlightened studies this paper rather aims at developing some ideas on how to put the port city into perspective. Beginning in the 17th century, the main actor on the Southeast Asian trading scene was the VOC. Under its influence new and old port settlements became "company towns", both functionally, with respect to a global network, and internally with respect to the physical and social structure (Blussé 1988, Sutherland 1989, Taylor 1983). Looking on these settlements from the Company's perspective they were, or should have been, made *locational* problems: Where and why capital should be invested in the material infrastructure of a particular place? How much and at what time should it be invested? How could the investments' returns be measured? Which effects were expected? Which effects, both predicted and unexpected, actually occurred?

The port city was the target of heavy investment during the 17th century and that is now what makes it a valuable target for research. This is especially true when an attempt is made to uncover the decision making processes which, finally, resulted in a particular network structure of emporia. If investigations proceed further and look on the port city, and its development, as being instrumental to the overall expansion process, then attention is inevitably turned to the *organizational level* of those main actors who inscribed their projects into the Southeast Asian trade history during the 17th and early 18th century. As is well-known, these actors were the big trading companies, foremost the VOC itself. The VOC investment policy, therefore, seems to be an appropriate first step in researching the impact infrastructure has had on the course of the European expansion in the Southeast Asian archipelago.

The question of infrastructural development should be embedded within a comprehensive approach. At least *three analytical levels* could be identified: Firstly, the concrete materialization of ideas, plans, strategies are faced when there is concern only with infrastructural development. There are two dimensions connected with this: Besides having real effects on the process (effects which have to be studied), infrastructure also contains an epistemological component, because it provides a unique kind of source material, which despite of the large number of detailed studies on port settlements, could perhaps be more comprehensively used for analysing the expansion process. Secondly, as mentioned above, the measures taken on the level of the concrete settlement raise questions about the institutional setting lying behind those measures; it is the decision making process with respect to investment behaviour which is of a particular importance. Thirdly, since there is a difference between a plan, its realization and its effects on the given socio-economic environment, the region itself, its response to extra-regional claims and opportunities brought in by the new powers have to be systematically included within an analytical framework. The functional description of Southeast Asia as *external arena* in the sense introduced by Wallerstein (1974, 1980) seems to be a promising starting point to combine the two mentioned analytical levels - the concrete location-centered and the organizational level (which delivers the *relevant description* of the main actors) - with a broader spatial and historical perspective on the socio-economic development of the maritime world of Southeast Asia in the period of European expansion.

In the remaining pages of this article the aforementioned research framework will be presented in greater detail. However it should be added, that research on this matter is just beginning and that data collection has not yet taken place. What is presented here, therefore, is nothing more than the theoretical view of the problem. Its empirical productivity still waits to be demonstrated.

## 2.   Trade and Infrastructure

This section begins with a passage from Meilink-Roelofsz' classical study on trade in the Indonesian archipelago. She writes:

> Unlike the Dutch ... the British did not have to reckon with the heavy cost of military actions, punitive expeditions against the natives, and *the upkeep of forts*, ships and *garrisons*. They had to pay a much higher purchase price for the cloves in Macassar and Bantam, of course, yet they, too, were able to make high profits (1962 pp.203; emphasis added).

Obviously, a fundamental difference existed between two modes of trade organization: On the one hand a method involving a heavy burden of *overhead costs*, the "Dutch method"; on the other hand the British (and other European and Asian trading powers) who capitalized on a given logistical infrastructure, perhaps adding one or the other element to it but not changing it fundamentally.

At this point a second statement concerning infrastructure should be quoted. This time it is from one of the leading contemporary historians on Indonesia, M.C.Ricklefs, who raises a further point when he speaks about a difference between the retreating Portuguese and the expanding Dutch power in the 17th century:

> "... the Dutch did one thing the Portuguese had not done: they established a *permanent foothold* in Java. This was to make their involvement fundamentally different from the Portuguese, and was to lead ultimately to the Dutch becoming a *land-based colonial power* in Java (1981, p.23; emphasis added)."

Ricklefs stresses two points: The effectiveness of establishing permanent footholds in order to gain a leading position in trade *and* the *long-term result* of such action which finally ended in becoming a "land-based power". To be accurate, the second point is not the necessary outcome of the first one. Rather it asks for an explanation in its own. There is a connection between both but not a simple prolongation like 'permanent footholds first, becoming a land-based power second'.

Taking both cited statements together the following rough and tentative *classification* can be derived concerning the interrelation of trade and infrastructure:

1. Connecting to, and utilizing of, the pre-existing infrastructure of the regional trading world,

2. reorganization of the trade infrastructure, in order to support a certain strategy of trade organization,

3. becoming a land-based power with growing political involvement in internal affairs including a growing portion of non-trade infrastructural activities.

The distinction which will be stressed in this paper is the difference between the first and second type of the interrelationship between infrastructure and trade. The second level was reached, or implemented, by the VOC during the 17th century, the period when the archipelago still may be classified as *external arena*. It is felt that there might be a connection between the transgression from the second to the third level of this classification on the one side and the transformation of the region from *external arena* to the *periphery* of the emerging *world system* on the other. But this is still not more than a passing remark which will be returned to at the end of this article.

Focussing on the first two levels of the above classification the regional dynamics may be approached by referring to a pair of concepts which A.Lipietz (1980) uses for analyzing the *structuration of space* under conditions of a developed capitalist world. He speaks of *inherited* and *projected space*. Within the context of the structural Marxist inspired regulation theory, he assigns the state the intervening regulatory task to organize "... the substitution of projected space for present space...", when the "... 'projected space' comes into more or less violent conflict with 'inherited space'" (p.74). Territorial development in his view is a matter of "redeployment of ... social relationships" against constraints imposed by the "already existing concrete space" (p.61).

If the heavy burden of structuralist regulation theory is, for the moment, cut back, both terms, *inherited* and *projected space*, could be incorporated into the present context of a changing 17th century maritime trading world. If focussing on the economic space, a particular spatial division of labour, of resources, of commodities, and of trading networks and trading points is to be identified. The observed massive restructuring of trade was a redeployment of given social relationships in the broadest sense. It had to work against *well-developed* trading networks as well as political, legal and cultural structures (which is rightly captured within the *external arena* concept).[1] For centuries the region had far-reaching trading contacts with many parts of the world. To be sure, compared with the early trading nations of that region (Indians, Chinese, Persians and others), which more or less unconsciously produced an inheritance of an economic space, the Portuguese were the first who tried to implement a comprehensive and systematic project on this space beginning in the early 16th century. But in great difference to the later Dutch project the Portuguese aimed not at mere economical ends, but rather it was the crown and the church which provided for the ultimate rationale of Portuguese action. This was quite the opposite with the Dutch. For the first time an economic space being created solely *as* an economic space can be seen. Principles of capital accumulation, and the yield from investment activities were the ultimate action-guiding formal rationale. A mapping of this difference in motivation onto the physical infrastructure is to be found in the following quotation:

> "The Dutch in the East were not, on the whole, such great builders of castles
> and fortifications as were their Portuguese predecessors, whose strongholds
> they usually reduced in size once they had captured them, with the view to
> *economizing on the garrison* and cannon necessary to defend their walls
> (Boxer 1965, pp.206-207; emphasis added)."

The Dutch started creating a modern "operational space" of global dimension which was based on an economic rationale. This, therefore, allows for the first time the development of a theoretical understanding of a concrete space's principles of creation: As far as these were economical rational they can be modelled with well-known tools of economic theory (or subsets of these tools). For any such theory, however, the actual investment activities in material infrastructure could serve as one of the major empirical testbeds.

To illuminate the basic idea of the proposed approach a closer look should be taken at what characterized the two first levels of infrastructural and economic access to the region.

## 3.    The Inheritance: Trade in the Archipelago before the VOC

From works of authors such as Bronson (1977), K.R.Hall (1985), van Leur (1955), Reid (1980), Wheatley (1983), Whitmore (1977), Wisseman (1977), Wolters (1967) and many others who used archaeological and (Chinese, Javanese) written source material many details are known about the early trade organization in the archipelago. K.R.Hall, for instance, tried to develop a systematic view on the trade systems of pre-16th century Southeast Asia and investigated their impact on early state formation. One type of trade system he calls the *riverine network system* (see also Bronson 1977). The legendary empire of *Srivijaya* (c.600-1200) with its center in South Sumatra serves as its paradigmatic illustration.

The importance of Srivijaya was based primarily on its intermediate location as a "guardian" between the trading poles China and India (and further west Persia and the Arab world). Spatially, a network of coastal and riverine centers was the system's characteristic feature. The underlying political structures, interconnecting chiefdoms and the paramount ruler, were based on mutual interest and interdependency as well as on mutual validation of status; relations of tribute and protection gave substance to this system.[2] Owing to political and economic reasons the trading points' locations were often shifted. The logistical infrastructure of trade, therefore, was of a most *flexible* type, even if the trading activity was backed up by an elaborated political system.[3]

Reid (1980) gives a vivid description of the early Southeast Asian city in connection with its trading point function. Remembering that even for the 16th and 17th centuries a total population of about 8 million people for the whole of Southeast Asia and a density of 3.7 persons per square kilometre is estimated (Reid, 1980, p.239), there was, of course, enough space and enough jungle, in which to retreat in times of hostilities and warfare. It was not the buildings, the physical infrastructure and the productive land which constituted the critical medium of accumulation and the basis of power: "The really important resource of the rich and powerful was their manpower, whether we call them slaves, bondsmen, clients, or subjects" (ibid., p.243). Houses could be rebuilt, land could be recultivated in short time. Therefore, the strategy of *spatial flexibility* was quite adequate.

The European powers could, like the Chinese, Indians, Persians and Arabs, capitalize on this kind of logistical infrastructure. They could realize external savings on transaction costs (i.e. costs of getting information about trade goods and price-differences, costs of assuring contracts) when utilizing this ready-made commercial network. Even if they added one or the other single stronghold to the overall pattern like the Portuguese who took Melaka in 1511 and fortified it - which caused other groups of merchants to shift their centre from Melaka to Makassar on Sulawesi - they still relied on this kind of network-structure.

Though European stone-built strongholds had some impact on the regional infrastructural setting in a way that the flexibility was successively given up (especially the early Portuguese fortifications at Melaka and Maluku exemplified how power and infrastructural equipment could be effectively interrelated (Reid 1980, p.246)), the local rulers, nevertheless, relied on a wide range of merchant groups living and trading in their places. To the latter, in turn, these places and the authority of the rulers were part of the protective and organizational infrastructure. On the other hand, these new infrastructural developments also made the indigenous port cities a target for the blockade-technique which was very effective because they usually were dependent on rice supply from the exterior (ibid., p.247).

The most lively center of Southeast Asian trade from the early 15th until the 17th centuries was *Melaka*. It was situated in a strategic position. It is Braudel where the following can be found:

> "The concentrated gratis energy generated from the monsoon spontaneously organized the sailing ships' journeys, and the *meeting* of the merchants, with a degree of certainty which is usually absent in sea-transport at that time (1986a, p.125; author's translation)."

Fig. 1 shows Melaka's geographical location as really being a kind of "natural centre" given the trade-flows, the seasonally changing winds and the stage of sailing technique.[4]

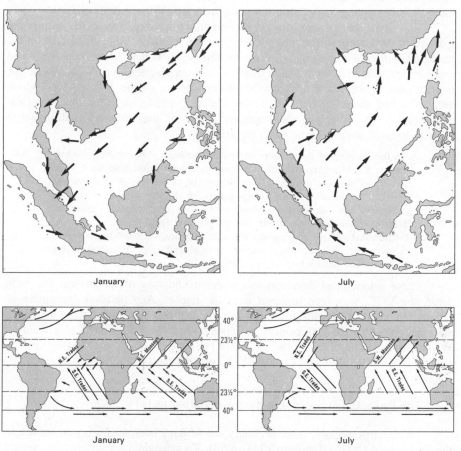

*Fig.1: Prevailing winds in January and July*
Sources: DALE (1953), BRUIJN et al. (1987)

Because waiting for the proper winds was a must, long journeys like from India to the Banda Islands in the eastern part of the archipelago, or from India to China were undertaken only rarely. Usually it was a *chain-like trading pattern* in which the trading points, especially Melaka, played the part of interconnecting different trading networks (Reid 1980, p.237). When, later, the Dutch came into the region their fleets, which were using the "roaring forties", were exactly scheduled with respect to this monsoon dependent regional trading pattern (Boxer 1965, pp.197-198; also Bruijn et al. 1987).

From a very early stage these nodes became true international places, where internationality was really a phenomenon of co-presence with a wide range of nations. The beginning of the European contact with the Southeast Asian *external arena* was essentially furthered by an infrastructural setting which was characterized by its openness to foreign traders instead of being instrumental to monopolize trade, and the exclusion of single trader communities or trading networks from taking their chance (see, for instance, Villiers 1990, p.154).

The small indigenous *port-states* emerging during the 17th century in reaction to the European growing influence and dominance, were still based on the traditional entrepôt function, but were combining it with a tighter relationship to some production area as its hinterland. For example, this has been shown for Banten, one of the major indigenous trading centers of the 17th century which eventually had to be handed over to the Dutch in 1684. As in his other articles, Kathirithamby-Wells grasps the changing nature of the inherited trading world from an internal point of view when he writes:

> "In the tradition of the seventeenth century maritime states of Aceh, Johor, Ternate and Makassar, Banten's pre-eminence did not derive from the relative security of regional trade as known during the earlier age of Srivijaya and Melaka. The foremost task of the new generation of port-states was to keep the indigenous network of trade alive in the face of European challenges. Moreover, the maritime capital no longer functioned exclusively as an international entrepôt or a regional commercial centre, but found it essential to combine this with the complementary role of cash-crop production in the hinterland and surrounding regions (1990, p.120)."

These autonomous developments stood in the way of the project the main actor, the VOC, was keen to cover over the region. Any detailed description of the internal con- and reconfiguration of group interests and power relations which were so typical for the 17th century archipelago must be left aside (Kathirithamby-Wells 1986a, 1986b, 1987, 1990, Ricklefs 1981, pp.29-46). Instead attention will now be turned to a closer look on the VOC-project itself.

## 4.    The Project: Forging a one-purpose trading world

"Without monopoly, there probably would have been no European empire in the East before 1800" (Hamilton 1948, p.53). To substantiate this statement would be somewhat difficult, but, nevertheless, it points to the fact that market conditions were of prime importance for the European expansion. On the other hand, these conditions alone surely will not explain the whole process. What was of even a greater importance was the *institutional innovation* brought on stage by the VOC. It was the 1602 founded Dutch United East India Company which made the archipelago's infrastructural setting a subset of an overall commercial strategy. Again, instead of going through single historical events the basic principles should be elucidated which led to the functional refitting of infrastructure during the 17th century.

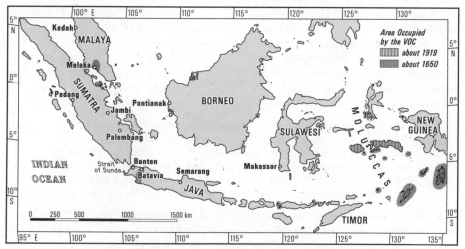

*Fig.2: Area occupied by the VOC around 1619 and 1650*
Source:*MEYN et al.(1987)*

From Fig. 2 it is evident that besides the small spice islands of the Moluccas in the eastern part of the archipelago, there were only a few centres like Batavia (the hub of the system) and later on Melaka, Makassar and some smaller secondary trading points under the authority of the Dutch. Through this regional network of forts and factories the archipelago's trade activities were controlled and the trading goods from and to Europe, the Indian subcontinent and East Asia were channeled through and connected to a globally extending network of Dutch footholds. What was the institutional background to this spatial pattern?

A huge body of literature on the VOC indicates that a quite distinct kind of entrepreneurial institution is being dealt with. Perhaps most convincingly this has been demonstrated by N. Steensgaard who even speaks of an "Asian trade revolution" which was pioneered by company trade. Given the non-transparency of the overseas markets he stresses two innovations which were made by the VOC: First of all, *protection costs* were internalized thereby making protection foreseeable and a matter of mercantile calculation. This, secondly, was part and parcel of aiming at, and maintaining of, a quasi-monopoly position which generated *market transparency* for the company and supported a *long-run-calculus*: "The Companies could safeguard themselves not only against the unpredictable protection costs, but also against the unforeseeable price fluctuations that the pedlar had to take into account in his calculation" (Steensgaard 1974, p.152).

The *peddling trade*, so vividly described by the doyen of modern studies on Indonesian trade, van Leur (1955), did not come to a standstill when the VOC began to operate. However, the VOC revolutionized the trading world and it was in this respect that it was also quite different to, for instance, the English East Indian Company as far as the first half of the 17th century is concerned. About 60 years ago E. Heckscher, the Swedish historical economist, pointed out, that the essential feature of the VOC lies in its nature as a *joint-stock company*. This type of enterprise guaranteed a *permanent stock of capital* and enabled a long-run investment policy independent from personal preferences. Trading partners (shareholders), for instance, could not demand to be paid out after certain ventures were finished as was ususal with its institutional predecessors (like the *vóórcompagnieën* or the English *regulated company*, Heckscher 1932, p.347)[5]:

> "The active partners of the *vóórcompagnieën*, the *bewindhebbers*, established themselves as a managerial group, and with the active support of the authorities of the republic, they could ignore the expectations and protests of the ordinary participants. The sums they had forwarded for a ten year investment were appropriated into a permanent, anonymous capital in the second and third decade of the 17th century (Steensgaard 1981, p.249).

The managers' interests and those of the government went hand-in-hand, but, following Heckscher (pp.347-348) again, it was even under violation of the charter that the distribution of dividends was retained by the management advancing the burden of a high share of fixed capital (i.e. the built environment, the ships etc.) as an argument. Adding to this the moderate governmental fiscal policy on the Company's activities all necessary pre-conditions were especially fulfilled for the VOC to develop:

> "... an attitude towards investments that differed from that of the traditional merchant entrepreneur. The speedy return of capital was no longer a major strategic aim. The companies like the Portuguese Crown, could invest in conquests and fortresses, and they did so to a certain extent. But their most important investments were in their inventories and their organization (ibid., p.254)."

For sure, a merchant enterprise is still dealt with, for whom the development of "backward people" was not a task and target but rather the search for profit. What is of interest in the present context is that the VOC *could* invest in physical infrastructure (that its investment behaviour actually differed in a significant way from that of the Portuguese was already mentioned above). With reference to the "necessities of these long distance trade relations" (trade with Africa, Asia and America) P.W. Klein, an expert on VOC-history, points out, that all the personal and material infrastructural investments were "essential but expensive requirements, that had to be subjected to *long term management* and control" and he continues: "It is evident that no single, arbitrary group of individuals would have been *able* or *willing* to engage in this comprehensive, hazardous and difficult business, without guarantee of security and profitability" (1981, p.24; emphases added).

In summary, therefore, a twofold distinction may be detected between the ordinary merchant and the company: The first difference concerns the institutional, or social, pre-conditions as there was the amount and the anonymity of capital, the management problem etc., in short, the *ability* to do something. The second difference refers to the pattern of expenditure, a problem of the *will* to invest in one or the other object. And it is in this respect that the analogy of two "hearts" beating in the VOC's chest may be used, something which Steensgaard describes as "a tension between a traditional and an inventory point of view" (1981, p.250): The merchant's *short term* against the manager's and government's *long-term* economical goals.

Nevertheless, the in-built long run perspective of corporate capital accumulation, the predictability of trade volume and prices as well as the long-run profit calculation were necessary for approaching the infrastructural refit in the Asian trading world and were essentially backed-up by this new infrastructural environment. It was Jan P. Coen, "the founder of the Dutch empire of the east Indies" (D.G.E.Hall 1981, p.336), who, on these grounds, persistently, and with murder and cruelty, implemented the appropriate plan:

> "According to his plans, Batavia was to be the centre of a great commercial empire based upon complete control of the sea. He did not envisage any wide extension of territorial power and was not interested in the political affairs of the interior of Java. The territories which, in his view, the V.O.C. should have in actual possession were small islands such as Amboina and the Bandas. The remainder of the empire should consist of strongly fortified trading settlements closely linked and protected by invincible sea-power. Nor would it be confined to Indonesia: its forts and trading stations should be far-flung over the whole of the East (ibid.)."

The pre-conditions and rationale of this plan, quoted at length from Hall, should be quite clear by now (see also Steensgaard 1974, pp.406-407). However there was also quite an objective check to this plan: "His [Coen's] warlike measures vastly increased the Company's expenses..."(ibid.)! Establishing permanent footholds and the more or less continuing warfare with actual and potential rivals (Melaka, Makassar, Banten...) was a costly undertaking which could be compensated only in the long run and, probably, never to the whole. That there *should* be a return from these investments follows from the incorporation of protection costs into the Company's cost-benefit-structure (Steensgaard's thesis). On the other hand, even if a comprehensive strategy existed and even if protection costs were part of the company action (and not the action of the king or the state) which led to their efficient allocation, this kind of investment could hardly co-exist with the mercantile part of the Company's nature. This is due to the fact that it fixed a substantial part of capital thereby slowing down its average turn-over time and hence its ability to generate profit. This tendency of reduction of the return to investment had to be checked and could only be accepted (on pure economical grounds), if today's postponed returns could have been balanced or even surpassed by expected future returns. The possibility to make "waiting" a rational choice implied a certain degree of independence from competitive market forces, or stated positively: It implied the possibility of *strategic action*. The realization of this possibility is given by a monopolistic, or at least an oligopolistic, market position as was actually held by the VOC on both sides of the market: The side of supply with Asian products and the side of selling them on the European market.[6]

Following these lines of the argument it is possible to see how the development of the port city became interconnected with that what occurred on the European scene. The internal physical and social structures of the archipelago's emporia, their growth and contraction, were effects of distant events as filtered through the economical rationale of the main actors, the companies. Adopting, now, a methodological perspective, it follows that if the ups and downs of the built environment can be related not only to the ever changing market conditions but also to the VOC's internal structural tension between the merchant's and the manager's interests, then some new insights into the archipelago's port city history, as well as into the ambivalent nature of the institution *company* itself could be achieved. At this point an open question is touched upon which calls for thorough investigation. However, what is evident so far is that an adequate assessment of the changing functional role of the port city, as it was incorporated into a certain project, obviously has to focus on those internal structures which steared the process of fixed capital formation.

Concrete knowledge on the investment behaviour is scanty. What, for instance, were the deeper reasons for the wasteful investment into "the castle with the golden walls" at "Negapatam (Coromandel), which cost the Company about a million and a half guilders"? (Boxer 1965, p.207) A counter-example:

> "... the Directors in the Netherlands, who strove for a maximum return on their investment, observed with distress the capital-consuming character of their headquarters in Asia ... By 1756 the Gentlemen XVII [the directorial board of the VOC] felt that most of the profits from the China trade were disappearing into the upkeep of Batavia. They decided to deal with the China trade themselves and left it to the Batavian High Government to solve the town's economic woes in an Asian trading world that was being transformed into a more open international market... and this in turn speeded up the transformation of Batavia from its position as a maritime trade entrepôt into the position of capital to a territorial hinterland (Blussé 1988, p.154)."

The reported decision on Batavia is all the more interesting as beforehand it was exactly the efficiency of the Chinese junk trade centred on Batavia which influenced the VOC-management's decision to stay away from directly trading with China at an earlier stage (ibid.). That, in general, a cost-benefit consideration was undertaken, should not be argued about. However, whether the actual resulting infrastructural investment activities were always in accordance with such considerations is another matter.

What is observed in maritime Southeast Asia when the infrastructural investment activities are viewed is not just a different "method" of trading, a method containing a greater share of overhead costs for example. The significance that method had for the course of the expansion process is that it led to an increasing *immobilization* of future activities caused by the spatial distribution of past activities. It is not that the decision makers were so irrational that they deliberately dug their own grave. It rather seems to be the consequence of a formal rationality working in that way that growing portions of capital had to be implanted into spatially fixed infrastructure in order to support the monopoly position. The monopoly strategy is the ultimate mechanism which accounts for the spatial allocation of infrastructural investment. It follows that as far as this strategy was undermined - as it actually was the case when trade flows changed - the capital invested in the physical network was depreciated, too.

It is Harvey (1982) who gives a convincing analysis of this process under conditions of capitalist accumulation. "The built environment", he writes, "has to be regarded, then, as a geographically ordered, complex, composite commodity" (p.233), a commodity, which reveals some special problems of economic depreciation, "... because the different elements have different physical lifetimes and wear out at different rates" (p.234). The allocation of the company capital was led by the monopolizing strategy, though the allocation principles still remain to be extracted from the historical source material. Hence viewing the Asian footholds as a kind of "productive commodity" follows not only from the simple fact that the Company was a commercial enterprise, it is substantiated, too, by empirical evidence like that quoted, for instance, from Blussé. The consumption of capital for "productive" trade purposes gradually went over to reproductive "upkeep" consumption needs thereby reducing the mercantile use-value of the footholds.

The transformation from the "British" to the "Dutch method" may be formally described by referring to a concept which, though only in passing, is used by Braudel. When he wonders how it came that single European city centred economies were capable of managing huge empires he sees a part of the explanation in the fact that the world economy itself was still a *weak network* which, in case of getting ripped up, could have been repaired without causing greater problems (1986b, p.93). The transformation of the infrastructural pattern, then, may be described as something which took place along a continuum starting with a weak network structure and ending in a structure which accordingly may be called a *rigid network*. Therefore, the process of European expansion, in terms of the subset *infrastructure*, may be described as a process of growing rigidity of the trade network; a growing "rigidization" as it was caused by fixed capital formation in "concrete space". This process may be grasped empirically by concentrating on the nodes, the port cities, and their potential to attract capital or, on the other hand, by looking at the connecting flows of goods, money or information (i.e. the command-structure), taking them as indicators of formalization and rigidization. A short illustration of the first type of description may be added.

Pertaining to the port city as a part of a complex control-structure H.Sutherland (1989) gives a detailed description of the economy and social life in one of the more important indigenous emporia of the archipelago, the town of Makassar (South Sulawesi). As a major competing centre Makassar eventually was seized by the Dutch in 1667 and was made a *company town*. Sutherland not only gives a detailed picture of the different trading nations centered on Makassar but she also accentuates the function of ethnical segregation which, to her, besides being the outcome of a particular idea of natural order, directly aimed at controlling the different ethnical groups according to their presumed loyalty. Even a map from the end of the 19th century shows more or less the same ethnical segregation she describes for 17th-century Makassar (see Fig. 3).

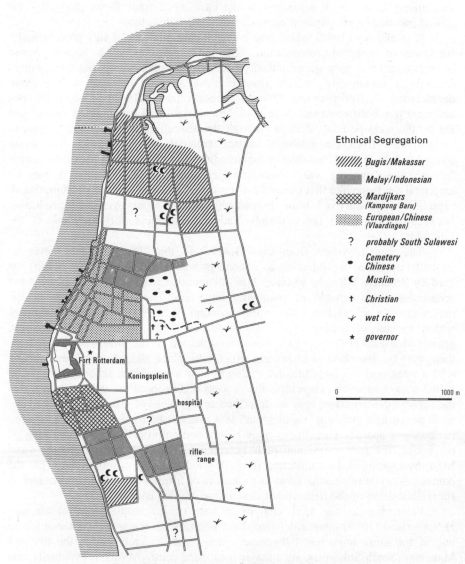

*Fig. 3: Ethnic Segregation in Makassar (19 th century)*

*Sources: SUTHERLAND (1989), Atlas van Nederlandsch-Indie (further bibliographical data not available)*

A kind of a socio-ecological *gradient of loyalty* is observed with the Europeans, Chinese and a Portuguese-Indian-Mestizo population, the so-called *mardijkers*, living in the middle of the town around the fort. These groups and the presumed "hot-blooded" and "hostile" South Sulawesian population of Bugis and Makassar were punctuated by a kind of sanitary belt of "docile" people from other islands of the archipelago (Javanese, Ambonese, Malays for example).

The ethnic compartmentalization of the emporia was well-known in the archipelago even before the VOC-pattern. The different trading communities living together in a settlement did have their own administrative privileges and duties; single persons fulfilled important administrative functions (harbourmaster for example). All this is also true for pre-VOC Makassar (ibid.; Reid 1983, Villiers 1990). The difference between the meaning of the traditional spatial pattern and that of the company town is that the former was a method of organizing an open, decentralized international "urban society" whereas the latter was a method of better controlling that society by dividing it up. However, this method of control had inherent difficulties, too: The necessary funds for a growing body of officials who run a whole urban administration were, besides the problem to recruit the people for these jobs, not easy to raise. Under a mercantile perspective they were unproductive but necessary costs. The example of Makassar, thus, gives a rough illustration of the growing portion of unproductive capital consumption. It would be a central research question whether this shift from productive to unproductive capital consumption is measurable in quantitative terms and how far these matters were included consciously within the decision-making process.[7]

If infrastructure did make sense only on the background of a monopoly-position in trade, because only then the investment could be compensated, then every risk of *not fully* realizing and preserving this position did mean a *proportional devaluation* of the capital investment. Even Coen's vision to attract settlers and private traders (!) from Holland to the East Indies did not succeed because, if realized, it would have been like an assault on the backbone of the whole VOC-system, namely the trade monopoly and the market transparency thereby generated.

The strategy caused positive feedbacks: The more investment had been made in preceding periods the more follow-up investments were necessary. This mechanism was part, if not the driving principle, of becoming increasingly involved in the internal affairs of the region: The VOC's approaching of the third level of the infrastructural setting was neither the mere outcome of a straight-forward strategy of occupation (against this stands the merchant half of the Company's heart) nor the only result of accidental action of particular officials of the VOC in the East Indies. Instead, it was in some ways the result of a self-enhancing and self-contradictory process as the concomitant of the "all-or-nothing-strategy" forced by the monopoly target. One hindrance to this plan was the very fact that local and other European trading networks were a continual, and perhaps a steadily growing, threat against it which, accordingly, continually forced a, perhaps steadily growing, portion of capital to be invested in regulation activities.[8]

In summary, the VOC's project led to a new articulation of "modes of trade" and certainly implemented its own mode as the dominant one. But the in-built contradictions enforced the involvement of the legal and whole political sphere of social life in the archipelago into the project which then gradually shifted away from a mercantile one eventually ending in the social formation of colonialism.

There exists a whole range of accounts as to how the VOC gradually became a land-based power (see, for instance, D.G.E.Hall 1981, pp.336-365, or M.C.Ricklefs 1981, 1986). This process concentrated on Java but also the strategic points of the other islands were put under the increasingly tighter grip of VOC-authority during the course of the late 17th and 18th centuries. When studying this process a type of explanation seems necessary which, more than it is true for the two, so far treated levels of the European access to the archipelago, needs a close interconnecting of both the organizational development of the expanding powers and the evolving local dynamics. However then the problem arises, that it is possible to become lost in single historical events or, on the other hand, that a too smooth theory is used, which interprets all the frictions of the process away. It is the second type of problem to which, finally, some short remarks will be devoted.

## 5.    A Note on Southeast Asia as an External Arena

The following attaches to just one very limited aspect of Wallerstein's world systems project, his discussion of the *external arena*. For the period under study here, according to Wallerstein, Southeast Asia may be classified as an element of the *external arena* of the emerging European world-economy. The nature of this type of region Wallerstein defines as follows:

> "The external arena of a world-economy consists of those other world-systems with which a given world-economy has some kind of trade relationship, based primarily on the exchange of preciosities, what was sometimes called the 'rich trades' (Wallerstein 1974, p.302)."

Southeast Asia fulfils these conditions: For a long period it was part of one of the, if not *the*, largest international trade systems of the world and it surely needs no further comment that from the beginning of the European intrusion until the end of the 17th century trade relationships (including products like pepper, cloves, nutmeg and mace) were based on luxuries or "rich trade".

Like all types of the Wallerstein-regions the external arena is defined in relation to the core-region of the European world-system. Nevertheless, he also joins to it a particular *internal structure* which was characterized by economic and political forces which defended their claims with considerable strength. This was the reason why the Europeans had to use gold and silver in order to come to terms with those forces. An elaboration of this point will not be attempted here, but it may be said, that the internal as well as the world-system related properties (being an element of another world-system) were in a sense the causes of the "superficial" economic relationship which existed between the core-region and the external arena.

Wallerstein's relational definition of the system of regions is, at the same time, of a spatial and socio-economic nature. Therefore, if socio-economic development is observed at all, then it has to be expected that the objects will shift from one regional class to another. In particular it is historical fact that what was an external arena first eventually became an internal one: To be incorporated into the *periphery* of the European world-system was the ultimate fate of most parts of the former *external arena*. The question Wallerstein put on the agenda, then, is *'why* the external arena gradually transgressed into the periphery?'. An interesting and important question, indeed. However, the answer Wallerstein gives, though only as a rough sketch as will be obvious from the quotation below, is typically functionalist and falls under the general critique on Wallerstein's epistemology which has been correctly advanced in the past.[9] He writes:

> "It was a matter of what made most sense to capitalist entrepreneurs in the short run - the profits of exploitation or the profits of speculation. In the short run, those who were in favor of speculation prevailed; but in the long run ... the profits of productive exploitation are the only solid base on which to stay ahead in the capitalist world-economy. The core powers ... launched in the eighteenth century the peripheralization of the Indian Ocean arena, which really took root after 1750 (1980, p.48)."

The empirical evidences the quotation contains will not be queried. However what is not demonstrated by Wallerstein is *how* that, which worked, and was apparently rational in the *short run*, stopped working and was no longer rational after some time! Questions immediately arising are: How can we grasp the structural change in a coherent way? When and for what reason did the actors (to a considerable extent self-interested, short-term thinking merchants!) change their principles of action? What was the part the regional "inheritance" added to the story? If an answer to these questions is really sought after, it should be based on the *mechanisms* which root within the concrete historical principles of action; be it the normal merchant enterprise or the local ruler experiencing a changing configuration of forces and chances; be it the management of a joint stock company or its officials in the East Indies who's action may be driven by "statesman visions". Such search for motives and mechanisms implies detailed research of source material but the alternative, a functionalist or a mere structuralist type of explanation, is too wide a by-pass to the problem of structural change. The conviction put forward in this paper is, that the companies, especially the VOC, would be the most rewarding object to begin with, because they, as an institutional innovation, brought with themselves far-reaching innovations into the region.

It is the Polish historian W.Kula (1976) who introduces the idea of a *short-term* and a *long-term dynamic* as basic for a complete historical research on economic change. Both types of dynamic and change have their own internal paths and descriptions. What becomes analytically more tangible then is the interface between both of these levels. Wallerstein, instead of adhering to such an approach, has no problems in reducing the historical process to a mere object of an objective long-run rationale. He remains completely unspecific as to *how* this rationale was implemented and *who* was the social actor acting it out. The notion of vagueness of the interrelationship between regions of different world-systems, which Wallerstein introduces as a defining property of the external-arena-relationship, is dissolved in a functionalist manner by him in favour of the *later* winner.

Similarly it is true that Wallerstein's explanation reduces the aforementioned difference between the "British" and the "Dutch" mode of trade organization, as previously termed, to nil. To put both under the heading "in favor of speculation", as he does in the above quotation, is completely missing the chance to theorize the whole process by referring to the short-term dynamic as is shown by the actors; for instance the companies. It is quite evident that the mode of incorporation of the archipelago into the European world-system worked at least partly through the penetrating restructuring of the already existing commercial nets and structures of which in this paper the infrastructral subset was central.

Starting with the reconstruction of the *short-term dynamic* as based on the rationale of the expanding enterprises, a kind of a *model of mercantile accumulation*, future research should, on the other hand, take the term *external arena* as seriously as possible, in the sense that this notion refers to a deeply pre-structured part of the world which had (due to the very fact of being part of another world-system) a particular kind of *inheritance* to offer to any mercantile *project*. What is presented in this paper, as already mentioned above, is a sketchy framework which has to, and will, be furtherly refined. The author is convinced that by interconnecting all three levels of analysis, the concrete materialization of strategic action, the institutional and rational background of that action and the region's response to it, a distinct and promising view on the process of European expansion will be yielded.

## Notes

1.   This statement will be given more substance below; but here the reference to a recent illuminating inquiry into the traditional *legal space* of the region provided by C.-B.Kaehlig (1986) is in order.

2.   Some insights into this early political structure and its transformation is provided in two recent articles by Kathirithamby-Wells (1986a, 1987); the classic study on Srivijaya is Wolters (1967).

3.   Both the flexibility and the elaborated political system is also true for the 17th century when local states like Aceh, Johor, Banten, and Makassar emerged. Except Makassar these states were located in the western part of the archipelago.

4.   Of course, the "exact position" of a trading center is not determined in this way. But obviously there should be a center in this part of the archipelago. See also de Josselin de Jong/van Wijk 1960, pp.21-23.

5.   "Common to all these transitional organizations is their tendency whenever possible to give up their economic co-operation and to revert to traditional entrepreneurial forms" (Steensgaard 1981, p.249).

6.   Concerning the possibility of *formal rational* action - the well-known Weberian topic! - a technical problem should be mentioned, too: The system of *accountancy*! According to

7.     Heckscher (1932, p.348), the system employed by the VOC did not, for instance, really fit into the division between private property and the entire enterprise's capital, thereby obscuring the decision-making process' principles.

7.     In this respect the nature of bookkeeping as it is reported by v.Leur: "The balance thus did not include the capitalized value of fixed properties" (1955, p.228) was, of course, a hindrance to formal rationality!

8.     Concerning these trade networks refer to the cited literature on trade; Evers (1984) gives a sketchy outline of these local networks, Dick (1975) is concerned with *perahu* shipping today but organization principles probably could be written back to the period under study; see also Needham (1983), Sutherland (1983), Taylor (1983) concerning slave trade which was of a particular importance and a domain of those networks, a further example is Andaya (1989) on cloth imports from India to Sumatra.

9.     A short but convincing discussion of this point is presented in Corbridge (1986).

## References

Andaya, B.W. (1989): The Cloth Trade in Jambi and Palembang Society during the Seventeenth and Eighteenth Centuries. Indonesia 48, pp.26-46

Arasaratnam, S. (1984): European Port-Settlements in the Coromandel Commercial System, 1650-1740. Paper presented to the Second International Conference on Indian Ocean Studies. Perth

Asian Business, June 1989, pp.16-23

Bassett, D.K. (1963): European Influence in South-East Asia, c. 1500-1630. Journal of Southeast Asian History 4, pp.173-209

Blussé, L. (1988): Strange Company. Chinese Settlers, Mestizo Women and the Dutch in VOC Batavia. (Verhandelingen v.h. Koninklijk Instituut v. Taal-, Land- en Volkenkunde 122) Dordrecht

Blussé, L. and F. Gaastra (eds.) (1981): Companies and Trade. Essays on Overseas Trading Companies during the *Ancien Régime*. (Comparative Studies in Overseas History 3, Leiden Centre for the History of European Expansion) Den Haag

Boxer, C.R. (1965): The Dutch Seaborne Empire, 1600-1800. London

Braudel, F. (1986a): Sozialgeschichte des 15.-18. Jahrhunderts, Bd.2: Der Handel. München

Braudel, F. (1986b): Sozialgeschichte des 15.-18. Jahrhunderts, Bd.3: Aufbruch zur Weltwirtschaft. München

Broeze, F.J.A. (ed.) (1989): Brides of the Sea: Port Cities of Asia from the 16th-20th Centuries. Honolulu

Bronson, B. (1977): Exchange at the Upstream and Downstream Ends: Notes Towards a Functional Model of the Coastal State in Southeast Asia. In: K.L.Hutterer (ed.): Economic Exchange and Social Interaction in Southeast Asia: Perspectives from Prehistory, History, and Ethnography. (Michigan Papers on South and Southeast Asia 13) pp.39-52, Ann Arbor

Bruijn, J.R., F.S.Gaastra and I.Schöffer (1987): Dutch-Asiatic Shipping in the 17th and 18th Centuries. Vol.I: Introductory Volume. (Rijks Geschiedkundige Publicatiën, Grote Serie 165) The Hague

Corbridge, S. (1986): Capitalist World Development. A Critique of Radical Development Geography. Houndmills

Dale, W.L. (1956): Wind and Drift Currents in the South China Sea. Journal of Tropical Geography 8, pp.1-31

Dick, H.W. (1975): Prahu Shipping in Eastern Indonesia. Part I. Bulletin of Indonesian Economic Studies 11, pp.69-107

Evers, H.-D. (1984): Traditional Trading Networks of Southeast Asia. (Paper read at the Fifth Bielefeld Coll. on Southeast Asia "Trade and State in Southeast Asia") Bielefeld

Glamann, K. (1958): Dutch-Asiatic Trade, 1620-1740. Copenhagen/The Hague

Hall, D.G.E. (1981): A History od South-East Asia. (4th edition) Houndmills

Hall, K.R. (1985): Maritime Trade and State Development in Early Southeast Asia. Honolulu

Hall, K.R. and J.K. Whitmore (eds.) (1976): Explorations in the Early Southeast Asian History: The Origins of Southeast Asian Statecraft. (Michigan Papers on South and Southeast Asia 11) Ann Arbor

Hamilton, E.J. (1948): The Role of Monopoly in the Overseas Expansion and Colonial Trade of Europe before 1800. American Economic Review, Papers and Proceedings 38, pp.33-53

Harvey, D. (1982): The Limits to Capital. Oxford

Heckscher, E.F. (1932): Der Merkantilismus. Jena

de Josselin de Jong, P.E. and H.L.A. van Wijk (1960): The Malacca Sultanate. An account from a hitherto untranslated Portuguese source. Journal of Southeast Asian History 1, pp.20-29

Kaehlig, C.-B. (1986): Gesellschaftsrecht in Indonesien. Autonome und nationale Gesellschaftsformen. (Mitteilungen des Instituts für Asienkunde 151) Hamburg

Kathirithamby-Wells, J. (1986a): Royal Authority and the Orang Kaya in the Western Archipelago, Circa 1500-1800. Journal of Southeast Asian Studies 17, pp.256-267

Kathirithamby-Wells, J. (1986b): The Islamic City: Melaka to Jogjakarta, c. 1500-1800. Modern Asian Studies 20, pp.333-351

Kathirithamby-Wells, J. (1987): Forces of Regional and State Integration in the Western Archipelago, c. 1500-1700. Journal of Southeast Asian Studies 18, pp.24-44

Kathirithamby-Wells, J. (1990): Banten: A West Indonesian Port and Polity During the Sixteenth and Seventeenth Centuries. In: Kathirithamby-Wells/Villiers (eds.), pp.106-125

Kathirithamby-Wells, J. and J. Villiers (eds.) (1990): The Southeast Asian Port and Polity. Rise and Demise. Singapore

Klein, P.W. (1981): The Origins of Trading Companies. In: Blussé, L. and F. Gaastra (eds.), pp.17-28

Kula, W. (1976): An Economic Theory of the Feudal System. Towards a model of the Polish Economy, 1500-1800. London

Leur, J.C.v. (1955): Indonesian Trade and Society. Essays in Asian Society and History. Den Haag

Lipietz, A. (1980): The Structuration of Space, the Problem of Land, and Spatial Policy. In: J.Carney, R.Hudson and J.Lewis (eds.), Regions in Crisis. New Perspectives in European Regional Theory, pp.60-75, New York

Macknight, C.C. (1973): The Nature of Early Maritime Trade: Some Points of Analogy from the Eastern Part of the Indonesian Archipelago. World Archaeology 5, pp.198-208

Manguin, P.-Y. (1988): Of Fortresses and Galleys. The 1568 Acehnese Siege of Melaka, after a Contemporary Bird's-eye View. Modern Asian Studies 22, pp. 607-628

Meilink-Roelofsz, M.A.P. (1962): Asian Trade and European Influence in the Indonesian Archipelago Between 1500 and About 1630. Den Haag

Meyn, M., M.Mimmler, A.Partenheimer-Bein et al. (1987): Der Aufbau der Kolonialreiche. (Dokumente zur Geschichte der europäischen Expansion Vol.3, edited by E.Schmitt) München

Needham, R. (1983): Sumba and the Slave Trade. (Working Paper 31, Centre of Southeast Asian Studies, Monash University) Melbourne

Pearson, M.N. (1988): Brokers in Western Indian Port Cities. Their Role in Servicing Foreign Merchants. Modern Asian Studies 22, pp. 455-472

Reid, A. (1980): The Structure of Cities in Southeast Asia, Fifteenth to Seventeenth Centuries. Journal of Southeast Asian Studies 11, pp. 235-250

Reid, A. (1983): The Rise of Makassar. Review of Indonesian and Malaysian Affairs 17, pp. 117-160

Reid, A. (1988): Southeast Asia in the Age of Commerce, c. 1450-1680. Vol.I: The Lands Below the Winds. New Haven

Reid, A. and L.Castles (eds.) (1975): Pre-Colonial State Systems in Southeast Asia. (Monographs of the Malaysian Branch of the Royal Asiatic Society 6) Kuala Lumpur

Ricklefs, M.C. (1981): A History of Modern Indonesia, c. 1300 to the Present. Houndmills

Ricklefs, M.C. (1986): Some Statistical Evidence on Javanese Social, Economic and Demographic History in the Later Seventeenth and Eighteenth Centuries. Modern Asian Studies 20, pp.1-32

Steensgaard, N. (1974): The Asian Trade Revolution of the Seventeenth Century. The East India Companies and the Decline of the Caravan Trade. Chicago

Steensgaard, N. (1981): The Companies as a Specific Institution in the History of European Expansion. In: Blussé, L. and F. Gaastra (eds.), pp.245-264

Subrahmanyam, S. (1988): Commerce and Conflict: Two Views of Portuguese Melaka in the 1620s. Journal of Southeast Asian Studies 19, pp.62-79

Sutherland, H.A. (1983): Slavery and the Slave Trade in South Sulawesi, 1660s-1800s. In: A.Reid (ed.): Slavery, Bondage and Dependency in Southeast Asia, pp.263-285, St.Lucia, London

Sutherland, H.A. (1989): Eastern Emporium and Company Town: Trade and Society in Eighteenth-Century Makassar. In: F.J.A.Broeze (ed.), Brides of the Sea: Port Cities of Asia from the 16th-20th Centuries. pp.97-128, Honolulu

Sutherland, H.A. and D.S.Bree (1984): The Harbourmaster's Specification: A Pilot Study in Computer-Aided Identification of Regional Trade Patterns in V.O.C. Asia. Paper presented to the Second International Conference on Indian Ocean Studies. Perth

Taylor, J.G. (1983): The Social World of Batavia. European and Eurasian in Dutch Asia. Madison

Villiers, J. (1981): Trade and Society in the Banda islands in the Sixteenth Century. Modern Asian Studies 15,pp. 723-750

Villiers, J. (1990): Makassar: The Rise and Fall of an East Indonesian Maritime Trading State, 1512-1669. In: Kathirithamby-Wells, J. and J. Villiers (eds.), pp.143-159

Wallerstein, I. (1974): The Modern World-System I. Capitalist Agriculture and the Origins of the European World-Economy in the Sixteenth Century. New York etc.

Wallerstein, I. (1980): The Modern World-System II. Mercantilism and the Consolidation of the European World-Economy, 1600-1750. New York etc.

Wheatley, P. (1983): Nagara and Commandery. Origins of the Southeast Asian Urban Traditions. (Research Paper 207/208, University of Chicago, Dep. of Geography) Chicago

Whitmore, J.K. (1977): The Opening of Southeast Asia, Trading Patterns through the Centuries. In: K.L.Hutterer (ed.): Economic Exchange and Social Interaction in Southeast Asia: Perspectives from Prehistory, History, and Ethnography. (Michigan Papers on South and Southeast Asia 13) pp.139-153, Ann Arbor

Wisseman, J. (1977): Markets and Trade in Pre-Majapahit Java. In: K.L.Hutterer (ed.): Economic Exchange and Social Interaction in Southeast Asia: Perspectives from Prehistory, History, and Ethnography. (Michigan Papers on South and Southeast Asia 13) pp.197-212, Ann Arbor

Wolters, O.W. (1967): Early Indonesian Commerce: A Study of the Origins of Srivijaya. Ithaca

# ERDKUNDLICHES WISSEN
Schriftenreihe für Forschung und Praxis.
Herausgegeben von Gerd Kohlhepp in Verbindung mit Adolf Leidlmair und Fred Scholz